CW00971991

SYSTEMS THEORY APPLIED TO AGRICULTURE AND THE FOOD CHAIN

SYSTEMS THEORY APPLIED TO AGRICULTURE AND THE FOOD CHAIN

Edited by

J. G. W. JONES

and

P. R. STREET

Department of Agriculture,
University of Reading, UK

ELSEVIER APPLIED SCIENCE
LONDON and NEW YORK

ELSEVIER SCIENCE PUBLISHERS LTD
Crown House, Linton Road, Barking, Essex IG11 8JU, England

Sole Distributor in the USA and Canada
ELSEVIER SCIENCE PUBLISHING CO., INC.
655 Avenue of the Americas, New York, NY 10010, USA

WITH 21 TABLES AND 34 ILLUSTRATIONS

British Library Cataloguing in Publication Data

Systems theory applied to agriculture and the food chain.
1. Agricultural industries. Applications of systems theory
I. Jones, J. G. W. II. Street, P. R.
338.1011

ISBN 1-85166-510-2

Library of Congress Cataloging-in-Publication Data Applied For

Typeset and Printed in Northern Ireland by The Universities Press (Belfast) Ltd.

Foreword

COLIN SPEDDING

Greatness attaches to those who perceive the commonplace in a new (and different) light from their more earthbound wayfaring companions. Colin Raymond William Spedding arrived at the (then) Grassland Research Station at Drayton near Stratford-upon-Avon in 1949 and such a light was kindled then and has shone forth for the 40 years which have since elapsed. His contributions have been in parasitology, sheep production, the ecology of grassland, agricultural systems and strategy, and in the administration of research and university education.

Born into the manse, his early life gave him a wide experience of the world, experience from which he has rarely failed to profit subsequently. After (in his own words) school left him at the age of 14 and describing himself as not educated, he was employed by Ilford Ltd before entering the Royal Navy in 1943, in which he served for three years on the lower deck and in the wardroom. The anecdotes of this period continue to enliven any conversation with him. The following three years included service as a mountain guide in Snowdonia and employment as a laboratory technician with Allen and Hanburys.

Upon arrival at the Grassland Research Station he set about acquiring at two-year intervals the degrees of Bachelor of Science, Master of Science and Doctor of Philosophy of the University of London by the arduous route of external study. During this period, however, he still found time to marry his charming botanist colleague, Betsy George. After a decent interval he was admitted to the degree of Doctor of Science of the same university in 1967. In the meantime, in 1954, he had moved with the rest of the (by this time) Grassland Research Institute to Hurley in Berkshire.

The early work which brought him wide recognition was in the field of parasitology from which there emerged the sideways creep-grazing system for ewes and lambs. From these beginnings he went on to study the production of sheep and the interactions between sheep and the swards they grazed. During all his work over this period he was greatly supported by the founder Director of the Grassland Research Institute (Dr William Davies) and others but Spedding regarded his investigations as only part of a much larger whole and in 1965 he persuaded the (then) Director of the Institute (Dr E. K. Woodford) to establish a Systems Synthesis Department within his (Spedding's) Ecology Division of the Institute. This developed into a substantial modelling activity in the Biomathematics Division of the Institute which was established later.

It was soon after he had played a leading role in organising a seminal symposium at Hurley early in 1969 on *The Use of Modelling in Agricultural and Biological Research* that he was offered the Chair of Agricultural Systems at the University of Reading. This chair had been created some three years earlier but never filled. He accepted the chair in 1970 on a part-time basis and continued his research at the Grassland Research Institute of which he became Deputy Director.

Finally, in 1975 he resigned from the staff of the Grassland Research Institute and occupied the chair on a full-time basis. There followed a decade and a half in which Colin Spedding achieved the full flowering of his career. From his tenures of office as Head of the (then) Department of Agriculture and Horticulture for eight years, Dean of the Faculty of Agriculture and Food and Pro-Vice-Chancellor, the University has reaped the benefit of his leadership and administrative abilities. In 1981 he also became Director of the Centre for Agricultural Strategy at a critical time in its financing and history. He succeeded not only in ensuring its survival but also in expanding its activities and enhancing its reputation and authority. All these activities have been superimposed on his continuing pursuit of scholarship through his teaching and research.

Outside the University he has been much sought after to perform duties which demanded clarity of thought and expression, and his activities have been multifarious both in the UK and abroad. He has, among other tasks, served as Adviser to the House of Commons Select

Committee on Agriculture; edited *Fream's Agriculture* for the Royal Agricultural Society of England; been President of the British Society of Animal Production; founded the journal, *Agricultural Systems,* which he edited for 12 years; and become Chairman of the Farm Animal Welfare Council, the Board of the United Kingdom Register of Organic Food Standards and the Apple and Pear Research Council. Invitations to travel and lecture overseas poured in; for a number of years he was Chairman of the Programme Committee of the International Livestock Centre for Africa, one of the institutes of the Consultative Group for International Agricultural Research, and in 1977 he was the Commonwealth Prestige Fellow of the New Zealand University Grants Committee.

Apart from the editorial activities already mentioned, Colin Spedding has written or edited eight books and published some 200 papers. Notable amongst these are a series on the study of sub-clinical worm infestation in sheep (with T. H. Brown, R. V. Large and I. A. N. Wilson) in the *Journal of Agricultural Science, Cambridge,* between 1957 and 1965, papers on the physiological basis of grazing management published in both the *Journal of the British Grassland Society* and the *Proceedings of the Nutrition Society* in 1965 and on agricultural systems and grazing experiments (with F. H. W. Morley) in *Herbage Abstracts* in 1968, a steady stream of papers since 1970 on the biological efficiency of agriculture, agricultural systems and agricultural policy, and a book on *Biological Efficiency in Agriculture* (with J. M. Walsingham and A. M. Hoxey).

Honours have inevitably flowed from such a variety of activities, culminating in June 1988 in his appointment by Her Majesty The Queen to be a Commander of the Order of the British Empire.

This volume, while marking his formal retirement from the University, celebrates but a milestone in a career which promises to develop further in different directions. It is to be hoped that his enthusiasm and perspicacity which have illumined the way for so many involved in agriculture for the past 40 years will continue to do so for many years to come.

J. G. W. JONES

Preface

Agricultural systems has developed in a variety of directions from being a research tool in limited biological sectors of agriculture. Most importantly, it has come to encompass the whole food chain, the ecosystem with which agriculture interacts and the economics of agriculture, and has extended into the fields of education and extension to a remarkable degree. Over the 25 years or so during which these developments have occurred the name of Colin Spedding has been associated with many of them, mainly in philosophical terms. Some of his colleagues considered that his impending retirement provided a suitable opportunity to review these developments as well as to honour a leader in the subject.

Accordingly, 13 eminent colleagues were invited to contribute papers to a symposium and this volume contains those papers. The papers have been made available to participants at the symposium during which discussions were opened by the contributors. No attempt was made to present the papers at the symposium and no record of the discussions has been published.

Each contributor was briefed to review the current state of the sector in which his expertise resided, to speculate on the future development of that sector insofar as he felt able to and to draw attention to the contributions made by Colin Spedding wherever relevant. Both the time to complete the task and the space each had at his disposal were limited and it is inevitable that exhaustive coverage of the brief was rarely possible. Nevertheless, the Editors take the view that the volume represents a useful review of many aspects of agricultural and food chain systems and a worthy tribute to an outstanding pioneer of the subject.

The chapters of this volume progressively cover the application of systems theory at various levels of resolution of the agricultural ecosystem and the food chain. The first two contributions by Baldwin and Hanigan, and Maxwell strike at the origins of the application of the systems approach in agriculture and its use in the development of an understanding of the biological and physiological systems which underlie enterprise systems in agriculture. Baldwin and Hanigan explore the application of systems theory in animal sciences. Maxwell reviews applications to the understanding of plant–animal interactions.

There then follow four chapters which cover the exploitation of the systems approach to resource allocation and system improvement at the enterprise and farm levels. Bywater and Doyle review the application of systems theory in the design of agricultural enterprises whilst Dent considers the application of the approach to the optimisation of the enterprise mix at the farming system level. Hildebrand's chapter is concerned with the application of farming systems research in technology transfer and farming system development.

The two subsequent papers are devoted to the application of systems thinking in the understanding and development of food markets. Gray's chapter is specifically concerned with food security planning and hence mainly with applications of modelling in food policy development and with institutional market intervention systems. Street's contribution covers the systems interactions between the various players in the food chain from producers through marketing agents, processors, retailers to consumers, and the consequential development of commercial supply and marketing links.

The discussion then moves to even broader applications. Three contributors review the application of systems theory to the development of agricultural strategy in its widest context. These chapters cover interactions between the agricultural system, the ecosystem and the social system in agricultural development. Conway reviews the interactions between agriculture and the ecosystem, De Wit the problems of understanding and managing future changes in agriculture and Harvey the issues of sector modelling for policy development.

The final section of this volume comprises two chapters which are concerned with the application of systems thinking to new methods of transferring information and technology in agriculture. Bawden deals with the development of the approach in institutional education and Gartner in relation to extension education.

The committee responsible for organising the symposium consisted of Professor T. R. Morris (Head of the Department of Agriculture), Professor E. H. Roberts (Dean of the Faculty of Agriculture and Food), Professor P. R. Street (Professor of Agricultural Systems), Professor M. Upton (Department of Agricultural Economics and Management), Dr S. P. Carruthers and Dr J. G. W. Jones. Mrs M. Newbery acted as Secretary to the Committee and to the Symposium.

In the economic climate in which the universities find themselves, the mounting of a symposium such as this has to be a self-financing operation. In those circumstances the cost to participants can become unacceptably high and to keep this cost within bounds the Committee sought sponsorship from commercial and other organisations. Since Colin Spedding is well known to many of these, the task proved an easier one than the Committee had anticipated and the philanthropic generosity of the following organisations is gratefully acknowledged:

Bayer UK Ltd
Imperial Chemical Industries PLC
The Nuffield Foundation
Royal Agricultural Society of England
Royal Bank of Scotland
Unilever PLC

Many other people and organisations have contributed to the mounting of the Symposium and their contributions have been much appreciated. In particular, the Institute of Biology, of which Professor Spedding is currently a Vice-President, subsidised in considerable measure the distribution of the Symposium brochure in its own periodical, *The Biologist*.

Finally, the Committee acknowledges with gratitude the contributions of the authors without which it is hardly necessary to say that there would have been no Symposium.

P. R. Street and J. G. W. Jones

Contents

List of Contributors

R. L. Baldwin

Department of Animal Science, University of California, Davis, Calfornia 95616, USA.

R. J. Bawden

Faculty of Agriculture and Rural Development, University of Western Sydney, Hawkesbury, Bourke Street, Richmond, New South Wales 2754, Australia.

A. C. Bywater

Department of Farm Management, PO Box 84, Lincoln University, Canterbury, New Zealand.

G. R. Conway

The Ford Foundation, 55 Lodi Estate, New Delhi 110003, India.

J. B. Dent

Division of Rural Resource Management, The Scottish Agricultural College, West Mains Road, Edinburgh EH9 3JG, UK.

C. J. Doyle

The Scottish Agricultural College, Auchincruive, Ayr KA6 5HW, UK.

J. A. Gartner

Division of Agricultural and Food Engineering, Asian Institute of Technology, GPO Box 2754, Bangkok 10501, Thailand.

J. G. GRAY

Food Studies Group, International Development Centre, Queen Elizabeth House, University of Oxford, 21 St Giles, Oxford OX1 3LA , UK.

M. D. HANIGAN

Department of Animal Science, University of California, Davis, California 95616, USA.

D. R. HARVEY

Department of Agricultural Economics and Food Marketing, and Centre for Land Use and Water Resource Research, University of Newcastle upon Tyne, Newcastle upon Tyne NE1 7RU, UK.

P. E. HILDEBRAND

Food and Resource Economics Department, Institute of Food and Agricultural Sciences, University of Florida, Gainesville, Florida 32611, USA.

T. J. MAXWELL

The Macaulay Land Use Research Institute, Craigiebuckler, Aberdeen AB9 2QJ, UK.

P. R. STREET

Department of Agriculture, University of Reading, Earley Gate, PO Box 236, Reading RG6 2AT, UK.

C. T. DE WIT

Department of Theoretical Production Ecology, Agricultural University, PO Box 430, 6700 AK Wageningen, The Netherlands.

1
Biological and Physiological Systems: Animal Sciences

R. L. BALDWIN and M. D. HANIGAN
Department of Animal Science, University of California, Davis, USA

EARLY AND CURRENT MODELS USED IN FEEDING SYSTEMS

The application of models in animal agriculture dates back to the turn of the century and resulted from the development by a number of workers, including Von Voit, Armsby, Atwater, Kellner and Rubner (Lusk, 1926), of methods of evaluation of foods and feeds in terms of metabolizable and net energy values. Metabolizable energy was defined by Armsby (1917) as 'the gross energy in the feed minus the gross energy of the excreta' and 'energy capable of transformation in the body'. Net energy values of feeds for maintenance were measured as the amount of body energy loss prevented by an increment of feed, while net energy values of feeds for productive processes were measured as amounts of energy stored in gain (protein and/or fat) or recovered in a product such as milk per increment of feed provided above maintenance. These enabled the development of models to predict animal energy and feed requirements based upon their expected productivity or, conversely, to predict animal productivity based upon feed intake.

A number of systems generally based upon estimates of feed values and animal requirements in terms of metabolizable energy were developed. These included the Physiological Fuel Value (PFV) system of Atwater, which is still in use by human and laboratory animal dieticians, and the Total Digestible Nutrients (TDN) system for which data are still included in US National Research Council (NRC) publications on the nutrient requirements for several livestock species. Several alternative systems, of which the Starch Equivalent System (SES) was the most prominent and widely adopted example, were based upon the net energy concept. For simplicity in application, TDN values of feeds and animal requirements were expressed in pounds where the relative values assigned to digestible protein, carbohydrate and fat were 1, 1 and 2·25 lb/lb. TDN requirements of animals were estimated by simply summing tabular values of TDN required for maintenance expressed as a function of body weight, TDN required for a given rate of gain, TDN required for production of milk expressed as a function of fat content, TDN required for activity, pregnancy and so on, or:

$$\mathrm{TDN}_{\mathrm{required}} = \mathrm{TDN}_{\mathrm{maintenance}} + \mathrm{TDN}_{\mathrm{growth}} + \mathrm{TDN}_{\mathrm{lactation}} + \cdots \quad (1)$$

Similarly, for simplicity in application of the SES, feed values were measured in terms of amount of energy stored in fat per kilogram of feed provided above maintenance in adult animals as compared to (divided by) the energy retained in fat per kilogram starch equivalent (SE) provided above maintenance to the same animal. Animal requirements were, again, a simple sum:

$$\mathrm{SE}_{\mathrm{requirement}} = \mathrm{SE}_{\mathrm{maintenance}} + \mathrm{SE}_{\mathrm{growth}} + \mathrm{SE}_{\mathrm{lactation}} + \cdots \quad (2)$$

These early models incorporated a number of systematic errors which were addressed in the development of newer systems or models currently in use by animal nutritionists. Two problems with the TDN system were recognized early. One was that maintenance requirements of animals were estimated based upon the observation of Rubner that heat production at maintenance varied across species as a function of surface area or body weight (W) raised to the two-thirds power ($W^{0\cdot67}$). Brody (1945) and Kleiber (1932) developed interspecific relationships between fasting heat production (FHP) and body weight which led to adoption of the well-known allometric equation

$$\mathrm{FHP} = a(\mathrm{kg}\ W)^{0\cdot75} \quad (3)$$

for estimating FHP based upon body weight. For standardizing the expression of FHP data for young adults, the value of the coefficient, a, was 0·29 MJ (kg W)$^{-0\cdot75}$. It was also noted early that application of the concept of metabolic body weight (MBW = (kg W)$^{0\cdot75}$) within a species led to considerable variance in the coefficient dependent upon age, previous plane of nutrition and physiological state (growing, lactating, pregnant).

Several workers utilizing statistical fitting routines, including Thonney *et al.* (1976), have shown that superior fits for the relationship between FHP and body weight within species are obtained when both the coefficient and exponent are allowed to be varied by statistical fitting routines. This approach reduces variance attributable to age but not to previous plane of nutrition or physiological state and, as a result, has not been adopted. Most current feeding systems accommodate variations in FHP and apparent maintenance requirements by adjusting the coefficient dependent upon age and physiological state. However, a number of current modelling analyses are directed to the development of improved, largely mechanistic models which address this problem (see below).

A second major problem with the TDN system was that TDN values from forages and concentrates fed to ruminants were not additive, a unit of TDN from forage having a lesser feeding value than a unit from a concentrate feed, and further that the relative values of forage and concentrate varied for different animal functions (maintenance, growth, lactation). This was found to be due, in part, to the fact that relative energy losses in methane from forage are greater than those for concentrates. This error is accommodated in current systems by definition of metabolizable energy (ME) as energy in feed minus energy losses in excreta and combustible gases. Other components of the non-additivity of forage and concentrate TDN values are attributable to differences in heat losses during fermentation, differential changes in digestibility associated with intake, and differences in products of fermentation and their efficiencies of utilization by the animal (see below). These were found to vary with ME density of the diet. Thus, most current ME-based feeding systems incorporate a correction factor based upon ME density of the feed. A generalization of most current ME-based models is that the ME requirement of an animal is calculated as the sum of ME values required for maintenance, growth, lactation and other activities. The ME requirement for maintenance (ME_m) would be calculated as:

$$ME_m = a(\text{kg } W)^{0\cdot75}\, k_m^{-1} \tag{4}$$

where the allometric coefficient, a, might vary from 0·25 to 0·5 for animals on a low plane of nutrition or lactating, respectively; and k_m or the efficiency of ME utilization for maintenance might vary from 0·65 to 0·80 dependent upon ME density of the feed. Similarly, the ME requirement for gain (ME_g) would be calculated as:

$$ME_g = \text{energy in gain}\ k_g^{-1} \tag{5}$$

where k_g or efficiency of ME utilization for gain might vary from 0·33 to 0·65 depending upon the composition of expected gain and ME density of the diet. Alternatively, energy gain in fat and protein may be calculated separately using estimates of their respective efficiencies which would vary dependent upon ME density of the diet. The latter approach might be judged superior on academic grounds, but estimates of the partial efficiencies of fat and protein gain are not very stable from experiment to experiment nor over time (age). This appears to be due to the fact that in conventional energy balance experiments we cannot satisfy the underlying assumption of parameter independence essential to estimation of generally applicable, partial efficiencies for protein and fat gain when using multiple linear regression techniques (Roux and Meissner, 1984; Bernier *et al.*, 1986).

This problem, along with the problem of variation in apparent maintenance requirements noted above, suggests that current feeding systems models have reached a level of complexity which cannot be surpassed using classical input–output, empirical data to estimate required parameter values.

The first net energy (NE)-based model used as the basis for feeding standards was the SES mentioned briefly above. While this system was adopted widely and was effective, its use in the UK and Western Europe has been discontinued in favor of ME-based models analogous to the generalized model discussed above. The major problem with the SES is the same one encountered with the TDN system, namely, that relative values assigned to feeds are not additive across functions. This is due to the fact that feeds are evaluated solely on their efficiency of use for fat gain and relative values of feeds for fattening differ from their relative values for support of, for example, maintenance. An example would be the comparison of diet fat with carbohydrate fermented in the rumen where the products of digestion (fatty acids) and fermentation (acetate) are used to synthesize body fat. Approximate efficiencies of these processes are 94–97 per cent and 60–65 per cent, respectively, resulting in a relative energetic efficiency of starch to fat of about 0·5. When diet fat and starch are used to support maintenance, the relative value of starch to fat is about 0·85. Thus, in the SES the value of starch for support of maintenance is significantly underestimated relative to fat. A similar undervaluation of forage relative to corn for support of maintenance is also observed.

The US NRC adopted a net energy model for use in the beef cattle industry. In this, feedstuffs are assigned separate net energy values for maintenance (NE_m) and gain (NE_g). Then, animal requirements for maintenance and gain are computed separately. Feed NE_m values are used to compute the feed requirement for maintenance and feed NE_g values are used to compute the feed required to support a given rate of gain. This approach overcomes the major problems with additivity of feed values discussed above for the TDN system and the SES. The system yields excellent predictions for British breeds of cattle since their maintenance requirements and composition of gain are similar throughout the standard feeding period. Some adjustments for 'frame size' were recently included to accommodate differences in composition of gain at a given weight in medium vs. large breeds of cattle but these corrections are inadequate (see below). A solution would be to incorporate separate estimates of net energy values of feeds for protein and fat accretion, but this step, as discussed above with regard to current ME-based models, goes beyond our ability to obtain valid, generally applicable estimates of efficiency parameters using the classical input–output, empirical approach.

MECHANISTIC ELEMENTS OF MODELS FOR USE IN FEEDING SYSTEMS

As discussed above, there are several areas where problems or limitations exist with regard to the application of classical (only input–output data used to parameterize models) approaches in improving our feeding systems. These include problems in accommodating variance in apparent maintenance requirements and in defining costs of productive functions due to differences in product composition or the efficiencies of protein and fat accretion. Additional problems which must be addressed include effects of previous plane of nutrition on current and future performance and optimization of protein availability to producing animals. It appears that resolution of these problems will require greater utilization of our knowledge of underlying functions in our models. This approach is called mechanistic modelling (Thornley and France, 1984) and requires that equations used to simulate, in this case, animal performance be based on concepts derived from studies conducted at the tissue, cellular and organelle levels and, further, that parameters used to implement the equations be derived from studies at these lower levels. Significant progress toward this end has been achieved in recent years and it appears that mechanistic elements are being or will be introduced into various feeding system models to increase accuracy of prediction and general applicability. In this section, progress in development of some appropriate mechanistic elements will be discussed.

Apparent Maintenance Requirements

As noted above, most systems for estimating the maintenance requirements of livestock are based upon the allometric equation of Kleiber (1932) and Brody (1945). For example, in the current NRC system for beef cattle the maintenance requirement in net energy for maintenance (NE_m) in megajoules is:

$$NE_m = 0{\cdot}322(\text{kg } W)^{0{\cdot}75} \quad (6)$$

In the current NRC system for dairy cattle (1989) the coefficient used is 0·335 MJ $(\text{kg } W)^{-0{\cdot}75}$. This estimate is adjusted upward 20 per cent for first lactation heifers and 10 per cent for second lactation cows. Koong *et al.* (1985) proposed that fasting heat production (kJ) for growing rats is best estimated using the equation

$$\text{FHP} = 445(\text{kg } W)^{(0{\cdot}686+0{\cdot}165\ \text{ADG})} \quad (7)$$

where ADG is average daily gain in grams implying that true metabolic body size is a variable dependent upon rate of gain. A similar equation form was utilized in a model developed for the pig (Black *et al.*, 1986), wherein metabolic body size was based upon body protein with additive adjustments for rate of gain and thermal environment. In the model of Fox *et al.* (1988), a more complex equation is used:

$$NE_m = [(a_1 + a_2)(\text{kg } W)^{0{\cdot}75}(\text{AX})(\text{BE})(L)(NE_{MHS})] + NE_{MCS} \quad (8)$$

where a_1 is 0·332 MJ $(\text{kg } W)^{-0{\cdot}75}$, a_2 is an acclimatization factor dependent upon previous temperature, AX is a grazing adjustment factor, BE is a breed factor from a table, L is a tabular factor for lactation, and NE_{MHS} and NE_{MCS} are adjustments for heat (panting) and cold stress, respectively. Graham *et al.* (1976) further modified the equation for use in their sheep model by incorporating empirically deduced maintenance adjustments for milk production, age, rate of gain, endogenous urinary energy loss, work associated with food intake, and digestible energy (DE) intake.

All of these represent attempts to identify factors which contribute to variation in apparent maintenance requirements and to accommodate these using equation forms and parameter values deduced from large input–output data sets and multiple linear regression techniques. Obviously, there is a limit to progress or advantages that can be gained using this approach. In order better to address this variation, several groups have developed mechanistic models which attempt to capture cause and effect relationships underlying maintenance energy expenditure as they vary across physiological states, environmental conditions, breed and other factors. A number of physiological/metabolic functions which contribute to variance in apparent maintenance requirements have been identified; these functions have been characterized, at least partially, using mechanistic models.

Physiological Service Functions

Baldwin *et al.* (1980) divided the basal energy expenditures of animals into two types: those associated with service functions performed by specific tissues in support of the whole animal; and costs required to maintain a particular cell or tissue. Some service functions including those associated with nerve function and Na^+ reabsorption by kidney are relatively constant functions of body weight and can be represented in aggregate (Baldwin *et al.*, 1987*c*). Others, including heart work and work of the muscles of respiration, vary dependent upon feed intake and animal energy expenditures and can be adequately represented as dependent variables. It is noteworthy and that these vary continuously in animals on low to high planes of nutrition and manifest themselves as components of both maintenance energy expenditures and costs of production. This is also true of costs of protein synthesis and ion transport (see below). As a result, mechanistic models are increasingly deviating from use of the classical concept of depicting costs of maintenance and production separately. Rather, energy expenditures in relatively constant processes are represented in aggregate (as above) and summed with costs associated with physiological work, protein synthesis, ion transport and other functions to obtain estimates of total animal energy expenditures to be satisfied through oxidation of available nutrients.

Relative Organ Weights

A second source of variance in apparent energy expenditures which has been investigated is differences in the relative weights (percentage of body weight) of organs and tissues. These vary with age and previous plane of nutrition. Koong *et al.* (1985), among others, have evaluated these effects in some detail. For example, in animals subjected to prolonged undernutrition relative liver weights are low (1·0–1·3 per cent of body weight) while in lactating animals they are high (1·7–1·9 per cent), as might be expected, due to the higher demands placed upon the livers of lactating animals. Energy requirements or expenditures per unit mass in liver are 25 times higher than in muscle or carcass (Baldwin *et al.*, 1985). Thus, when relative liver weight increases from 1·3 to 1·8 per cent of body weight and muscle weight decreases from 42·0 to 41·5 per cent of body weight in a lactating as compared to a non-lactating, non-pregnant cow, the increase in highly active metabolic tissue (liver) results in a 10 per cent increase in energy requirements. Similar changes in relative organ weights occur with age. Attempts to accommodate these sources of variance by incorporation of provisions for differential tissue growth rates will be discussed in a subsequent section.

Cell Maintenance Costs

Major variable components of energy expenditures associated with the maintenance of cells and tissues are energy expenditures in ion pumping (transport) and protein turnover. Together, these can comprise 40–55 per cent of basal energy expenditures. Milligan and McBride (1985), Summers *et al.* (1988) and Baldwin *et al.* (1980) discussed sources of variation in these which include age, plane of nutrition, physiological state and, for ion transport, cold adaptation. Baldwin *et al.* (1987*c*) and DiMarco and Baldwin (1989) included provisions for variation in these functions in dynamic models of lactating dairy cows and growing steers and found that such provisions helped simulate total body energy expenditures. However, very limited data were available to support parameterization of the equations used. Gill *et al.* (1989*a, b*) undertook comprehensive data and simulation analyses of energy costs associated with protein turnover and Na^+ and K^+ transport (ion pumping) in ten tissues in growing lambs. Protein synthesis in the several tissues was represented using equations of the form:

$$\text{rate} = K_{\text{XaXp}} Q_{\text{Xp}} (1{\cdot}0 + K_{\text{Xa,XaXp}} / C_{\text{Xa}})^{-1} \qquad (9)$$

where $K_{\text{XaXp}} Q_{\text{Xp}}$ is the maximum rate of conversion of amino acids into protein as a function of the quantity of protein (Q_{Xp}) in that tissue; $K_{\text{Xa,XaXp}}$ is affinity of the protein synthesis system for intracellular amino acids; and C_{Xa} is the intracellular concentration of amino acids. This equation, derived from the enzyme kinetic arguments of Michaelis and Menten, is commonly used in mechanistic models of animal metabolism and can be parameterized using tissue- or cell-level data. Equations for protein degradation were of the form:

$$\text{rate} = K_{\text{XpXa}} Q_{\text{Xp}} (1{\cdot}0 + C_{\text{Xa}} / J_{\text{Xa,XpXa}})^{-1} \qquad (10)$$

where the maximum rate is defined by a constant (K_{XpXa}) and quantity of protein (Q_{Xp}); and $J_{\text{Xa,XpXa}}$ is an inhibition constant for an effect of intracellular concentrations of amino acids on degradation rate. Transfer of amino acids into the tissue from blood is represented using a mass action equation. The equation for transfer of amino acids out of the cell has the same form as equation (10) with the concentration of amino acids in blood acting to inhibit efflux. Equations for ion transport provide for a basal rate and a variable rate which increases with feed intake. Simulation analyses of growing lambs fed at two intakes indicated that protein turnover accounted for 19 per cent of the increment in energy expenditure due to increased nutrient input while increased costs in Na^+ and K^+ transport accounted for 39 per cent and fat turnover and accretion for 25 per cent, leaving 17 per cent of the increment in energy expenditure unaccounted for.

Chemical Composition of Rations

Another source of variance in energy expenditures in maintenance and production is attributable to differences in diet and products of digestion absorbed and actually available to the animal. A number of examples illustrating this source of variance were developed by Baldwin *et al.* (1980). Products from fermentation of soluble carbohydrates, starch, hemicellulose and cellulose differ considerably as they are fermented by different microbes (Murphy *et al.*, 1982). In general, greater doses of energy in the form of methane and heat are associated with fermentation of the principal components (fiber) of forages as compared with concentrates (starch). Also, the fermentation products available to the animal from forages as compared to concentrates are used for maintenance and production at lesser efficiencies by the animal (see example comparisons below). In order to compensate for these, dietary energy density adjustment factors were adopted for use in both ME and NE systems as discussed above. This approach helped a great deal but does not completely eliminate the problem, leading Baldwin *et al.* (1977, 1987*b*) to suggest that the use of mechanistic elements which explicitly describe specific chemical entities in diets, the biochemistry of their fermentation and pathways of utilization of the several fermentation products by the animal would better accommodate these sources of variance.

An example of this approach is the following comparison of the use of glucose vs. acetate as a source of ATP for use in maintenance. Because the biochemical efficiency of trapping acetate energy in ATP is less than that for glucose, heat production when acetate is used as an energy source is 118 per cent of the heat produced when the same amount of ATP is produced for physiological work functions by the oxidation of glucose. As previously discussed, an extreme example of differences in energy losses in nutrient use by the animal is that dietary fat is incorporated into body or milk fat at an efficiency of 94–97 per cent while the efficiency of conversion of dietary carbohydrate, via fermentation to acetate, to body fat is only 60–65 per cent. The importance of this difference is becoming more prominent as the use of protected fats and whole oilseeds in cattle rations has increased markedly in recent years. The models of Gill *et al.* (1989*a, b*) and Baldwin *et al.* (1987*a, b*) are examples of the application of this concept and illustrate the utility of using our knowledge of underlying function, intermediary metabolism in this case, to address problems encountered with models used in support of animal agriculture.

Models for Meeting the Protein Requirements of Ruminants

As for energy, early models used to calculate the protein requirements of ruminants were based upon regression equations of the form:

$$\mathrm{DP}(\mathrm{kg}\,W)^{-0{\cdot}75} = 1{\cdot}6 + 5{\cdot}2G \tag{11}$$

where DP denotes digestible protein (g) required per day per kg $W^{0{\cdot}75}$ and G is the rate of gain (kg day^{-1}) (Preston, 1982). Depending upon the selection of data sets used for evaluation, R^2 values for such equations range from 0·48 to 0·94. Important sources of variance noted were DP/DE ratio in the diet and age or composition of gain. Therefore, appropriate adjustments were made when developing table recommendations for DP or CP (crude protein) in cattle rations.

Subsequent to this, it was recognized that the types of protein supplements used in ruminant rations can have marked effects upon performance. Specifically, it was noted that microbial growth in the rumen is not adequate to support maximal productivity. Thus, diets containing only proteins which are highly available and do not escape degradation in the rumen support neither maximum growth rates nor maximum milk production. Many conventional diets have this property such that delivery of proteins or amino acids to the lower gut to supplement those available from the rumen microbes can increase production 20–30 per cent. Thus, it became apparent that the physical properties of dietary proteins must be balanced to provide adequate nitrogen and protein (or amino acids and peptides) to support optimal microbial growth and, at the same time, adequate escape of diet protein from the rumen to support optimal or maximal rates of growth and milk production.

Widespread recognition of this problem led, in the 1970s and early 1980s, to extensive research and the development of at least ten models of the static, factorial type and several dynamic models directed to the study and formulation of feeding recommendations for protein (Owens, 1982). Most of these static, factorial models were based upon the set of concepts depicted in Fig. 1. Dietary crude protein is partitioned into true protein that is degraded to ammonia and fatty acids by the rumen microbes, true protein which escapes rumen degradation and non-protein nitrogen (including urea) which acts as a source of ammonia for rumen microbial growth. Microbial growth was usually calculated as a constant function of DE or ME intake or organic matter digestion in the rumen. In some cases, provisions for salivary protein and urea entry to the rumen were included.

In these models, microbial protein and dietary protein which escapes from the rumen are digested at differing efficiencies in the lower gut yielding an estimate of amino acid availability to the animal. Absorbed amino acids are in turn partitioned to support maintenance and productive functions including metabolic fecal protein losses, protein accretion, milk production and amino acid degradation and gluconeogenesis. Parameter values used to implement the equations vary somewhat among models, but these differences tend to cancel out because of the factorial nature of the computation and the fact that nitrogen balance data were used to adjust many of the key parameters in the models (Owens, 1982). Objectives in these models are to ensure that the nitrogen requirements of the rumen microbes are met and, further, that the amino acids requirements of the animal are met economically by a combination of rumen microbial growth and protein escaping rumen degradation. Key elements of these systems are accurate definition of the portions of rumen degradable and undegradable protein in feedstuffs and, particularly, protein supplements and estimates of the extent of degradation of limiting amino acids by the animal.

Rumen:

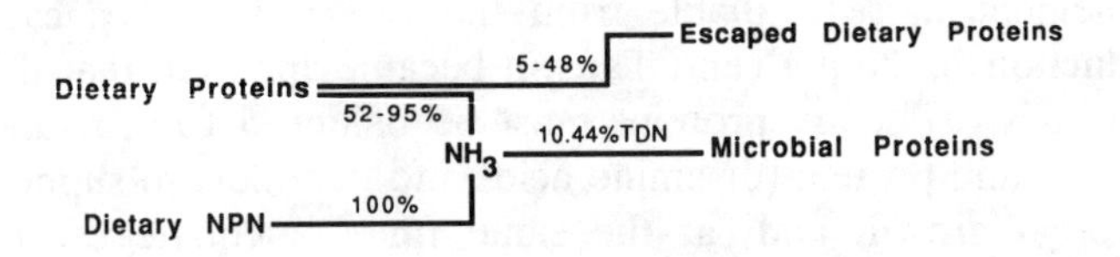

Intestine:

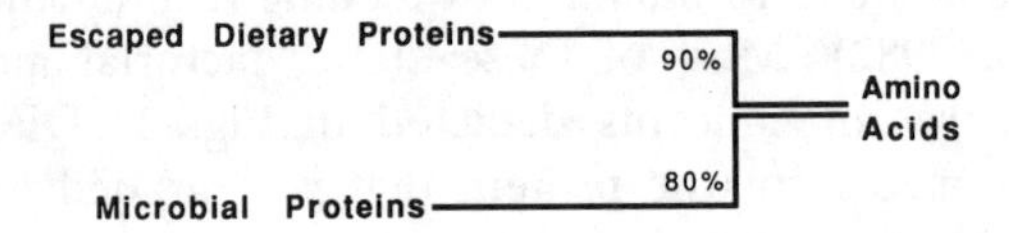

Metabolism:

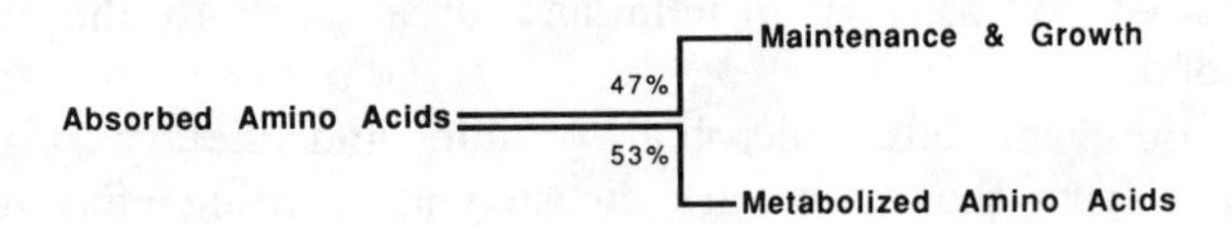

Fig. 1. Diagram of the metabolizable protein system. (Adapted from Trenkle, 1982.)

These early modelling efforts led to the development of a sophisticated static, factorial model by Fox *et al.* (1988) which is used along with the current NRC system for computation of the nutrient requirements of dairy cattle. Separate provisions for the digestion of soluble carbohydrates, starch and fiber, and the microbes fermenting each type of carbohydrate are included in the model. Microbial growth yields are allowed to vary as a function of growth rate (driven by carbohydrate availability), microbial maintenance requirements, theoretical maximum microbial growth yield and pH. Also, growth rates or yields of microbes fermenting non-structural carbohydrates are a function of the availability of peptides and amino acids arising from the degradation of true protein in the rumen. In our view, addition of provisions to accommodate variance in microbial protein availability due to diet composition, fermentation rates and protein availability in the rumen adds considerable utility to this model.

At least two dynamic models which contain many of the same provisions as the static model of Fox *et al.* (1988) are available (Black *et al.*, 1980–1; Baldwin *et al.* 1977, 1987*b*, *c*). These offer the advantage of considering the effects of changes which occur over time in composition of gain, e.g. protein requirements, rates of nitrogen cycling from the animal to the rumen and variations in rumen microbial growth yields.

Growth Elements

Several problems discussed above indicate a need for improved representations of growth processes. These include effects of previous and current plane of nutrition on relative organ weights and apparent maintenance requirements, composition and, as a result, efficiency of gain and effects of breed on fasting heat production and composition of gain at a given weight. In formulating tabular values for the current NRC system for beef cattle, separate regression equations for large- and medium-framed bulls, steers and heifers were used. Also, effects of rate of gain on composition of gain and of previous plane of nutrition were accommodated in the tables. Fox and Black (1984) introduced empirically derived correction factors for environment and breed (dairy vs. beef breeds) effects upon FHP, corrections for mature body weight or frame size (breed), corrections for sex effects upon composition of gain, and effects of antibiotics, monensin and lasalocid upon NE_m and NE_g values assigned to feeds. Many of these were incorporated into the Fox *et al.* (1988) and current NRC models (1984).

Oltjen *et al.* (1986*a*) observed that these adjustment factors are adequate in well-defined systems similar to those from which the data used to formulate the equations were derived but that significant errors were associated with application of these to cattle fed low-quality diets, cattle varying in condition scores when entering the feeding period and exotic breeds of cattle whose use in the beef industry is increasing. Based upon these observations, Oltjen *et al.* (1986*a*) developed a dynamic, largely mechanistic growth element for use along with the NRC system to improve estimates of rate and composition of gain. The mechanistic growth element was based upon three concepts regarding growth in mammalian systems which had been evaluated for adequacy in simulations of the growth of nine tissue and organ systems in mice, rats, sheep and pigs by Baldwin and Black (1978). The concepts were: (1) the primary genetic determinant of organ size is the final DNA content of the organ in mature, normally grown individuals of the species; (2) each unit of DNA specifies, on a genetically defined basis for each tissue and each species, the ultimate formation of a specific amount of cell material, and nutritional and physiological status determines whether this target is achieved; and (3) the specific activities of enzymes responsible for tissue growth vary exponentially with organ size and their kinetic properties are relatively constant across species.

Dynamic equations depicting rates of accretion of body DNA (dDNA/d*t*) in grams per day and protein (dPROT/d*t*) in kilograms per day derived by Oltjen *et al.* (1986*a*) from these concepts were:

$$\mathrm{dDNA/d}t = K1(\mathrm{DNAMX} - \mathrm{DNA})\mathrm{NUT1} \quad (12)$$

$$\mathrm{dPROT/d}t = \mathrm{SYNTHESIS} - \mathrm{DEGRADATION} \quad (13)$$

$$\mathrm{SYNTHESIS} = K2(\mathrm{DNA}^{0\cdot 73})\mathrm{NUT2} \quad (14)$$

$$\mathrm{DEGRADATION} = K3(\mathrm{PROT}^{0\cdot 73}) \quad (15)$$

where $K1$, $K2$ and $K3$ are rate constants, NUT1 and NUT2 are nutritional constants set to 1 to simulate normal ME intakes over time, and DNAMX is the normal DNA content of a steer with a mature body weight of 750 kg. For animals of different mature sizes, rate constants are adjusted by the size scaling factor of Taylor (1980); DNAMX values were adjusted to account for differences in mature body weights; and $K2$ was increased 4 per cent for cattle with hormone adjuvants.

The model has been evaluated first with respect to its ability to predict growth and composition of steers as affected by nutrition, initial condition, frame size and use of growth promotants. Second, its performance was compared against other systems including the original NRC (California) net energy and Fox and Black (1984) systems described above. Using two independent data sets, the model predicted empty body weight and fat content with standard deviations of 14 and 10 kg, respectively (Oltjen *et al.*, 1986*b*), a result considerably better than that obtained using the other systems (Table 1); body weight estimates were less biased and standard errors of prediction were as good or better for both data used in model development and the independent data sets (Oltjen *et al.*, 1986*a*). In particular, frame size effects were best accounted for by the model. This demonstrates

TABLE 1
Residual Final Body Weight (RBW) and its Standard Deviation (SD) in a Comparison of Four Feeding Systems

Data	*System*[a]							
	Model		*NRC*		*FOX*		*ARC*	
	RBW[b]	*SD*	*RBW*	*SD*	*RBW*	*SD*	*RBW*	*SD*
EBWM[b] > 751	−1·2	22·8	34·0	27·9	−24·5	40·9	−82·3	21·6
EBWM < 751	−0·3	16·7	22·4	17·3	2·8	31·4	−31·3	14·7
−Implant[c]	−2·9	14·3	−10·4	18·6	−22·1	25·2	−31·3	14·7
+Implant[d]	1·6	18·2	22·4	17·3	20·4	22·1	−10·3	19·7

[a] Systems compared are those of Oltjen *et al.* (1986*b*) (*Model*), the National Research Council (1984) (*NRC*), Fox and Black (1977) (*FOX*) and the Agriculture Research Council (1980) (*ARC*).
[b] RBW is residual body weight (observed–predicted). EBWM is empty body weight at maturity. Data for steers fed monensin are not included in NRC or ARC comparisons.
[c] Steers were not implanted.
[d] Steers were implanted.

that a mechanistic model based on fundamental biological concepts can accommodate greater variance observed in production situations and possesses greater predictive accuracy and wider application than empirically derived models such as those currently available and in use.

In practice, the Oltjen *et al.* (1986*b*) dynamic model is especially useful in situations where ongoing collection of feed intake and/or body weight data is routine, such as in a feedlot. The incoming data can be used to recalibrate model predictions using adaptive filtering techniques, allowing increasingly more precise projections of performance as a feeding period progresses (Oltjen and Owens, 1987). Also, the model is sensitive to conditions which affect composition of gain, thereby providing more robust predictions of gain. This is particularly true over short feeding intervals, whereas the previously described systems require longer feeding periods for systematic errors to cancel out (Oltjen and Garrett, 1988).

Similar concepts were incorporated into a pig growth model (Black *et al.*, 1986); a steer growth model which incorporated additional mechanistic elements to accommodate and study variance due to digestive and animal metabolic effects on growth and energy expenditures (DiMarco and Baldwin, 1989; DiMarco *et al.*, 1989); and a lactating dairy cow model (Baldwin *et al.*, 1987*c*) with encouraging results. In the pig model, a term for maximum body nitrogen at maturity replaced the DNAMX term.

Lactation

Probably the most common equation form used to simulate lactation curves of dairy cattle is the gamma function (Wood, 1969):

$$\gamma_n = an^b E^{-cn} \tag{16}$$

where γ_n is expected milk yield in the nth week of lactation and a, b and c are statistically deduced parameters. Experience has shown that this equation can accurately define the lactation curves of groups of cattle. However, significant errors are evident, particularly during early lactation, when lactation curves are fitted to data on individual cattle (Broster and Thomas, 1984). The gamma function is used in several feeding systems to anticipate milk yields and by dairy herd improvement associations to compute full lactation performance estimates based upon incomplete records. Broster and Thomas (1984) identified a number of limitations in this approach including the fact that it is 'retrospective rather than predictive' and that there are no provisions for accommodation of differing planes of nutrition, particularly 'the cumulative and residual effects from periods of over or underfeeding'.

In recognition of these limitations, Neal and Thornley (1983) developed a mechanistic model of the mammary gland to simulate the lactation curves of cattle. This model included provisions for specification of an individual animal's genetic (mammary cell numbers) capacity to produce milk, for hormonal effects upon mammary development and function, for variations in nutrient availability to the mammary glands and for feedback inhibition due to milk retained in the udder. Baldwin *et al.* (1987*c*) included these mammary elements along with mechanistic elements describing digestion, metabolism and growth in a model of lactating dairy cows. Inputs to the digestion element are feed intake, the chemical and physical properties of the feeds and nitrogen cycling into the rumen from the animal. Outputs are rates of absorption of nutrients by the animal. Provisions are made for variation in rates of protein degradation in the rumen and effects of ammonia, amino acids, peptides and maintenance requirements on microbial growth. The metabolic elements include functions for regulation of patterns of nutrient utilization among tissues, e.g. rates of nutrient oxidation and use for biosynthesis within tissues. Requirements for nutrient oxidation are computed as the sum of energy requirements for specific functions (such as ion transport, protein synthesis, service functions, metabolism) rather than the sum of estimated maintenance plus productive energy needs. The growth and lactation elements provide driving functions for the partitioning of amino acids between body and milk protein synthesis and trace tissue reserves over time. This latter provision allows simulation of effects of the previous plane of nutrition upon current and subsequent performance.

Results of a model evaluation of effects of feeding strategy upon lactational performance are presented in Fig. 2. Two diets differing in CP content (15 per cent CP = –M and 18 per cent CP = –H), where the increment in protein content was provided as fishmeal, were fed at two levels (low = L– and high = H–). Treatments were changed after 12 weeks of lactation. Results compare very favorably with those presented by Broster and Broster (1984) and Broster and Thomas (1984) and address their comments (above) regarding the utility of

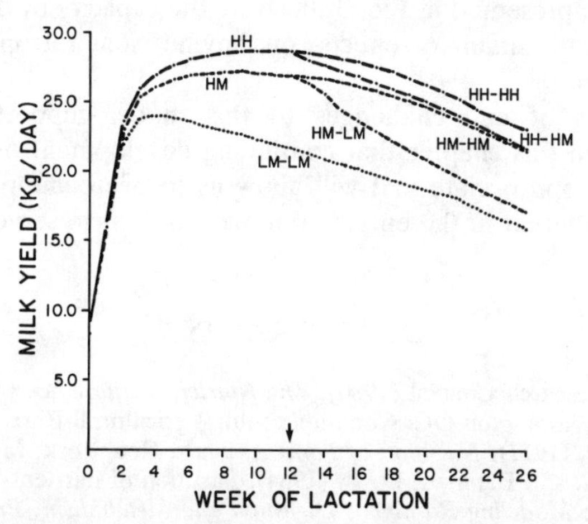

Fig. 2. Effect of different feeding strategies upon lactational performance. L indicates a feeding rate of 5 kg day^{-1} plus 1 kg feed per 3 kg milk averaged over the previous three weeks. H indicates a feeding rate of 8 kg feed day^{-1} plus 1 kg feed per 3 kg average daily milk yield. M indicates a standard forage:concentrate (50:50) ration at 15 per cent crude protein. H represents the standard ration adjusted to 18 per cent crude protein with fishmeal. Changeovers of diet and feeding strategy occurred at the twelfth week of lactation. (Adapted from Baldwin and Bauman, 1984.)

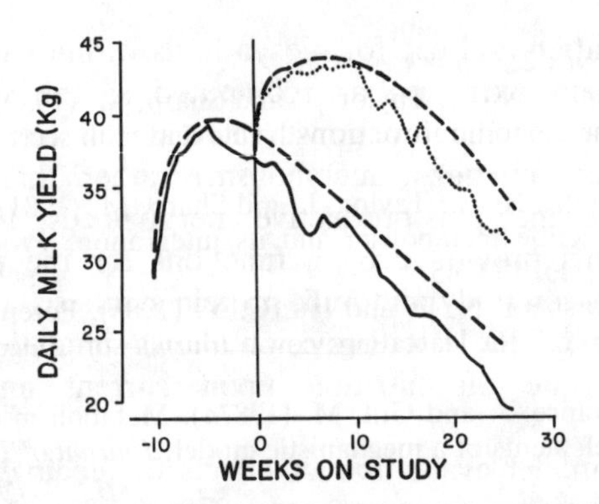

Fig. 3. Simulated response to bovine somatotrophin (BST) administration. Solid and dotted lines depict lactation curves of dairy cows treated with placebo and BST, respectively. Dashed lines are simulated responses. (Adapted from Baldwin and Bauman, 1984.)

previous, empirical systems. Since model outputs are very detailed, causal relationships producing these responses can be readily traced and quantified.

The results presented in Fig. 3 illustrate the capacity of the lactating cow model to simulate effects of bovine somatotrophin (BST) administration.

Many tests of and challenges to this model support the view emphasized in this chapter that continuing development of mechanistic models is appropriate and will allow us to overcome many of the limitations inherent in the empirical models that have served us in the past.

BIBLIOGRAPHY

Agriculture Research Council (1980). *The Nutrient Requirements of Ruminant Livestock*. Washington DC: Commonwealth Agricultural Bureaux, ARC.

Armsby, H. P. (1917). *Nutrition of Farm Animals*. New York: Macmillan.

Baldwin, R. L. and Bauman, D. E. (1984). Partition of nutrients in lactation, pp. 80–8 in: *Modeling Ruminant Digestion and Metabolism, Proceedings of the Second International Workshop,* eds R. L. Baldwin and A. C. Bywater. Davis: University of California.

Baldwin, R. L. and Black, J. L. (1978). Simulation of the effects of nutritional and physiological status on the growth of mammalian tissues: Description and evaluation of a computer program, *CSIRO Animal Research Laboratory Technology Paper* No. 6.

Baldwin, R. L., Koong, L. J. and Ulyatt, M. J. (1977). A dynamic model of ruminant digestion for evaluation of factors effecting nutritive value, *Agricultural Systems* **2,** 255–88.

Baldwin, R. L., Smith, N. E., Taylor, J. and Sharp, M. (1980). Manipulating metabolic parameters to improve growth rate and milk secretion, *Journal of Animal Science,* **51,** 1416–28.

Baldwin, R. L., Smith, N. E., Taylor, J. and Sharp, M. (1981). The synthesis of models to describe metabolism and its integration, *Proceedings of the Nutrition Society,* **40,** 139–45.

Baldwin, R. L., Forsberg, N. E. and Hu, C. Y. (1985). Potential for altering energy partition on the lactating cow, *Journal of Dairy Science,* **68,** 3394–402.

Baldwin, R. L., France, J. and Gill, M. (1987*a*). Metabolism of the lactating cow. I. Animal elements of a mechanistic model, *Journal of Dairy Research,* **54,** 77–105.

Baldwin, R. L., Thornley, J. H. M. and Beever, D. E. (1987*b*). Metabolism of the lactating cow. II. Digestive elements of a mechanistic model, *Journal of Dairy Research,* **54,** 107–31.

Baldwin, R. L., France, J., Beever, D. E., Gill, M. and Thornley, J. H. M. (1987*c*). Metabolism of the lactating cow. III. Properties of mechanistic models suitable for evaluation of energetic relationships and factors involved in the partition of nutrients, *Journal of Dairy Research,* **54,** 133–45.

Bernier, J. F., Calvert, C. C., Famula, T. R. and Baldwin, R. L. (1986). Maintenance energy requirements and net energetic efficiency in mice with a major gene for rapid postweaning growth, *Journal of Nutrition,* **116,** 419–28.

Black, J. L., Beever, D. E., Faichney, G. J., Howarth, B. R. and Graham, N. M. (1980–1). Simulation of the effects of rumen function on the flow of nutrients from the stomach of sheep: Part 1—Description of a computer program, *Agricultural Science,* **6,** 195–219.

Black, J. L., Campbell, R. G., Williams, I. H., James, K. J. and Davies, G. T. (1986). Simulation of energy and amino acid utilisation in the pig, *Research and Development in Agriculture,* **3**(3), 121–45.

Brody, S. (1945). *Bioenergetics and Growth.* New York: Reinhold Publishing.

Broster, W. H. and Broster, V. J. (1984). Reviews of the progress of dairy science: Long term effects of plane of nutrition performance of the dairy cow, *Journal of Dairy Research,* **51,** 149–96.

Broster, W. H. and Thomas, C. (1984). The influence of level and pattern of concentrate input on milk output, *Journal of Dairy Research,* **51,** 49–69.

DiMarco, O. N. and Baldwin, R. L. (1989). Implementation and evaluation of a steer growth model, *Agricultural Systems,* **29,** 247–65.

DiMarco, O. N., Baldwin, R. L. and Calvert, C. C. (1989). Simulation of DNA, protein and fat accretion in growing steers, *Agricultural Systems,* **29,** 21–34.

Fox, D. G. and Black, J. R. (1977). A system for predicting performance of growing and finishing beef cattle, *Michigan Agricultural Experiment Station Research Report,* **328,** 141.

Fox, D. G. and Black, J. R. (1984). A system for predicting body composition and performance of growing cattle, *Journal of Animal Science,* **58,** 725.

Fox, D. G., Sniffen, C. J., O'Connor, J. D., Russell, J. B. and Van Soest, P. J. (1988). *The Cornell Net Carbohydrate and Protein System for Evaluating Cattle Diets*. Ithaca, NY: New York State College of Agriculture and Life Sciences.

Gill, M., France, J., Summers, M., McBride, B. W. and Milligan, L. P. (1989*a*). Mathematical integration of protein metabolism in growing lambs, *Journal of Nutrition,* **119,** 1269–86.

Gill, M., France, J., Summers, M., McBride, B. W. and Milligan, L. P. (1989*b*). Simulation of the energy costs associated with protein turnover and Na^+K^+-transport in growing lambs, *Journal of Nutrition,* **119,** 1287–99.

Graham, N. M., Black, J. L., Faichney, G. J. and Arnold, G. W. (1976). Simulation of growth and production of sheep—model 1: A computer program to estimate energy and nitrogen utilisation, body composition and empty liveweight change, day by day for sheep at any age, *Agricultural Systems,* **1,** 113–38.

Kleiber, M. (1932). Body size and metabolism, *Hilgardia,* **6,** 315–53.

Koong, L. J., Ferrell, C. L. and Nienaber, J. A. (1985). Assessments of interrelationships among levels of intake and production, organ size and fasting heat production in growing animals, *Journal of Nutrition,* **115,** 1383–90.

Lusk, G. (1926). A history of metabolism, pp. 3–78 in: *Endocrinology and Metabolism, Vol. 3,* ed. L. F. Barker. New York: D. Appleton & Co.

Milligan, L. P. and McBride, B. W. (1985). Shifts in animal energy requirements across physiological and alimentation states, *Journal of Nutrition,* **115,** 1374–82.

Murphy, M. R., Baldwin, R. L. and Koong, L. J. (1982). Estimation of stoichiometric parameters for rumen fermentation of roughage and concentrate diets, *Journal of Animal Science,* **55,** 411–21.

National Research Council (1984). *Nutrition Requirements of Beef Cattle.* Washington, DC: National Academy Press.

National Research Council (1989). *Nutrition Requirements of Dairy Cattle.* Washington, DC: National Academy Press.

Neal, H. D. and Thornley, J. H. M. (1983). The lactation curve in cattle: A mathematical model of the mammary gland, *Journal of Agricultural Science, Cambridge,* **101,** 389–400.

Oltjen, J. W. and Garrett, W. N. (1988). Effects of body weight, frame size and rate of gain on the composition of gain of beef steers, *Journal of Animal Sciences,* **65,** 1732–8.

Oltjen, J. W. and Owens, F. N. (1987). Beef cattle feed intake and growth: Empirical Bayes derivation of the Kalman filter applied to a nonlinear dynamic model, *Journal of Animal Science,* **65,** 1362–70.

Oltjen, J. W., Bywater, A. C. and Baldwin, R. L. (1986*a*). Development of a dynamic model of beef cattle growth and composition, *Journal of Animal Science,* **62,** 86–97.

Oltjen, J. W., Bywater, A. C. and Baldwin, R. L. (1986*b*). Evaluation of a model of beef cattle growth and composition, *Journal of Animal Science,* **62,** 98–108.

Owens, F. N. (1982). *Protein Requirements for Cattle: Symposium*. Stillwater: Division of Agriculture, Oklahoma State University.

Preston, R. L. (1982). Empirical value of crude protein systems for feedlot cattle, pp. 201–17 in: *Protein Requirements for Cattle: Symposium,* ed. F. N. Owens. Stillwater: Division of Agriculture, Oklahoma State University.

Roux, C. Z. and Meissner, H. H. (1984). Growth and feed intake patterns: 1. The derived theory, pp. 672–90 in: *Herbivore Nutrition,* ed. F. M. C. Gilchrist and R. I. Mackie. Johannesburg, South Africa: The Science Press.

Summers, M., McBride, B. W. and Milligan, L. P. (1988). Components of basal energy expenditure, pp. 257–85 in: *Comparative Aspects of Physiology of Digestion in Ruminants,* eds A. Dobson and M. J. Dobson. Ithaca, NY: Comstock.

Taylor, St. C. S. (1980). Genetic size-scaling rules in animal growth, *Animal Production,* **30,** 161.

Thonney, M. L., Touchberry, R. W., Goodrich, R. D. and Meinske, J. C. (1976). Intraspecies relationship between fasting heat production and body weight: A re-evaluation of 75-W, *Journal of Animal Science,* **43,** 692–704.

Thornley, J. H. M. and France, J. (1984). Role of modelling in animal production research and extension work, pp. 4–9 in: *Modeling Ruminant Digestion and Metabolism, Proceedings of the Second International Workshop,* eds R. L. Baldwin and A. C. Bywater. Davis: University of California Press.

Trenkle, A. (1982). The metabolizable protein feeding standard, pp. 238–44 in: *Protein Requirements for Cattle: Symposium,* ed. F. N. Owens. Stillwater: Division of Agriculture, Oklahoma State University.

Wood, P. D. P. (1969). Factors affecting the shape of the lactation curve in cattle, *Animal Production,* **11,** 307–16.

2

Plant–Animal Interactions in Northern Temperate Sown Grasslands and Semi-natural Vegetation

T. J. MAXWELL

The Macaulay Land Use Research Institute, Aberdeen, UK

INTRODUCTION

The simplicity of the title of this paper belies the biological complexity of the grazing process and its effect on plant productivity and sward composition. The title is a short-hand way of describing the effects of defoliation, excretal return and treading by grazing animals on the physiology, morphology and chemistry of the plant, and the effect of available biomass, physical configuration and quality of the forage on the behaviour and forage intake of grazing animals.

In the preface to his book *Grassland Ecology,* Spedding (1971) explained some of the potential benefits afforded by adopting an ecological approach to an understanding of grassland. Tansley (1946) had pointed out that 'anything like a complete study of the ecology of a plant community necessarily includes a study of the animals living or feeding upon it', but, while ecological principles are required to elucidate the inter-relationships among climate, soils, plants and animals, Professor Spedding also recognized that a quantitative approach was indispensable to a fuller understanding of grassland ecology. He has devoted much of his subsequent professional effort to developing a systems and modelling approach to research on this subject.

In the UK, ruminant livestock production systems are based on the utilization of sown and indigenous grasslands, and research has been concerned primarily with understanding how the efficient use of these grazing resources can be achieved. The objective has been more often

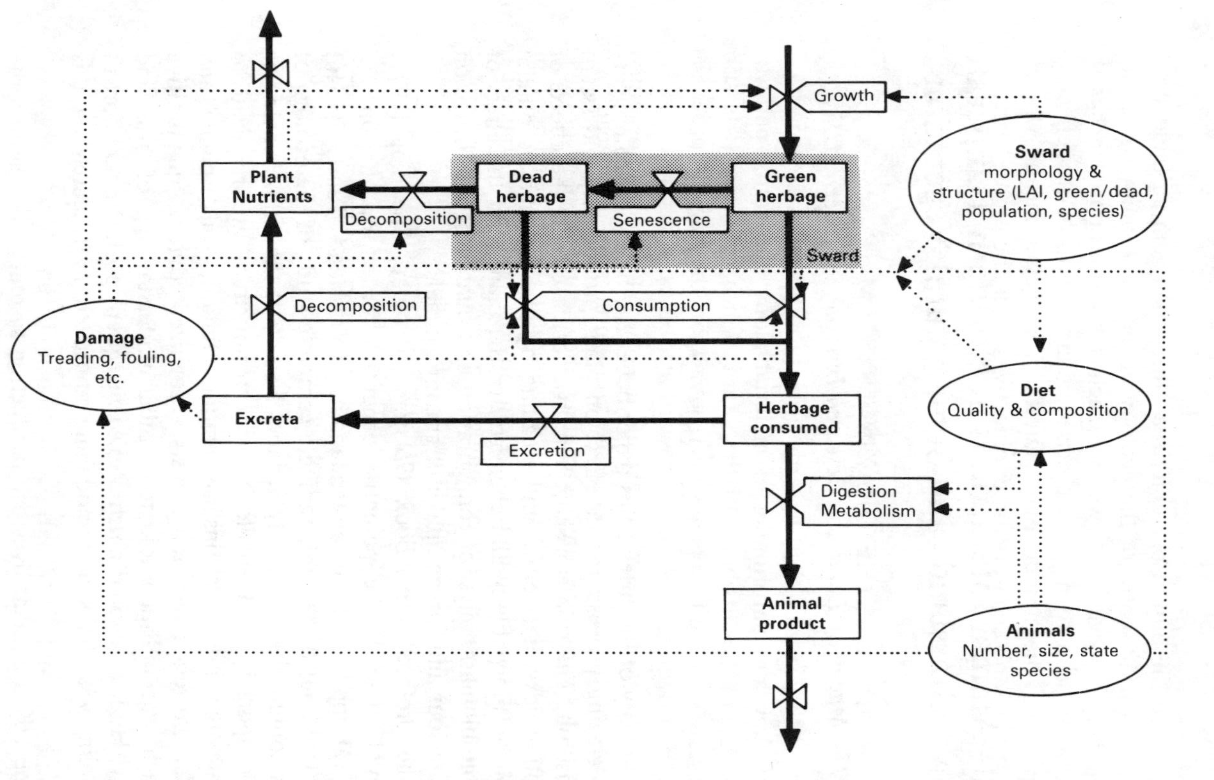

Fig. 1. Plant and animal inter-relationships in grazing systems (adapted from Hodgson, 1985). The bold lines and arrows indicate the flow and partitioning of material through the system; the factors which are altered by grazing management decisions and which influence the various rate processes (growth, consumption, etc.) are circled; dotted lines are used to indicate impacts and inter-relationships.

to maximize the output per unit area rather than gain an understanding, in a holistic sense, of the ecology of the plant–animal interface. This paper, however, reviews how such a holistic approach has the potential to lead to a more precise understanding of the effects of manipulating the animal–plant interface, and the means whereby agricultural and conservation objectives can be more precisely achieved.

THE GRAZING SYSTEM

Grant and Maxwell (1988) have described the grazing system in terms of the flow and partitioning of material along alternative pathways (Fig. 1), and the attributes of the sward, which collectively describe the sward state and have a direct impact on the various processes. Herbage growth is influenced by sward attributes, which in turn affect light interception, ability to expand leaves, and the photosynthetic activities of those leaves. Herbage consumption is influenced by a variety of sward attributes, including the nature, (green or dead, species composition, digestibility), the amount (height, biomass) and structure of the plant material. Demand for food is determined by several animal factors (stock numbers, species, size and physiological state of the animal) which interact with the sward in ways which influence diet selection and quality, and which together determine the rate of consumption and, thus, nutrient intake. The effect of stock numbers on herbage production and animal performance is largely mediated through change in sward state. Sward state is influenced by variation in soils, climate and topography. In semi-natural vegetation these local factors, together with management history, contribute to the unique character of the vegetation.

EXPERIMENTAL APPROACH

Historically, and particularly with respect to agricultural objectives, research of the grazing system has centred on the use of stocking rate as the determining treatment variable. In such experiments a set number of animals are equated with a unit area of land and effects are measured in terms of output per unit area of land. There are serious limitations to this approach in attempting to understand the underlying biology of the 'grazing system'; as Hodgson (1985) has pointed out, it tends to ignore the biological variation in space and time of grazing systems. For example, between-year variation in climate can result in a significant variation in herbage production of sown swards. It is virtually impossible in these circumstances to characterize and extrapolate the resulting relationships among stocking rate, fertilizer use and output where stocking rate has been used as a determining treatment variable. The problems are exacerbated on semi-natural vegetation where the unfenced hill grazings are vegetatively heterogeneous and unique in character (Miles, 1985). This is not to argue that stocking rate studies have no place; much can be learned if experiments are well-designed and properly monitored.

An alternative approach is based on determining the effect that sward conditions have on the components of the system. Grazing experiments in which treatments are based on the direct manipulation of sward characteristics have been found to lead to real progress in understanding (Hodgson and Grant, 1981). However, the main areas of research interest differ between intensive grazing systems on sown pastures and more extensive systems on semi-natural vegetation. In the former case food quality is invariably high and it is the quantity of herbage which determines the level of output. Consequently, interest has focused on the ways in which sward biomass affects the rate of herbage growth and consumption, losses to senescence and decomposition, and animal responses.

On semi-natural vegetation the overall quantity of herbage is rarely limiting; the factors which require consideration are life form, seasonality of growth, and quality and quantity of herbage produced by particular plant species. Variation in these attributes between species results in uneven distribution of grazing activities in time and space, which have significant consequences for vegetation dynamics (Miles, 1979). An ability to predict the consequences of altering grazing management for vegetation succession is, therefore, of primary importance. It is inevitable that those who use, or have an interest in, the semi-natural grazings of the UK have different views as to what constitutes a satisfactory species composition. There is, nevertheless, a common interest in acquiring sufficient understanding of plant–animal relationships for objective criteria to be drawn up to formulate management prescriptions which achieve specific management objectives (Sydes and Miller, 1988).

THE SOWN SWARD

Over the last 10 to 15 years, three objectives of research have been

(a) to obtain estimates of growth, senescence and net production from measurements of tiller population density and of tissue turnover on sample tillers and stolons,
(b) to measure tissue fluxes using carbon exchange procedures,
(c) to construct conceptual and mathematical models to explore the dynamics of herbage production and methods of utilization.

Growth

The annual variation in the temperature, rainfall and radiation are largely responsible for the seasonal pattern of growth and for setting the upper limit of annual grass production. These factors are largely outside the control of management.

In the spring, when the grass plant is in its reproductive phase of growth, it more effectively exploits its environment than later in the year when it becomes vegetative. Where nutrient supply is non-limiting, the calculated potential yield from theoretical models and extrapolation from field experiments estimate potential harvestable levels of production of between 27 and 30 t DM ha^{-1} (Leafe, 1978). The national average grass crop yield, excluding semi-natural grazings, is thought to be about 6 t DM ha^{-1} (Robson, 1981). However, measurement of yields under grazing conditions, where the defoliation of plant tissue has a feed back effect on subsequent growth and partitioning of nutrients, has required a more detailed approach and an acceptance of the principle that the grass crop represents the sum of the yields of individual plants; also it is the morphological and physiological characteristics of the individual plant which determines the way in which the grazed crop is managed to achieve desired rates of growth.

In this approach the tiller, composed of a tubular structure made of the leaf sheaths surmounted by the leaf blades, is adopted as the fundamental unit of the grass crop. Grass growth then can be regarded as the regular production of new leaves from a tiller, composed of three to four leaves, and the production of secondary tillers from its base. As a new leaf emerges from the tiller senescence begins at the tip of the oldest leaf, transfer of nutrients to other parts of the tiller takes place and the oldest leaf dies. The length of life of a leaf in the summer is about four weeks and it is twice as long in the winter.

The disposition of leaves on a tiller affects its ability to capture sunlight. Even though swards may have a similar leaf area index (LAI), those which have leaves which are long and erect will allow better light penetration, therefore more light is absorbed by the canopy and as a consequence there will be a greater capacity for growth than in a prostrate-leaved crop. However, the leaves of an intensively and continuously grazed crop are subject to frequent defoliation; only short leaves or partially defoliated leaves make up such a crop and their total area is relatively small. The significance of these observations is that the morphological characteristics of a crop relative to the purpose for which the grass is grown will determine its growth potential (Wilson, 1981).

The rate of tillering is much influenced by the light reaching the tiller base (Langer, 1963); it also differs between grass species. Tiller development can be stimulated by defoliation, though the number of new tillers initiated and the rate at which they develop are also likely to be affected by sward conditions both before and after defoliation. Grant *et al.* (1981) also found that the rate at which tiller populations diverged in swards maintained at different levels of herbage mass was much lower before the summer solstice, when light conditions were steadily improving, than after the solstice when they were steadily declining, though all the swards were tillering actively. The loss of tillers is increased by both severe shading and defoliation.

The types of plant required for grazed, sown pastures are those that have characteristics which, on the one hand, contribute to herbage production and sward longevity under various systems of grazing management and, on the other, contribute to high intake potential and efficient utilization (Hodgson, 1982). There are grass varieties which fulfil some of these requirements, but it is doubtful on the basis of present evidence whether there are any which fulfil them all. Nevertheless, while it may be true that present opportunities and selection criteria for breeding plants particularly suited to grazing are limited, it would be misleading to suggest that existing recommended varieties seriously constrain levels of output from systems of animal production, but many of the varieties have been developed with the assumption that plant nutrient inputs, particularly nitrogen, are non-limiting.

White Clover

With an increasing desire to limit nitrogen use there is much interest in finding ways in which clover can make a greater contribution to the nitrogen economy of the sward (Ryden, 1983) and directly in terms of the nutrition of the grazing animal (Thomson, 1984). The ability of white clover to remain in a grass sward is dependent upon its ability to compete with grass. In the early stages of establishment, before nodulation has taken place, clover will compete with grass for mineral nitrogen. It may continue to do so, but as nodulation and nitrogen fixation takes place the transfer of mineral nitrogen to the soil pool will occur. Under most conditions nitrogen transfer occurs through the grazing animal and the decay of white clover herbage, roots and/or nodules. The amount of nitrogen transferred from clover to associated grasses and its effect on growth varies with season, species of grass, the proportion of white clover in the sward, age of sward and type of management. It is generally accepted that applications of fertilizer nitrogen beyond 40 kg N ha^{-1} will reduce the proportion of clover in the sward. Ryegrass dominance in UK grass swards is partly explained by the regular use of fertilizer nitrogen and the balance between nitrogen fixation by the clover, subsequent decay of clover and release of nitrogen which is then taken up by ryegrass leading to increased dominance during these periods. Superimposed on these factors in grazed swards are the influence of grazing animals which exert their effects through excretal returns and treading. Large quantities of nitrogen (300–600 kg N ha^{-1}) are returned in urine patches and only a small proportion of the sward is affected at any one time; thus differences in the amounts of plant available nitrogen (3–120 kg NH_4^+-N ha^{-1}) are created within the sward. This results in the development of spatial heterogeneity of clover within the sward (i.e. the presence or absence of clover in areas of the sward) which has feed-back effects on diet selection by animals (Grant and Marriott, 1989).

Differences in foliage architecture between white clover and grass are also important. Haynes (1981) concluded that legumes are generally more prone to be shaded by competitors than grasses, and compete poorly for light because they generally possess horizontally-inclined leaves, while in grasses light is distributed more evenly throughout the canopy because of their more upright leaves. In grazed swards, however, these effects are likely to be modified by defoliation. As grazing takes place in the upper horizons of a sward a higher proportion of white clover leaf compared to grass leaf is removed (Milne *et al.*, 1982), which puts the clover plant at a relative disadvantage in capturing sunlight. To some extent the severity of the effect will depend on the variety of white clover used, i.e. whether it is an erect or prostrate type (Harkness *et al.*, 1970), and also on the relative vigour with which it is growing. Though the white clover plant replaces leaves relatively quickly, it is likely that the frequency of defoliation will have a significant impact on its ability to compete and, indeed, to survive in swards that are continuously grazed. Defoliation of clover by grazing temporarily reduces nitrogen fixation; Moustapha *et al.* (1969) and Ryle *et al.* (1985) have both shown that it takes at least six days before the original level of nitrogen fixation is re-attained after defoliation and the rate of photosynthesis is reduced by 90%. Curll and Wilkins (1982) have argued, therefore, that under repeated defoliation, as in high stocking rate continuous grazing systems in which the interval between defoliation is less than seven to eight days, nitrogen fixation and sward productivity and percentage clover content is likely to be considerably reduced unless a rest period is introduced.

Recent evidence (Grant and Marriott, 1989) also suggests that stolon burial during the growing season, as a result of earthworm activity, and poaching, particularly combined with leaf removal, adversely affects clover growth. The increase in branching rate on the free-growing tips of buried main stolons, though strategically beneficial, is too small to fully compensate for the reduced branching rate and increased losses by death of branch stolons.

A factor of key importance in determining the balance between perennial ryegrass and white clover is the response of the two species to temperature. Perennial ryegrass has a lower temperature threshold for growth than white clover (Williams, 1970), and thus has an advantage in growth over white clover both early and late in the season. Once white clover is actively growing however, its greater specific leaf area coupled with its pattern of leaf expansion compared with perennial ryegrass (Parsons *et al.*, 1990) means that the advantage in growth changes to white clover.

The early work of Jones (1933), together with more recent work (Laws and Newton, 1987; Grant and Barthram, 1988) indicates that the white clover content of grass clover swards is reduced in laxly compared with closely grazed swards in spring. Shade reduces clover leaf appearance and branching rates (Solangaarachichi and Harper, 1987; Davies and Evans, 1990). In spring the lower temperature threshold for grass allows grass height increment and shading to the detriment of white clover. Later in the year, however, the clover petioles are able to extend during periods of herbage accumulation and thus maintain the clover laminae near the surface of the sward (Dennis *et al.*, 1984; Grant and Barthram, 1990). Davies and Evans

(1990) have shown that if the clover laminae are intact and unshaded then clover branching rate is unaffected by shade at the level of the stolons. The provision of a rest period from grazing in late spring or summer, when clover has a growth advantage over grass, should lead to improvements in clover content and, indeed, incorporation of a conservation cut in grass–clover grazing systems has frequently been shown to lead to improvements in the clover content of the herbage (Wolton *et al.*, 1970; Curll *et al.*, 1985; Sheldrick *et al.*, 1987). Wolton *et al.*, (1970) showed that the increases in amounts of clover occurred both in the period of growth up to the date of harvest and in the subsequent regrowth period.

The evidence to date would suggest that, while the presence of clover in the sward will be influenced by soil nutrient status, temperature, moisture and direct trampling effects, these all interact with the effects of grazing and the severity of defoliation. Continuous grazing systems have been shown to be feasible in some circumstances (Davies *et al.*, 1989) and rotational systems in others (Newton *et al.*, 1985). The view that a rest period should be incorporated into a grass–clover system has a strong physiological basis and it is the timing of the rest period relative to the respective growth cycles of grass and clover that will be crucial in determining the maximum benefit achieved (Grant and Barthrum, 1990).

Effects of Grazing Animal on Sown Sward

Grazed perennial ryegrass-dominant swards maintained in a steady state develop characteristic differences in tiller size and number; at any one height, however, the tiller numbers vary in response to seasonal variation in incoming radiation and effects of management. Changing from tall to short swards, and *vice versa* (Bircham and Hodgson, 1981; Grant and King, 1984) have shown that the growth rates of the individual tillers adjust more rapidly to the new conditions than do the tiller population densities and that a reduction in tiller numbers as herbage accumulates is a more rapid process than the build-up of numbers following grazing down. The increase in growth rate per tiller and in canopy photosynthesis on short swards which are released from grazing is not matched by an immediate increase in rate of leaf death. This increase is delayed for approximately 3 leaf-appearance intervals which represents the time taken for the new larger leaves to reach the oldest (i.e. senescing) leaf category (Parsons and Robson, 1982; Grant and King, 1984).

Effects of management on gross photosynthesis of the leaf canopy mainly arise as a result of changes in LAI. However, changes in photosynthetic potential can arise also, either because of changes in canopy structure which affect light interception or because of changes in photosynthetic efficiency of the leaves. Parsons and Leafe (1981) have shown that the photosynthesis of expanding leaves and young fully expanded leaves is three to four times greater than old leaves, and some 76% of the total photosynthetic uptake of the canopy of an intensively grazed sward is contributed by the expanding leaves and fully expanded leaves, even though these leaf categories represent only 40% of the leaf and sheath areas. In a laxly grazed sward there is tissue loss due to death and decay and a considerable proportion of the photosynthetic uptake is inevitably lost. This does not occur to anything like the same extent in more intensively grazed swards because tissue is harvested before it reaches this stage. Also, the energy maintenance requirement of an intensively grazed sward maintained at low herbage mass is less. Theoretically, at least, efficient utilization by the grazing animal of the energy captured by the sward requires that leaves should not reach the stage of senescence since animals are likely to consume only small amounts of such material, which therefore inevitably dies and decays.

These results suggest that little advantage can be gained in rotational grazing systems by extending the complete cycle beyond 3 leaf-appearance intervals, and also that the grazing down period should be short relative to the regrowth period.

Maxwell *et al.* (1988) have drawn the conclusion based on field experiments (Grant *et al.*, 1988; King *et al.*, 1988) and modelling (Parsons and Penning, 1988; Parsons *et al.*, 1988) that intermittent grazing systems result in alternating periods of increased and decreased growth rates compared with continuous stocking systems and that the average net herbage production is similar for both systems, and for intermittent cycles is maximized when the average LAI for the complete cycle is of the same low order as that at which net production is optimized in continuous systems. Thus, provided continuous stocking systems incorporate a means of maintaining sward conditions within the optimum height range, there is little to be gained by using more labour-intensive and costly rotational grazing systems. These conclusions are in accord with the appraisal of Ernst *et al.* (1980), who conducted a critical review of the literature on animal and sward production under continuous and rotational managements, and also

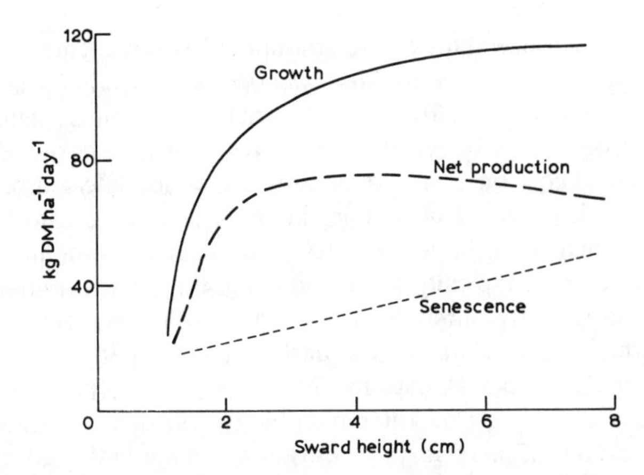

Fig. 2. The relationship between sward height and the rates of growth, senescence and net production (kg DM ha^{-1} day^{-1}) on a range of continuously stocked swards maintained at various heights (adapted from Bircham and Hodgson, 1983).

that of Lantinga (1985), who used a combination of field data and pre-predictive models of herbage production under grazing for his evaluation.

The importance of all of this with respect to the management of grazed sown swards is highlighted by the results of a series of studies where continuous variable stocking has been used to maintain a range of steady-state sward conditions on perennial ryegrass and on perennial ryegrass–white clover swards (Bircham and Hodgson, 1983; Grant *et al.*, 1983; Parsons *et al.*, 1983*a,b*). Gross tissue production has been shown to increase in a positive and curvilinear manner in relation to increasing sward surface height and leaf area index (LAI) while tissue death increases linearly. The net production, i.e. utilized herbage, though maximized on swards maintained at a sward surface height of around 3·5 cm (LAI 2–3), is relatively insensitive to variations in height between 2·5 cm and 8 cm (Fig. 2).

The important conclusion that can be drawn from these studies is that in general terms the control of growth and utilization of sown swards can largely be achieved through the control of sward height.

Distribution of Growth

The seasonal distribution of net herbage production under continuous grazing, where sward surface heights are maintained constant, as predicted using a simulation model (Lantinga, 1985), as estimated from direct measurements of the amounts of herbage eaten (Treacher, *et al.*, 1986), and by stock numbers needed to maintain constant sward conditions (HFRO, 1986), is characterized by a single mid-summer peak. Nitrogen can be used to increase growth without danger of deleterious effects on sward structure and hence late-season production potential, provided swards are maintained at a height of 4–5 cm.

The seasonal distribution of dry matter production in rotational grazing systems is typically skew, with highest net accumulation rates in late spring–early summer (Lantinga, 1985). The amount of dry matter which accumulates in the early regrowth periods has a major impact on this seasonal pattern (Morrison, 1977), presumably as a consequence of effects on tiller phenology and density. Clearly, care must be exercised in relation to duration of regrowth and pattern of nitrogen fertilizer application if the benefits of applying the nitrogen fertilizer are to be optimized and yet mid-summer depressions in herbage production avoided (Prins *et al.*, 1980; Morrison, 1980).

SEMI-NATURAL VEGETATION

The range of plants in semi-natural communities is very wide indeed and much has been written about the ways in which soils, climate and topography interact to influence botanical composition and plant succession (e.g. Miles, 1985). There is also a substantial body of information concerning the biomass production of semi-natural plant communities (Heal and Perkins, 1978; Hodgson and Grant, 1981).

The difficulties of collecting and interpreting herbage production data from an indigenous sward have been outlined by Grant and Hodgson (1986). Plants of different life form, for example, grasses, herbaceous perennials and dwarf shrubs, are often intermixed. The position of their growth points varies from ground level or below (grasses or sedges) to short tips (the heath species). Some species have a continuous turnover of tissue with growth and senescence proceeding simultaneously (e.g. many grasses and *Eriophorum vaginatum*), other species are more seasonal, leaf or shoot production being restricted to a few months in summer and separated in time from the period of senescence (e.g. *Molinia caerulea, Trichophorum ceaspitosum, Juncus acutiflorus*). Species also vary in the recycling of nutrients. The growth rate of those with no specialized storage organs is dependent upon leaf area and canopy photosynthesis, while plants with swollen shoot bases, extensive rhizomes or storage of carbohydrate in woody stems or roots have the ability to use these reserves as a contribution to leaf production; these reserves also affect the ability of the plant to withstand defoliation in one season and to survive into the next. These kinds of variation in the basic biology of species severely limits the usefulness of estimates of herbage production based on a sequence of quadrat cuts. Nevertheless, carefully interpreted information on changes in plant biomass can be useful, particularly in response curve experiments, where trends are followed over time and related to changes in plant cover consequent upon, for example, a stepwise increase in stocking rate (e.g. Grant *et al.*, 1985). Studies of this kind have been important in determining both a conceptual and practical approach to the management of semi-natural vegetation to meet agricultural objectives (Eadie and Maxwell, 1975).

However, extrapolation from these kinds of studies, except in a very general sense, is not easily achieved because at other sites the patterns and levels of utilization of particular species may differ because of differences in species composition and/or species balance, or because of differences in growth rate attributable to soils, climate and topography.

It is argued that to make progress in understanding the animal–plant interface within the context of semi-natural vegetation it is necessary to collect more precise quantitative data of plant responses to given levels of defoliation, bearing in mind that the value attached to the dry matter contribution of plants differs because of the widely differing life forms, seasonality of growth and nutritional value of the various hill species.

This approach has been used for heather (Grant *et al.*, 1978; Milne *et al.*, 1979; Grant *et al.*, 1982) and more recently for *Nardus* and *Molinia*. For heather it has been established that the level at which the current season's shoots are utilized is fundamental in determining the long term productivity of heather communities. It is also clear that damage by overgrazing varies with season and increases with age of stand. In general terms it has been shown that levels of utilization of 40% of the current season's shoots in the summer and autumn can be achieved without reducing plant productivity (Grant *et al.*, 1982).

The approach used to investigate effects of grazing on *Nardus* grassland is similar to that used for sown swards—that is, in relation to sward height. However, as *Nardus* is not a preferred species, it is the height of the preferred between-tussock grasses which, it has been found, can be used to predict the level of *Nardus* utilization, changes in sward morphology and species composition over time. Early results suggest (Grant, S. A., unpublished) that leaf extension rates of *Nardus* tillers are reduced by cattle and goat grazing at an intertussock grazing height of 4·5 cm with the tussocky nature of the sward disappearing within two years. However, the trends over time in plant species composition suggest that there is a limit to the extent to which *Nardus* will be reduced by cattle after an initial rapid decline in *Nardus* cover and an increase in broad-leaved grass cover. By contrast *Nardus* increased under sheep grazing.

The work on *Molinia* is concerned with establishing how much of the leaf of the plant can be used if *Molinia* dominance is to be retained. Cutting and grazing studies have shown that *Molinia* is very sensitive to defoliation, as little as 33% leaf removal leading to reduced early leaf extension the following year. The relative importance of recycled energy and nutrients is being investigated to identify amounts of annual leaf removal which would not depress the production capacity of the plant.

These examples illustrate the kind of information which it is necessary to acquire in order to find a simple criterion which has physiological relevance and which can be used to predict the level of utilization and consequently the level of productivity of the main plant species.

FORAGING BEHAVIOUR

The foraging strategy of a grazing herbivore defines the way in which food is selected. The community, plant or plant part actually selected is determined by an array of plant and animal factors but primarily depends on what is on offer and the requirements of the animal.

There are a number of animal factors which determine foraging strategy (Table 1) (Gordon and Iason, 1989). Many of these are adaptations to a particular diet which have arisen across evolutionary time and/or as a result of selection imposed by man. These factors need to be considered as constraints which limit how the grazing ruminant can be expected to perform or behave. Possibly the most all-pervasive variable is the body size of the ruminant. The absolute food requirement, the inability for fine-grained food selection and food retention time in the gut all increase with body weight (Gordon

TABLE 1
Characteristics of Animals with Different Foraging Strategies (Gordon and Iason, 1989)

Foraging strategy	*Grazer*	*Mixed (Preference for grass)*	*Mixed (Preference for browse)*	*Mixed*
Animal	Cow	Sheep	Goat	Deer
Muzzle width	Wide	Wide	Narrow	Intermediate
(relative to body size)	Flat	Flat	Pointed	Flat
Degree of selectivity	Low	High	High	Intermediate
Rumen size (relative to body size)	Large	Large	Small	Intermediate
Digestive Capability				
Fibrous grasses	High	High	Low	Intermediate
Trees, shrubs	Low	Intermediate	High	High

and Illius 1988; Illius and Gordon 1989). In between-species comparisons, larger animals ingest food of lower digestibility or protein content (Hodgson, 1981*b*; Owen-Smith, 1982). The size and shape of the animal's mouth also affects its ability to select discrete food items from a heterogeneous array of plant material (Gordon and Illius, 1988). More selective feeding habits are associated with a narrower and more pointed dental arcade which is better able to select food items from surrounding material of lower quality.

As well as having different abilities to select their diet, comparative studies between ruminant species show differences in their ability to digest the same forages. Large animals utilize poorer quality foods better because they can eat more and retain the cell wall fraction for more extensive digestion in their relatively larger guts. A computer model developed by Illius and Gordon (1990) predicts that small animals obtain about 80% of digested energy from the cell soluble fraction, compared with about 50% in the larger animals. It also quantifies the disadvantages of large body size under conditions of resource depletion, where low biomass limits intake rate. A model has also been developed which shows how dental morphology and vegetation structure interact to determine the weight and composition of plant material which can be removed by a single bite. The results indicate that short swards impose greater limitations on food intake by larger animals than smaller animals. Illius and Gordon (1990) have extended the use of the models to investigate the implications for the relative value of two alternative food patches, for example, a tall mature patch of herbage and a short vegetative sward. They conclude that choice not only depends on the properties of the food but also on the size of the animal confronted by the alternatives on offer. The concept of niche selection can be developed from these considerations.

In recent years there has been an increasing attempt to develop from these ideas a unifying theory to describe the foraging behaviour of ruminants. The evolutionary adaptation to a particular diet by animals has led to the development of the theory of optimal foraging which presupposes that animals make feeding choices in order to maximize their intakes of energy or some limiting nutrient while minimizing energetic costs of foraging and exposure to predators (Stephen and Krebs, 1987). Pyke (1984) and Stephens and Krebs (1987) both highlight the large number of assumptions underpinning optimal foraging theory for which validation is required. Gray (1986) has questioned the value of the optimality approach to the study of foraging behaviour and suggested that an epigenetic (i.e. development from the simple to the complex) approach integrating morphological, physiological and behavioural processes as being potentially more useful.

Illius and Gordon (1990) concluded after their review of some of the constraints on diet selection and foraging behaviour in mammalian herbivores that these largely arise from variables associated with body size and perceptual faculties. They propose that mammalian herbivores are sensitive to differences between homogeneous patches of food but within a patch show only limited ability in the short term to evaluate alternatives either prior to or following consumption. Without further understanding of these processes it would be premature to attempt a definition of optimal strategies for mammalian herbivores and they believe that the development of a 'procedural' theory of foraging offers a more realistic approach for the present. Grant and Maxwell (1988) came to a similar conclusion. Their approach was to suggest that it is first necessary to gain a sufficient understanding of the mechanics of diet selection and nutrient intake to predict how animals will respond in this highly heterogeneous environment.

Sward Factors Influencing Animal Herbage Intake—The Sown Sward

Evidence to date suggests that the individual performance of animals grazing a sown perennial ryegrass sward is determined by the quantity and quality of ingested herbage, with the rate of intake being determined primarily by sward canopy characteristics. On temperate swards, as the height of the sward being grazed decreases so also does the amount of herbage consumed per bite (Hodgson, 1982). Sheep respond to a low sward height by taking more bites per minute and grazing for longer, but these compensatory measures cannot counter the reduced intake per bite; the net effect is an overall fall in daily intake (Hodgson & Grant, 1985; Penning, 1986). The relationship between rate of food intake and food availability is known as the functional response of the herbivore (Noy-Meir, 1975), which in grazing mammals is usually an asymptotic relationship with intake declining once plant biomass falls below a critical level. Despite the general applicability of this relationship, there is always a large proportion of the total variability in intake, about 50%, which remains unexplained (Short, 1986; Milne *et al.*, 1988).

One of the reasons for this is that the distribution of plant biomass within the sward has a strong effect on the bite size and thus on intake rate. The depth of the leafy layer in the canopy and the bulk density within this are significant determinants of bite size (Burliston and Hodgson, 1985; Illius and Gordon, 1987). On sown swards containing little clover there is little evidence for differential diet selection in cattle and sheep although sheep tend to consume a diet containing a slightly higher level of organic matter digestibility (Forbes and Hodgson, 1985), but these swards have low plant heterogeneity and lack opportunities for differential diet selection or niche use by cattle and sheep.

Nevertheless, differences in animal trampling, dunging and urination patterns do give rise to a heterogeneous array of patches varying in height, nutrient content and digestibility and, in swards with higher clover content, species composition (Marriott *et al.*, 1987). This heterogeneity offers the opportunity for the animal to exercise some foraging choices to increase their relative nutrient intake above the mean value of the sward (Illius, 1986). On sown swards the composition of the diet is often considered to be a consequence of largely unselective foraging behaviour in relation to a stratified distribution of plant material in the sward. For example, when present in abundance the proportion of clover in the diet was found to be simply related to the proportion within the grazed horizon (Milne *et al.*, 1982). In general, the amount of clover in the diet is greater than that in the herbage available when there is an abundance of pasture, whereas on short swards the proportion of clover in the diet tends to be less (Milne *et al.*, 1982; Curll *et al.*, 1985). It is also possible that there is concentration of grazing on clover-rich areas of the sward (Clark and Harris, 1985).

The study of mixed grass–clover swards where there are patches differing in clover content, biomass and excretal contamination, and of varying size and distribution offers important opportunities of examining the significance of heterogeneity in influencing the choice of grazing site, bite selection and intake rate. In these relatively simple swards it is possible to examine more critically some aspects of foraging theory (Illius and Gordon, 1990) which can lead potentially to a better understanding of the ways in which animals graze mixed sown swards and heterogeneous semi-natural vegetation.

Sward Factors Influencing Animal Herbage Intake—Semi-natural Vegetation

The aim of studies of foraging behaviour on semi-natural grassland is to gain sufficient understanding of the mechanics of diet selection and nutrient intake to predict how animals will respond in a heterogeneous environment. Two phases of study are required. First, a detailed monitoring of grazing choice on sward conditions at a series of points in time which allows hypotheses to be established, is needed. Second, studies with experimentally manipulated swards are required in order to test hypotheses that have been developed. These latter studies need to explore single and mixed animal species, grazing on vegetation ranging from a single community type to combinations of more complex mixes.

Hunter (1954; 1962) took a descriptive approach and investigated the grazing behaviour of sheep in relation to their preference for particular vegetation communities. He examined the distribution of grazing sheep on selected parts of his study area in relation to the productive capacity of the vegetation and to its botanical and potential nutritive value. Grazing intensity was found to be most closely related to potential nutritive value. He differentiated between the preferred vegetation types, which occurred on the better or ‘mull’ soils, and the unpreferred types on the more acid, peaty or ‘mor’ soils. Heather communities belong to the latter group and have a characteristically low comparative grazing intensity, with the main period of use being in winter, but with a secondary peak around about July. It was suggested that availability of grazing on the preferred vegetation types was a major determinant of the pattern and level of use of the unpreferred types (Hunter 1954; 1962).

An experimental approach to testing hypotheses about foraging has been used to only a limited extent. With respect to different patches of heather, age of stand affected both the seasonal pattern and the overall level of use. Grant and Hunter (1968) studied interactions between grazing and burning, in which one quarter of each of a series of grazed plots was burned at two-yearly intervals to provide four ages of heather. Sheep preferentially grazed the younger heather stands at all times of the year. This preference for younger heather was more strongly developed at lower animal grazing pressures where old heather was able to increase rapidly in height and cover. At the highest grazing pressure all the heather was maintained short and there was little preference for age of stand.

Heather has poor feeding value and the sheep's diet must contain a proportion of grass to achieve acceptable production levels (Maxwell *et al.*, 1986). The factors influencing grazing choice between sown grass and heather have been studied in two experiments (Milne and Grant, 1987). In the first of these, on plots containing different proportions by area of grass, changes over time in grass biomass, heather utilization and grass:heather proportions in the diet were monitored during two- to four-week grazing periods in May, July and October. In a follow-up experiment, heather utilization during May to August was monitored on plots with 30% by area of grass, but with a grass biomass maintained at 400, 950 or 1500 kg DM ha^{-1} by adjusting stock numbers as necessary.

These studies (Milne and Grant, 1987) (Fig. 3) indicated that grass biomass was a major determinant of heather utilization thus confirming the suggestion of Hunter (1962). However, this relationship was modified by season, for example, sheep grazed less heather early in the season when grass, unlike heather, was growing rapidly. When grass biomass was controlled there was an inverse relationship between current growth of grass and heather utilization. The proportion of grass influenced first the number of grazing days and second the amount of grass remaining at the same level of heather utilization.

A further example illustrating the usefulness of treatments based on the control of a particular sward characteristic is given by comparison of different foraging strategies by different animal species on *Nardus stricta* grassland. Such studies have the added advantage that grazing choices of different animal species can be compared against a common

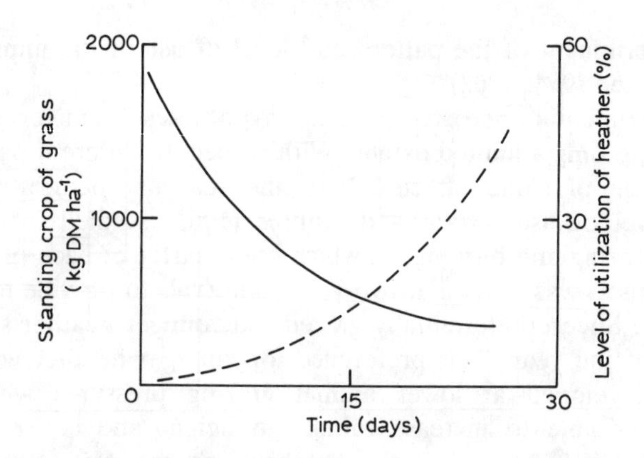

Fig. 3. Relationship between grass and heather in diet.

criterion. Cattle and goats consistently ingest more *Nardus* than do sheep and the proportions of *Nardus* in the diet of both sheep and cattle are inversely and curvilinearly related to the height or biomass of the preferred between-tussock grasses (Grant, S. A., unpublished).

Armstrong and Hodgson (1985) have reviewed the nutritional implications of grazing semi-natural vegetation and report on a series of indoor studies to establish relationships between herbage digestibility and herbage intake and field studies which have measured variations in diet quality, and on herbage intake due to animal species, plant community and season (Forbes, 1982). The results are summarized in Fig. 4 and demonstrate the annual range in herbage digestibility selected by sheep and cattle grazing five hill plant communities under sown swards. It shows the high potential digestibility and therefore the intake that can be obtained from the indigenous grass communities, and confirms in more general terms the importance of advancing maturity in limiting the herbage intake of animals grazing hill areas (Eadie, 1967). There are substantial differences in nutritive value and herbage intake between (a) the group of grass communities and (b) the dwarf shrub communities for both sheep and cattle. The very low levels of digestibility and intake recorded for cattle on the *Calluna* and blanket bog communities, coupled with observations of treading and grazing damage to dwarf shrubs in both cases, are indicative of the unsuitability of these communities for cattle grazing even at the relatively low levels of utilization.

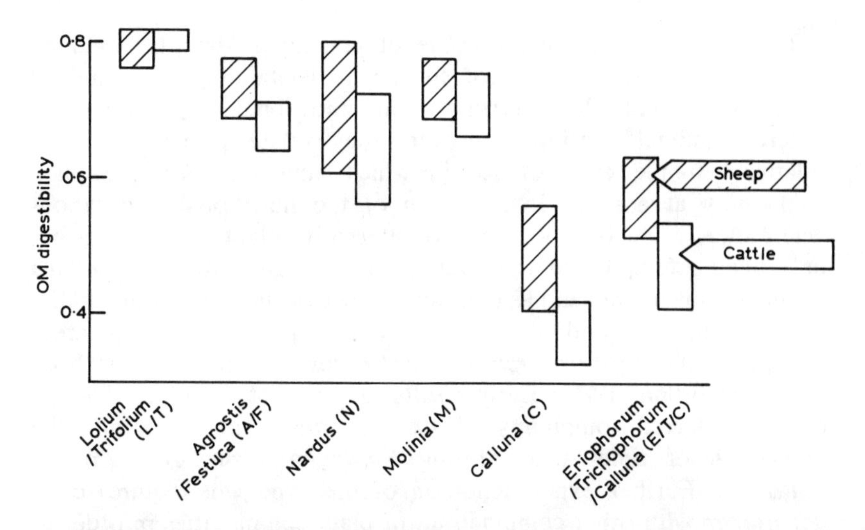

Fig. 4. The annual range in the digestibility of the diet (OMD) selected by sheep and cattle grazing five hill plant communities and a sown sward.

In general, sheep tend to maintain diet digestibility at the expense of rate of intake, whereas cattle tend to maintain rate of intake at the expense of digestibility. Cattle and sheep diets show greater differences in plant morphology than in plant species composition, the most characteristic difference being in the relative proportions of live leaf on the one hand and seed head and stem on the other (Grant and Hodgson, 1986). However, differences between animal species in diet digestibility were usually smaller than differences in botanical composition.

The contrasts between the morphological composition of sheep and cattle diets suggest the potential for using cattle to control excess herbage in summer and so improve grazing conditions for sheep, but this conclusion cannot necessarily be applied across all communities. The observations on the *Nardus*-dominated community provided the basis for suggesting that cattle are able to maintain nutrient intakes better than are sheep under comparable management constraints whilst at the same time exerting a greater impact on the undesirable *Nardus* component of the sward. This is in part a reflection of the fact that the digestibility of young *Nardus stricta* is roughly similar to that of other fine-leaved species, despite its unacceptability to sheep in particular (Armstrong and Hodgson, 1985).

These studies indicate the value of grazing treatments based on sward criteria as a means of further understanding plant–animal relationships and characterizing their nutritional implications for different animal species. However, they relate primarily to the relatively simple circumstance in which animals graze one plant community at a time. Investigation of the nutritional implications where there is a choice between two alternative plant communities has only been attempted in the context of associated areas of *Calluna* heath and sown grassland as indicated earlier (Milne and Grant, 1987). However, more recently the foraging behaviour of sheep grazing areas of contrasting *Agrostis–Festuca* and *Nardus* communities has been studied (Gordon, 1989). Early results suggest that the grazing height of the preferred community (*Agrostis–Festuca*) influences both the proportion of time spent grazing *Nardus* and the extent of its utilization. Further experimentation of this type will require to be undertaken with other combinations of plant communities in order to develop a body of information which should ultimately lead to an understanding of the criteria which determine the choice of grazing site and selection of the herbage within a site on semi-natural vegetation.

SYSTEMS IMPLICATIONS

Sown Swards

For the simple sown sward dominated by grass species, it has been possible, because of the close correlation that exists between sward characteristics and sward height, to use the latter parameter as a means of describing the response in herbage intake by the grazing animal to changing sward conditions. However, it is important to recognize that the level of intake at a given sward height varies with animal species. On grass swards the intake of ewes reaches a maximum at 5–6 cm while the intake of cows is maximized at 8–10 cm (Fig. 5). Grass–clover swards give a similar pattern of response (Orr *et al.*, 1987).

These relationships suggest that by managing sward conditions control can be exercized also over the level of animal production. At sward heights of 3, 6 and 9 cm, lactating ewes lost 1·2, 0·5 and 0·6 units of condition score during lactation (Orr *et al.*, 1987). Similar patterns of response have been observed in lambs with growth rates

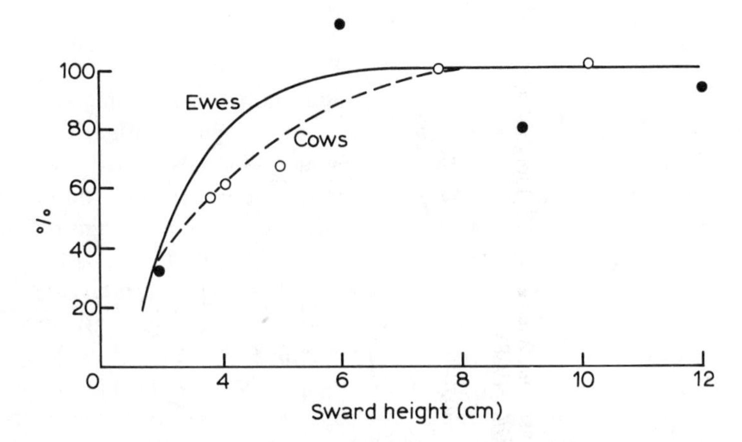

Fig. 5. Relative herbage intakes of lactating ewes and cows at different sward heights (adapted from Penning, 1986; Wright, 1986).

being depressed in swards with less than 6 cm. Recent evidence suggests that change in sward height may also be important particularly in weaned lambs. Compared to static sward heights of 3 or 6 cm, a sward increasing from 3 to 6 cm increased liveweight gain in weaned lambs four-fold, while a sward decreasing in height from 6 to 3 cm depressed liveweight gain (Doney *et al.*, 1987).

As with sheep, the liveweight gain of cows increases as the maintained sward height is increased, but reaches a maximum at 8 to 10 cm and thereafter shows a slight decline. Calf liveweight gain shows a similar pattern of response but as spring-born calves become older, and less dependent on milk and more dependent on herbage, their growth rates become more sensitive to sward height.

Under cattle grazing, when sward height increases beyond 8 cm grazing ceases to be uniformly distributed over the sward, but is concentrated on some areas with others becoming rejected. These ungrazed areas develop flower stems and digestibility declines. At 8 cm the area rejected soon levels off at less than 10%, while at 12 cm a large proportion of the area is rejected for much of the grazing season. Later in the season cattle are forced to graze this material of poor quality resulting in the decline of cow liveweight gain at sward heights in excess of 8–10 cm.

It follows from the analysis of animal–plant interactions of the sown sward that there is the potential for controlling both the growth and

TABLE 2
Output per Hectare and Proportion of Area Closed in Production Systems Where Sward Heights are Maintained Within Specified Limits (Maxwell *et al.*, 1988)

Year	*Weight of animal weaned* ($kg\ ha^{-1}$)			*Area closed for first silage cut* (%)			*Cattle turn-out date*
	Sheep[a]		*Cattle*[b]	*Sheep*[a]		*Cattle*[b]	
	Cheviot	*Beulah*		*Cheviot*	*Beulah*		
1985	630	635	538	38	33	50	16 May
1986	663	640	542	4	0	56	26 May
1987	618	609	582	42	42	18	8 May

[a] Sheep are stocked at 20 ewes ha^{-1} with 1·2 lambs ewe^{-1} and 200 kg N ha^{-1} $annum^{-1}$.
[b] Cattle are stocked at 2·5 cows and calves ha^{-1} and 250 kg N ha^{-1} $annum^{-1}$.

utilization of the herbage, and animal performance by managing swards to a specific sward height. Seasonal sward height profiles have been suggested for both sheep and cattle production systems (Maxwell and Treacher, 1987; Wright, 1988). Experiments to test their validity have shown that consistent levels of animal performance and output per unit area can be achieved from varying levels of herbage production (Table 2).

Sward height in these experiments has been controlled by closing a proportion of the grazed area for conservation. The area closed is inversely related to stocking rate and is subject to variation between years due to differences, in particular, in the time when growth starts in the spring. In the cattle experiments, turnout is delayed until swards reach their specified heights and so differences in spring growth are also reflected in turnout data and hence in winter feed requirements.

However, the relationship between stocking rate and the amount of grass conserved, for example, for sheep (Fig. 6) allows a choice of stocking rate (×) to be made at which it is possible to provide the feed required for the winter from conserved grass while maintaining consistent levels of animal performance and high levels of grass utilization during the grazing period. The variation in the position and slope of these relationships arises from different responses of herbage production to nitrogen fertilizer use. While experience suggests that it

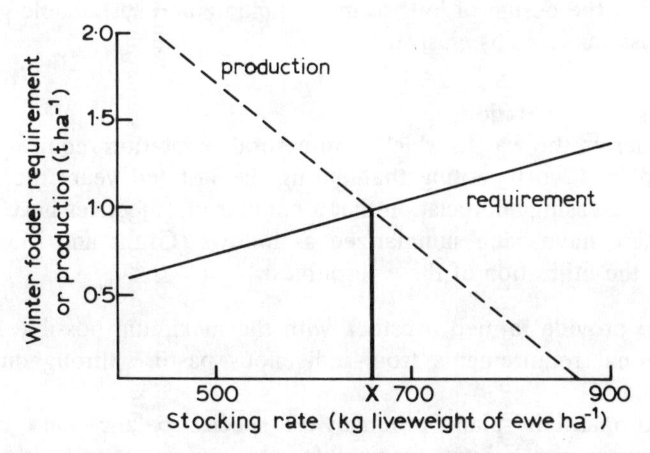

Fig. 6. Relationships between ewe stocking rate and winter fodder requirement (1·5 kg DM kg^{-1} liveweight) and winter fodder production. Based on Brecon Cheviot ewes with 100 kg N ha^{-1} $annum^{-1}$.

is possible to choose a stocking rate which usually provides the feed required for the winter, a greater degree of precision could be achieved if the amount of nitrogen supplied could be much more closely related to nitrogen demand. There is a real need for the identification and description of key soil and pasture parameters which would allow a more precise prediction of herbage production response to applied nitrogen (Whitehead, 1986).

In reviewing the use of white clover–perennial ryegrass mixtures for systems of sheep production, Newton and Davies (1987) conclude that herbage production on grass–clover swards with fertilizer nitrogen is similar to those of pure grass swards receiving up to 200 kg N ha^{-1} (Hoglund *et al.*, 1979; Morrison *et al.*, 1985) though there is a wide range of response depending upon soil, climate and amount of white clover present (Cowling, 1982). Carcass output (Newton *et al.*, 1985), and proportion of lambs finished (MLC, 1986) often favours rotational grazing systems but it is also acknowledged that continuous grazing systems in which clover content is maintained at a relatively high level are also successful. Davies *et al.*, (1989) showed that similar lamb production per hectare can be achieved from grass–clover swards and from grass-only swards grazed to the sown sward height profile but with a reduction in fertilizer input of 125 kg N ha^{-1}. The major limitation to the wider use of white clover-based systems is their apparent unpredictability and until the reasons for this are fully understood the design of long-term predictable and sustainable grass–clover systems remains elusive.

Semi-natural Vegetation

In considering the way in which semi-natural vegetation requires to be managed it is worth noting that during the last ten years there has been an increasing appreciation that a number of objectives have to be met. These have been summarized as follows (Grant and Maxwell, 1988) in the utilization of these resources:

(i) to provide farmed livestock with the maximum possible nutritional requirements from indigenous pasture throughout the year;
(ii) to maintain plant productivity, species balance and create appropriate habitats for wildlife;
(iii) to sustain or develop amenity areas and scenically attractive landscape.

Inherent in these objectives is the need to manage the grazing of these resources at levels of herbage utilization which either sustain the existing balance between species or are designed to manipulate a shift in that balance to one which is more appropriate. However, in the late 1960s and early 1970s the call for improved productivity and increased levels of production from hill and upland farms provided the impetus for a new approach to the management of hill grazings (Eadie and Cunningham, 1971). At that time an analysis of the nutritional requirements of the hill ewe on the one hand and the ability of hill grazings to support these requirements under traditional grazing systems on the other highlighted the relative inefficiencies inherent in the traditional system and also the opportunities that existed to bring about significant improvements (Eadie, 1971). This was particularly evident where the hill grazings were dominated by a grass heath in association with areas of high quality *Agrostis–Festuca* grassland.

By increasing the utilization of the existing hill vegetation during the spring, summer and autumn through grazing control and creating areas of better quality herbage, and improving the nutrition of stock in the winter by using bought-in feed, it was shown to be possible to bring about substantial increases in stock-carrying capacity, individual animal performance and total output (Armstrong *et al.*, 1984; Sibbald, 1990) (Table 3). However, the financial incentives to increase output have been largely removed. In an integrated approach to land management with forestry, however, there may still be worthwhile investment opportunities to be realized (Sibbald and Eadie, 1988).

Much more emphasis is now placed on ensuring that vegetation management practices are not wholly agriculturally orientated but also meet other objectives concerned with creating habitats for wildlife and developing attractive landscapes. In this paper it has been proposed that to achieve one, or any combination of these objectives under grazing management, it is necessary to find a simple criterion for each major vegetation type which, together with a knowledge of the physiology and seasonal patterns of growth of the main plant species, can be used to control the level of utilization and consequently the level of productivity of the vegetation. A number of examples have been quoted which indicate that some success can be achieved using this approach and it is possible to begin to quantify for specific sites the levels of stocking needed to meet the criteria set for some plant communities. This information can be used as a guide to what might be achieved at other sites; more precise extrapolation depends on

TABLE 3
Main Components of Physical Output from Traditional and Improved Sheep from Year Round Grazing Systems (Sibbald, 1990)

	Sourhope		*Glensaugh*		*Lephinmore*	
	Traditional	*Improved*	*Traditional*	*Improved*	*Traditional*	*Improved*
Ewes (ha^{-1})	1·41	2·35	0·98	1·23	0·76	1·02
Weaning (%)	90	128	70	96	85	92
Lamb ($kg\ ha^{-1}$)	27·5	76·5	17·9	34·1	17·2	23·8

acquiring better quantitative information that relates plant production to its site, soil and climatic environment. Even without this more precise information progress in developing management strategies can be made. Whatever level of utilization is chosen, the consequences for some plant communities can be estimated. Thus, if it is possible to define the conditions necessary to optimize the amenity or wildlife value of an area, then the effect of doing so on agricultural production can be assessed reasonably accurately.

Good progress has been made in developing management strategies for heather moorland, for example Maxwell *et al.* (1986) using quantitative information from research (Milne *et al.*, 1979; Grant *et al.*, 1982; Milne and Grant, 1987) has led to the construction of a computer-based management model (Sibbald *et al.*, 1988). The model can be used to:

(i) provide the management framework to maintain existing balances in *Calluna* content and distribution;
(ii) calculate the number of sheep likely to prevent overgrazing and to remove the risk of replacement of heather by native grass species;
(iii) avoid under-utilisation and prevent a degenerative change towards scrub regeneration and the loss of the agricultural resource.

The model was designed in the first instance to provide a basis for setting stocking rates for sheep and a relevant burning regime; it is now being extended further to include other vertebrate grazing species and a more detailed description of vegetation dynamics and foraging behaviour (Armstrong, 1990). It is important to extend this approach and to develop management decision rules for all our major indigenous vegetation types, which in principle need to be similar to those described for heather moorland.

As we acquire a more precise knowledge of the nutritional attributes of the major vegetation types and an increasing understanding of the factors which determine the choice of grazing site, there is the future prospect of being able to define more objectively the grazing area requirements for livestock. It should be possible to describe the proportion of each of the major vegetation types which would optimize levels of nutrition throughout the year from grazed herbage. This may require a significant adjustment of some boundaries of the existing grazings and the appropriate division of others. Conversely, given a defined grazing area, it should be possible to calculate much more precisely the sustainable numbers of livestock needed to obtain specific levels of animal performance and output, and at the same time to achieve defined amenity and wildlife objectives.

Much of the research which has been referred to in relation to developing an understanding which is necessary for the management of semi-natural vegetation has referred to domestic grazing animals. The approach, however, is equally valid for studies of the consequences of grazing for vegetation dynamics in the case of the major wild herbivores, such as red deer (*Cervis elaphus*) or rabbits (*Oryctologus cuniculus*). In the wider context, given that conservationists, or those with wildlife, sporting or amenity interest, can define the conditions they require (for example, Sydes and Miller, 1988; Hudson, 1988), it should be possible that the experimental approach described in this paper can be used to provide the understanding whereby these conditions could be predictably achieved using grazing by domestic and other animals. The objectives in managing vegetation for agricultural purposes may of course be quite different from those required for conservation, wildlife or other interests. If the grazing management of agricultural animals has to be modified to meet conservation or wildlife objectives it is likely that the economic performance of the agricultural enterprise will be adversely affected (Grant and Maxwell, 1988).

In determining future land use policies therefore for semi-natural grazings it is reasonable to evaluate the consequences of alternative policies in cost benefit terms. A more explicit and quantitative description of grazing management protocols for semi-natural vegetation provides a much more rational basis on which to carry out such an analysis. This does not imply, however, that a cost benefit analysis would be the only criteria determining land use, but it does provide an important reference point.

ACKNOWLEDGEMENT

Dr J. A. Milne, Miss S. A. Grant, Dr I. J. Gordon, Dr I. A. Wright and Mr A. R. Sibbald made significant and valuable contributions to this paper which the author gratefully acknowledges.

REFERENCES

Armstrong, R. H. (1990). Modelling the effects of vertebrate herbivore populations on heather moorland vegetation, in: *Modelling Heather Management,* Proceedings of Nickerson Foundation Workshop, ed. S. A. Grant and M. Whitby. (in press).

Armstrong, R. H. and Hodgson, J. (1985). Grazing behaviour and herbage intake in cattle and sheep grazing indigenous hill plant communities, pp. 211–18 in: *Grazing Research at Northern Latitudes,* NATO ASI Series A 108, ed. O. Gudmundsson. New York: Plenum Press.

Armstrong, R. H., Eadie, J. and Maxwell, T. J. (1984). Hill sheep production: A modified management system in practice, in: *Hill Land Symposium, Galway,* ed. M. A. O'Toole. Dublin: An Foras Taluntais.

Bircham, J. S. and Hodgson, J. (1981). The dynamics of herbage growth and senesence in a mixed-species temperate sward continuously grazed by sheep. Paper presented at the XIIIth International Grassland Conference, Lexington, USA.

Bircham, J. S. and Hodgson, J. (1983). The influence of sward conditions on rates of herbage growth and senescence under continuous stocking management, *Grass and Forage Science,* **38,** 323–31.

Burliston, A. J. and Hodgson, J. (1985). The influence of sward structure on the mechanics of the grazing process in sheep. *Animal Production,* **40,** 530–31.

Clark, D. A. and Harris, P. S. (1985). Composition of the diet of sheep grazing swards of differing white clover content and spatial distribution, *New Zealand Journal of Agricultural Research,* **28,** 233–40.

Cowling, D. W. (1982). Biological nitrogen fixation and grassland production in the United Kingdom, *Philosophical Transactions of the Royal Society,* B **296,** 397–404.

Curll, M. L. and Wilkins, R. J. (1982). Frequency and severity of defoliation of grass and clover by sheep at different stocking rates, *Grass and Forage Science,* **37,** 291–8.

Curll, M. L., Wilkins, R. J., Snaydon, R. W. and Sharmingalingham, V. S. (1985). The effect of stocking rate and nitrogen fertiliser on a perennial ryegrass-white clover sward. 1. Sward and sheep performance, *Grass and Forage Science,* **40,** 129–40.

Davies, A. and Evans, M. E. (1990). Auxiliary bud development in white clover in relation to defoliation and shading, *Annals of Botany* (in press).

Davies, D. A., Fotheringill, M. and Jones, D. (1989). Assessment of contrasting perennial ryegrasses, with and without white clover, under continuous sheep stocking in the uplands. 2. The value of white clover for lamb production, *Grass and Forage Science,* **44,** 441–50.

Dennis, W., Woledge, J., Culhane, K. and Stokes, J. (1984). Effect of clover morphology on growth and photosynthesis in mixed swards, p. 168 in: *Forage Legumes,* ed. D. J. Thomson. British Grassland Society Occasional Symposium No. 16.

Doney, J. M., Milne, J. A., White, I. R. and Colgrove, P. (1987). Liveweight gain of weaned lambs grazing swards increasing and decreasing or at constant surface heights, *Animal Production,* **44,** 471.

Eadie, J. (1967). The nutrition of hill sheep: Improved utilisation of hill pastures, pp. 38–45 in: Fourth Report 1964–7, Hill Farming Research Organisation, Edinburgh, UK: HFRO.

Eadie, J. (1971). Hill pastoral resources and sheep production. *Proceedings of Nutrition Society*, **30,** 204–10.

Eadie, J. and Cunningham, J. M. M. (1971). Efficiency of hill sheep production systems, pp. 239–49 in: *Potential Crop Production,* ed. P. F. Waring and J. P. Cooper. London: Heineman.

Eadie, J. and Maxwell, T. J. (1975). Systems research in hill sheep farming, pp. 395–414 in: *Study of Agricultural Systems,* ed. G. E. Dalton. London: Applied Science Publishers.

Ernst, P., Le Du, Y. L. P. and Carlier, L. (1980). Animal and sward production under rotational and continuous grazing management—A critical appraisal, pp. 119–26 in: *The Role of N in Intensive Grassland Production,* ed. W. H. Prins and G. W. Arnold. Wageningen, The Netherlands: EFG.

Forbes, T. D. A. (1982). Ingestive behaviour and diet selection in grazing cattle and sheep. PhD thesis, University of Edinburgh, UK.

Forbes, T. D. A. and Hodgson, J. (1985). Comparative studies of the influence of sward conditions in the ingested behaviour of cows and sheep, *Grass and Forage Science,* **40,** 69–77.

Gordon, I. J. (1989). Develop and test foraging strategy theories for ruminants grazing mixed indigenous communities, p. 65 in: Macaulay Land Use Research Institute (MLURI) Annual Report. Aberdeen, UK: MLURI.

Gordon, I. J. and Iason, G. R. (1989). Foraging strategy of ruminants: Its significance to vegetation utilisation and management, pp. 34–41 in: Macaulay Land Use Research Institute (MLURI) Annual Report 1988–9. Aberdeen, UK: MLURI.

Gordon, I. J. and Illius, A. W. (1988). Incisor arcade structure and diet selection in ruminants, *Functional Ecology,* **2,** 15–22.

Grant, S. A. and Barthram, G. T. (1988). Investigate the manipulation of clover content in grazed swards and the effect on herbage production, pp. 90–1 in: Macaulay Land Use Research Institute (MLURI) Annual Report 1987. Aberdeen, UK: MLURI.

Grant, S. A. and Barthram, G. T. (1990). The effects of contrasting cutting regimes on the components of clover and grass growth in microswards, *Grass and Forage Science* (in press).

Grant, S. A. and Hodgson, J. (1986). Grazing effect on species balance and herbage production in indigenous plant communities, pp. 69–77 in: *Grazing Research at Northern Latitudes,* NATO ASI Series A 108, ed. O. Gudmundsson. New York: Plenum Press.

Grant, S. A. and Hunter, R. F. (1968). Interactions of grazing and burning on heather moors and their implications for grazing management, *Journal of the British Grassland Society,* **23,** 285–93.

Grant, S. A. and King, J. (1984). Grazing management and pasture production: The importance of sward morphological adaptation and canopy photosynthesis, pp. 119–29 in: Hill Farming Research Organisation Biennial Report 1982–3. Edinburgh, UK: HFRO.

Grant, S. A. and Marriott, C. A. (1989). Some factors causing temporal and spatial variation in white clover performance in grazed swards, pp. 1041–2 in: *Proceedings of the XVI International Grassland Congress, Vol. II.* Nice, France: Association Francais pour la Production Forragere.

Grant, S. A. and Maxwell, T. J. (1988). Hill vegetation and grazing by domesticated herbivores: The biology and definition of management options, pp. 201–14 in: *Ecological Change in the Uplands,* Publication No. 7 of The British Ecological Society. London: Blackwell Publications.

Grant, S. A., Barthram, G. T., Lamb, W. I. C. and Milne, J. (1978). Effect of season and level of grazing on the utilisation of heather by sheep. 1. Responses of the sward, *Journal of the British Grassland Society,* **33,** 289–300.

Grant, S. A., King, J., Barthram, G. T. and Torvell, L. (1981). Responses of tiller populations to variation in grazing management on continuously stocked swards as affected by time of year, pp. 81–4 in: *Plant Physiology and Herbage Production,* ed. C. E. Wright. British Grassland Society Occasional Symposium No. 13.

Grant, S. A., Milne, J. A., Barthram, G. T. and Soutar, W. G. (1982). Effects of season and level of grazing on the utilisation of heather by sheep. 3. Longer-term responses and sward recovery, *Grass and Forage Science,* **37,** 311–20.

Grant, S. A., Barthram, G. T., Torvell, L., King, J. and Smith, H. K. (1983). Sward management, lamina turnover and tiller population density in continuously stocked *Lolium perenne*-dominated swards, *Grass and Forage Science,* **38,** 333–44.

Grant, S. A., Bolton, G. R. and Torvell, L. (1985). The response of blanket bog vegetation to controlled grazing by hill sheep, *Journal of Applied Ecology,* **22,** 739–51.

Grant, S. A., Barthram, G. T., Torvell, L., King, J. and Elston, D. A. (1988). Comparison of herbage production under continuous and intermittent grazing, *Grass and Forage Science,* **43,** 29–39.

Gray, R. (1986). Faith and foraging, pp. 69–140 in: *Foraging Behaviour,* eds. A. C. Kanul, J. R. Krebs and H. R. Pulliain. New York: Plenum Press.

Harkness, R. D., Hunt, I. V. and Frame, J. (1970). The effect of variety and companion grass on the productivity of white clover, pp. 175–86 in: *White Clover Research,* British Grassland Society Occasional Symposium No. 6.

Haynes, R. J. (1981). Competitive aspects of the grass–legume association, *Advances in Agronomy,* **22,** 227–261.

Heal, O. W. and Perkins, D. F. (1978). *Production Ecology of British Moors and Montane Grasslands.* New York: Springer.

Hill Farming Research Organization Annual Report (1986). Edinburgh, UK: HFRO.

Hodgson, J. (1981*a*). Influence of sward characteristics on diet selection and herbage intake by the grazing animal, pp. 153–66 in: *Proceedings of International Symposium on Nutritional Limits to Animal Production from Pastures,* ed. J. B. Hacker. Slough: Commonwealth Agricultural Bureaux.

Hodgson, J. (1981*b*). Variations in the surface characteristics of the sward and the short-term rate of herbage intake by calves and lambs, *Grass and Forage Science,* **36,** 49–57.

Hodgson, J. (1982). Influence of sward characteristics on diet selection and herbage intake by the grazing animal. Nutritional Limits to Animal Production from Pastures, ed. J. B. Hacker, pp. 153–66 in: *Proceedings International Symposium,* St Lucia, Queensland, Australia. Brisbane, Australia: CSIRO.

Hodgson, J. (1985). The significance of sward characteristics in the management of temperate sown pastures, pp. 63–7 in: *Proceedings of the XV International Grassland Congress,* Kyoto, Japan: Science Council of Japan and Japanese Society of Grassland Production.

Hodgson, J. and Grant, S. A. (1981). Grazing animals and forage resources in the hills and uplands, pp. 41–57 in: *The Effective Uses of Forage and Animal Resources in the Hills and Uplands,* ed. J. Frame. Maidenhead, Berkshire, UK: British Grassland Society Occasional Symposium No. 12.

Hodgson, J. and Grant, S. A. (1985). The grazing ecology of hill and upland swards, pp. 77–84 in: *Hill and Upland Livestock Production,* ed. T. J. Maxwell and R. G. Gunn. British Society of Animal Production Publication No. 10.

Hoglund, J. H., Crush, J. R., Brock, J. C., Ball, R. & Carron, R. A. (1979). Nitrogen fixation in pasture. XII General discussion, *New Zealand Journal of Experimental Agriculture,* **7,** 45–51.

Hudson, P. J. (1988). Spatial variations, patterns and management options in upland bent communities, pp. 381–97 in: *Ecological Change in the Uplands,* ed. M. B. Usher and D. B. A. Thompson. Special Publication No. 17 of British Ecological Society. London: Blackwell Publications.

Hunter, R. F. (1954). The grazing of hill pasture types, *Journal of the British Grassland Society,* **9,** 195–208.

Hunter, R. F. (1962). Hill sheep and their pastures: A study of sheep grazing in South-east Scotland, *Journal of Ecology,* **50,** 651–680.

Illius, A. W. (1986). Foraging behaviour and diet selection, pp. 227–36 in: *Grazing Research at Northern Latitudes,* NATO ASI Series A 108, ed. O. Gudmundsson. New York: Plenum Press. (In association with NATO Scientific Affairs Division, London).

Illius, A. W. and Gordon, I. J. (1987). The allometry of food intake in grazing ruminants, *Journal of Animal Ecology,* **56,** 989–1000.

Illius, A. W. and Gordon, I. J. (1989). Prediction of size-scaling effects on forage intake and digestion using a model of digestion flow in ruminants, *Animal Production,* **48,** 637.

Illius, A. W. and Gordon, I. J. (1990). Constraints on diet selection and foraging behaviour in mammalian herbivores, in: *Behavioural Mechanisms of Food Selection,* ed. R. N. Hughes. Berlin: Springer-Verlag.

Jones, M. G. (1933). Grassland management and its influence on the sward. II. The management of a clovery sward and its effects, *Empire Journal of Experimental Agriculture,* **1,** 122–8.

King, J., Sim, E., Barthram, G. T., Grant, S. A. and Torvell, L. (1988). Photosynthetic potential of ryegrass pastures when released from continuous stocking management, *Grass and Forage Science,* **43,** 41–8.

Langer, R. H. M. (1963). Tillering in herbage grasses. *Herbage Abstracts,* **33,** 141–8.

Lantinga, E. A. (1985). Productivity of grasslands under continuous and

rotational grazing. PhD thesis, Agricultural University, Wageningen, The Netherlands.
Laws, J. A. and Newton, J. E. (1987). The effect of early defoliation in the spring by sheep on the proportion of clover in a grass/white clover sward, pp. 203–5 in: *Efficient Sheep Production from Grass,* ed. G. E. Pollott. British Grassland Society Occasional Symposium No. 21.
Leafe, E. L. (1978). Physiological, environmental and management factors of importance to maximise yield of the grass crop, pp. 37–49 in: *Maximising Yields of Crops,* London: HMSO.
Marriott, C. A., Smith, M. A. and Baird, M. A. (1987). The effect of urine on clover performance in a grazed upland sward, *Journal of Agricultural Science,* **109,** 177–85.
Maxwell, T. J. and Treacher, T. T. (1987). Decision Rules for Grazing Management, p. 67 in: *Efficient Sheep Production from Grass,* ed. G. E. Pollott. British Grassland Society Occasional Symposium No. 21.
Maxwell, T. J., Grant, S. A., Milne, J. A. and Sibbald, A. R. (1986). Systems of sheep production on heather moorland, pp. 188–211 in: *Hill Land Symposium, Galway,* ed. M. A. O'Toole. Dublin: An Foras Taluntais.
Maxwell, T. J., Grant, S. A. and Wright, I. A. (1988). Maximising the role of grass and forage systems in beef and sheep production, pp. 84–97 in: *Proceedings of the 12th General Meeting of the European Grassland Federation,* Dublin. Galway: Irish Grassland Society.
Meat and Livestock Commission (1986). Sheep Improvement Services, South-West Regional Flockplan Report 1985. Milton Keynes, UK: MLC.
Miles, J. (1979). *Vegetation Dynamics.* London: Chapman and Hall.
Miles, J. (1985). The ecological background to vegetation management, pp. 3–20 in: *Vegetation Management in Northern Britain,* Monograph No. 30, ed. E. B. Murray. Croydon, UK: British Crop Protection Council.
Milne, J. A. and Grant, S. A. (1987). Sheep management on heather moor, pp. 165–7 in: *Efficient Sheep Production from Grass,* ed. G. E. Pollott. British Grassland Society Occasional Symposium No. 21.
Milne, J. A., Bagley, L. P. and Grant, S. A. (1979). Effect of season and level of grazing on the utilization of heather by sheep. 2. Diet Selection and Intake, *Grass and Forage Science,* **34,** 45–53.
Milne, J. A., Hodgson, J., Thompson, R., Souter, W. G. and Barthram, G. T. (1982). The diet ingested by sheep grazing swards differing in white clover and perennial ryegrass content, *Grass and Forage Science,* **37,** 209–18.
Milne, J. A., Colgrove, P. M., Kerr, W. G. and Elston, D. A. (1988). Herbage intake of ewes in the autumn as influenced by sward herbage, type and amount of supplement and fatness of the ewe, in: *British Grassland Society Research Meeting No. 1,* Session VI, Paper 3. Hurley, UK: British Grassland Society.
Morrison, J. (1977). The growth of *Lolium perenne* and response to fertiliser N in relation to season and management, pp. 108–14 in: *Proceedings 13th International Congress,* Leipzig, Section 7. Berlin: Akademie Verlag.
Morrison, J. (1980). The influence of climate and soil on the yield of grass and its response to fertiliser nitrogen, pp. 51–7 in: *The Role of N in Intensive*

Grassland Production, eds. W. H. Prins and G. W. Arnold. Wageningen, The Netherlands: EFG.

Morrison, J., Newton, J. E. and Sheldrick, R. D. (1985). Management and utilization of white clover. Information Leaflet No. 14. Hurley, UK: The Animal and Grassland Research Institute.

Moustapha, E., Ball, R. and Field, T. R. O. (1969). The use of acetylene reduction to study the effect of nitrogen fertilisers and defoliation on nitrogen fixation by field-grown white clover, *New Zealand Journal of Agricultural Research,* **12,** 691–6.

Newton, J. E. and Davies, D. A. (1987). White clover and sheep production, p. 79 in: *Efficient Sheep Production from Grass,* ed. G. E. Pollott. British Grassland Society Occasional Symposium No. 21.

Newton, J. E., Wilde, R. M. and Betts, J. E. (1985). Lamb production from perennial ryegrass and perennial ryegrass–white clover swards using set-stocking on rotational grazing, *Research and Development in Agriculture,* **2,** 1–6.

Noy-Meir, I. (1975). Stability of grazing systems: An application of predator-prey graphs, *Journal of Ecology,* **63,** 459–81.

Orr, R. J., Penning, P. D., Parsons, A. J. P. and Treacher, T. T. (1987). Effect of sward surface height of mixed swards of ryegrass/white clover on the intake and performance of ewes and their twin lambs, *Animal Production,* **44,** 470.

Owen-Smith, N. (1982). Factors influencing the consumption of plant products by large herbivores, pp. 359–404 in: *The Ecology of Tropical Savannas,* Ecological Studies Volume 42, ed. B. J. Hursilley and B. H. Walker. Berlin: Springer-Verlag.

Parsons, A. J. P. and Leafe, E. L. (1981). Photosynthesis and carbon balance of a grazed sward, pp. 69–72 in: *Plant Physiology and Herbage Production,* ed. C. E. Wright. British Grassland Society Occasional Symposium No. 13.

Parsons, A. J. P. and Penning, P. D. (1988). The effect of duration of regrowth on photosynthesis, leaf death and average rate of growth in a rotationally grazed sward, *Grass and Forage Science,* **43,** 15–27.

Parsons, A. J. P. and Robson, M. J. (1982). Seasonal changes in the physiology of S24 perennial ryegrass (*Lolium perenne* L). 4. Comparison of the carbon balance of the reproductive crop in spring and the vegetation crop in autumn, *Annals of Botany,* **50,** 167–77.

Parsons, A. J. P., Leafe, E. L., Collett, B. and Stiles, W. (1983*a*). The physiology of grass production under grazing. 1. Characteristics of leaf and canopy photosynthesis of continuously grazed swards, *Journal of Applied Ecology,* **20,** 117–26.

Parsons, A. J. P., Leafe, E. L., Collett, B., Penning, P. D. and Lewis, J. (1983*b*). The physiology of grass production under grazing. 2. Photosynthesis, crop growth and animal intake of continuously grazed swards, *Journal of Applied Ecology,* **20,** 127–39.

Parsons, A. J. P., Johnson, I. R. and Harvey, A. (1988). The use of a model to optimise the interaction between frequency and severity of intermittent defoliation and to provide a fundamental comparison of the continuous and intermittent defoliation of grass, *Grass and Forage Science,* **43,** 49–59.

Parsons, A. J. P., Harvey, A. and Woledge, J. (1990). Plant/animal interactions in a continuously grazed mixture. 1. Differences in the physiology of leaf expansion and the fate of leaves of grass and clover, *Journal of Applied Ecology* (in press).

Penning, P. D. (1986). Some effects of sward conditions on grazing behaviour and intake by sheep, pp. 219–26 in: *Grazing Research at Northern Latitudes,* NATO ASI Series A 108, ed. O. Gudmundsson. New York: Plenum Press. (In association with NATO Scientific Affairs Division, London.)

Prins, W. H., van Burg, P. F. J. and Wieling, H. (1980). The seasonal response of grassland to nitrogen at different intensities of nitrogen fertilisation with special reference to methods of response measurement, pp. 35–49 in: *The Role of N in Intensive Grassland Production,* ed. W. H. Prins and G. W. Arnold. Wageningen, The Netherlands: EFG.

Pyke, G. H. (1984). Optimal foraging theory: A critical review, *Annual Review of Ecological Systems,* **15,** 523–75.

Robson, M. J. (1981). Potential production—What is it and can we increase it?, pp. 5–18 in: *Plant Physiology and Herbage Production,* ed. C. E. Wright. British Grassland Society Occasional Symposium No. 13.

Ryden, J. C. (1983). The nitrogen cycle in grassland—a case for studies in grazed pastures, pp. 150–66 in: Annual Report 1982. Hurley, UK: The Grassland Research Institute.

Ryle, G. J. A., Powell, C. E. and Gordon, A. J. (1985). Defoliation in white clover: re-growth, photosynthesis and N_2-fixation, *Annals of Botany,* **56,** 9–18.

Sheldrick, R. D., Lavender, R. H. and Parkinson, A. E. (1987). The effect of subsequent management on the success of introducing white clover into an existing sward, *Grass and Forage Science,* **42,** 359–71.

Short, J. (1986). The effect of pasture availability on food intake, species composition and grazing behaviour of kangaroos, *Journal of Applied Ecology,* **23,** 559–71.

Sibbald, A. R. (1990). Biological and economic assessment of potential for sheep production on heather moorland, in: *Modelling Heather Management,* ed. S. A. Grant and M. Whitby. Proceedings of Nickerson Foundation Workshop (in press).

Sibbald, A. R. and Eadie, J. (1988). Optimum allocation of land between the farming and forestry enterprises, pp. 67–78 in: *Farming and Forestry*: Proceedings of a Conference held at Loughborough University, ed. G. R. Hatfield. Alice Holt, UK: Forestry Commission Occasional Paper 17.

Sibbald, A. R., Grant, S. A., Milne, J. A. and Maxwell, T. J. (1988). Heather moorland management—A model, pp. 107–8 in: *Agriculture and Conservation in the Hills and Uplands,* ed. M. Bell and R. G. H. Bruce. Grange-over-Sands: Institute of Terrestrial Ecology.

Solangaarachichi, S. M. and Harper, J. L. (1987). The effect of canopy filtered light on the growth of white clover (*Trifolium repens*), *Oecologia,* **72,** 372–6.

Spedding, C. R. W. (1971). *Grassland Ecology.* London: Oxford University Press.

Stephens, D. W. and Krebs, J. R. (1987). *Foraging Theory*. Princeton, NJ: Princeton University Press.

Sydes, C. and Miller, G. R. (1988). Range management and nature conservation in the British Uplands, in: *Ecological Change in the Uplands*, ed. M. B. Usher and D. B. A. Thompson. Special Publication No. 17 of British Ecological Society. London: Blackwell Publications.

Tansley, A. G. (1946). *Introduction to Plant Ecology*. London: Allen and Unwin.

Thompson, D. J. (1984). The nutritive value of white clover, pp. 78–92 in: *Forage Legumes*, ed. D. J. Thomson. British Grassland Society Occasional Symposium No. 1.

Treacher, T. T., Orr, R. J. and Parsons, A. J. P. (1986). Direct measurement of the seasonal pattern of production on continuously-stocked swards, pp. 204–5 in: *Grazing*, ed. J. Frame. Hurley, UK: British Grassland Society.

Whitehead, D. C. (1986). Sources and transformation of organic nitrogen in intensively managed grassland soils, pp. 47–58 in: *N fluxes in Intensive Grassland Systems*, ed. H. G. van der Meir, J. C. Ryden and G. G. Eurik. Dordrecht: Nijhoff.

Williams, W. (1970). White clover in British agriculture, pp. 1–10 in: *British Grassland Society Occasional Symposium No. 6.*

Wilson, D. (1981). The role of physiology in breeding herbage cultivars adapted to their environment, pp. 95–108 in: *Plant Physiology and Herbage Production*, ed. C. E. Wright. British Grassland Society Occasional Symposium No. 13.

Wolton, K. M., Brockman, J. S. and Shaw, P. G. (1970). The effect of stage of growth at defoliation on white clover in mixed swards, *Journal of the British Grassland Society*, **25**, 113–18.

Wright, I. A. (1988). Suckler beef production, pp. 51–64. in: *Efficient Beef Production from Grass*, ed. J. Frame. British Grassland Society Occasional Symposium No. 22.

3

Exploitation of the Systems Approach in Technical Design of Agricultural Enterprises

A. C. BYWATER

Department of Farm Management, Lincoln University, Canterbury, New Zealand

INTRODUCTION

The intention of this chapter is to examine approaches which have been used in the analysis and improvement of production systems on farms. It is not intended to ignore economic considerations in the design of agricultural enterprises; the distinction between this chapter and the next by Professor Doyle is based primarily on a hierarchical difference in the target of analysis. Here the perspective is of a whole farm or whole enterprise including technical, economic and production management considerations. The following chapter will concentrate more on issues concerning resource allocation within enterprises.

Formal and Informal Systems Approaches

Any analytical or management approach which attempts to accommodate the basic notions of general systems theory might be termed 'a systems approach'. Unfortunately, a number of different approaches and techniques have been used under this heading and the meaning of the term has become somewhat confused. It may be better in fact to do away with the term altogether, but if we are to use it perhaps it would be helpful to distinguish between formal and informal applications of the concepts.

The central ideas regarding the characteristics and behaviour of systems include the notion that systems contain components which interact with one another over time and that such interaction is a primary determinant of system behaviour. The human mind is incapable of tracing the impact of interactions between more than two or three components within a system over any reasonable time period even in a qualitative, let alone quantitative way. Formal systems approaches are defined here as those employing techniques which explicitly quantify the effects of dynamic interactions within the system being analyzed. Informal systems approaches are those which, while cognisant of dynamic interactions within systems, attempt to accommodate them primarily through descriptive or intuitive means rather than by rigorous quantification.

Any attempt to trace interactions quantitatively is almost certain to require the assistance of a numeric processor. Formal systems approaches therefore are likely to involve models solved by computers. Informal systems approaches may or may not involve computer models (budgets, for example, may be computerised; but they do not explicitly quantify the dynamics of the systems they represent, only its expected outcomes).

INFORMAL SYSTEMS APPROACHES

The use of budgets and other farm management techniques is an informal application of systems thinking. Farm management both as an academic focus area and a professional discipline must take a systems view of the farm if it is to have any validity. From a practical standpoint, the traditional farm management techniques are probably the most widespread systems management tools used for analysis and development of farming systems. Bio-economic budgeting can become quite sophisticated. The advent of the spreadsheet has caused an explosion in the range and types of whole farm or whole enterprise planning and decision aids available for use by farmers, consultants and researchers. Spreadsheets have been adopted to a much greater extent than, and with an enthusiasm that was noticeably lacking with, earlier quantitative management techniques.

However, the efficacy of these planning devices is absolutely dependent on the values assigned by the user to the expected response of the system or its components to changes in its inputs or management. It is this which distinguishes budgeting systems (either biological or economic) from simulation models; the response of the system to change is exogenously determined with the former and endogenously generated with the latter. This is not to deny the usefulness of budgeting and other similar techniques; it simply recognises the fact that they are informal techniques which depend on the practical experience and knowledge of the individual using them.

The other example of informal systems approaches considered here is the use of systems trials or farmlet studies. These occur both in developed countries and in less developed countries where such trials form part of the so called Farming Systems Research process. The rationale for, and process of Farming Systems Research *per se* are described in the chapter by Hildebrand (pp. 131–43). In this chapter, the term Farming Systems Development (FSD; FAO, 1989) is preferred as being more descriptive and perhaps more appropriate to the overall activity which combines and integrates a strong process consulting element with varying amounts and types of what has traditionally been thought of as research. In FSD the process of consultation and farmer involvement has been formalized, rather than the research process *per se*.

The consulting elements of FSD are very similar to those used in Farm Management and have been applied successfully in New Zealand, where they have been enjoyed (endured?) by generations of Lincoln students in their Farm Management courses, since the late 1930s and early 1940s. In our case, the context is one of large, commercial farms rather than small, subsistence or semi-subsistence farms, but the holistic whole-farm viewpoint and the sequence of data gathering, current system description, and farm system development and evaluation applied to large commercial farms closely parallels the process applied to groups of small farmers in FSD. The process is a continuous and usually incremental one in both cases and the mix of techniques is similar even though the objective criteria may be different. Suggestions that FSD is quite different from Farm Management (Byerlee *et al.*, 1982) may be a reflection of an academic rather than professional orientation to Farm Management teaching in the US. The need for professional training of field officers in New Zealand arose because of a need to administer a number of government initiatives to support farmers (such as the Mortgagors and Lessees Rehabilitation Act of 1936), much the same sort of reason for the development of FSD in developing countries. A strong, case-study based professional training has been employed since then.

The research process within FSD varies as much as research in any other context; in many respects it offers the same potential and faces similar methodological problems as its counterparts elsewhere. FSD recognises the importance of interactions in the farming system and the impact of the environment and attempts to accommodate these through systems trials on smallholders' own farms. A distinction is made here between systems trials which are concerned with the whole production system (or a major subsystem) of the farm and single factor experiments carried out on-farm rather than on-station. In some cases, the research is largely descriptive and prescriptive, involving intuitive reasoning based on some understanding of existing farming systems and farmers' circumstances (Collinson, 1980). In such cases it does not have the rigour of a formal systems approach. In other circumstances some attempt to use formal systems approaches has been made in the research phase (e.g. see Lambourne 1986).

There are obviously strong parallels between the systems trials of FSD and systems trials in developed countries. The fact that the former are conducted on smallholder farms may overcome some of the problems of research station trials, but it will not solve all of them.

Where significant gains in output are possible with changes in technology or management and these changes are assimilable, systems trials based on a combination of previous research evidence and intuition can be very successful in changing the way farmers produce. Examples in the New Zealand context include the early grazing management trials with dairy cows (e.g. McMeekan and Walshe, 1963), and the introduction and management of lucerne on dryland sheep properties (e.g. Stewart, 1967). Successful examples of the generation of new technology through FSD are cited by Norman (1986).

On the other hand, production systems are complex and it is rarely possible to account for all interactions and environmental effects. The New Zealand trials cited above, although very successful at the time, provide reason for caution in this respect. As we have learned more about the underlying biological systems in question, the superiority of the management systems proposed in these earlier trials has become much less certain (e.g. Bryant and Holmes 1985). Barlow *et al.* (1986) also provide several examples where gains from farming systems trials in an FSD context have been short lived.

One of the perpetual dilemmas of systems trials anywhere is to clearly distinguish between at least three possible objectives for such trials. Failure to do so invariably creates problems. Menz (1980) describes conflicts between two objectives in an FSD context—designing new systems or combinations of technologies and evaluating systems which have already been designed. To these may be added a third objective which is attempting to characterise and understand interactions at the production system level. These three objectives are usually incompatible since they ask three quite different questions: 'How does the system work?', 'What is the best or better system to meet some set of objectives?' and 'Does this system work under these particular conditions and constraints?' (the last often implies some demonstration role if the answer is yes).

None of these questions, with the possible exception of the last, is particularly easy to answer. Where the objective is to design better systems, there is no guarantee that, even with trials conducted on farmers' own properties, the results will be beneficial in other areas and other years or to other farmers (e.g. Menz and Knipscheer, 1981). The permutations of different elements within even relatively simple systems can be quite daunting, although, where significant numbers of farmers participate the opportunity to test a number of systems and relationships clearly exists (e.g. Garrity *et al.*, 1981).

The problem of transferring systems technology to other areas, other years and other managers is obviously compounded if it is not possible to provide hard, objective and quantitative explanations of why the system worked or did not work in the trial. This clearly requires an understanding of the interactions occurring at the production system level. As noted above, such interactions are often extremely complex, particularly in pastoral livestock systems. Bryant and Holmes (1985) have stated that in their opinion the science of whole-farm experimentation is 'woefully undeveloped'. They base this contention on a failure to provide objective decision rules to help in reaching realistic decisions about critical areas of [pastoral dairy farm] management. In general, systems trials have failed to characterise the dynamic interaction of components within systems with sufficient rigour to provide an adequate explanation of systems behaviour.

An understanding of interactions at a production system level is likely to be achieved in much the same way as an understanding of the components of such systems, i.e. by the formal application of the scientific method of hypothesis generation, testing and rejection. Where the target of such enquiry is dynamic interactions within a system, then it follows that hypotheses must be stated in quantitative and dynamic terms. It is likely to be only through the use of quantitative dynamic models in conjunction with systems trials that progress in understanding systems level interactions will be made.

In summary, farm systems trials, both in developed countries and within the context of FSD, are informal systems approaches because in themselves they do not quantify the dynamic interactions characteristic of agricultural production systems. The process of consulting elements of FSD, which are similar to professional approaches used elsewhere, are likely to yield short-term benefits on their own (an important consideration in many circumstances), but the use of systems trials, either on research farms or on farmers' own farms, is not a panacea. The methodology of systems trials is not well developed. The unfortunate tendency for new technical problems or environmental constraints to arise and thereby diminish previous gains both reinforces the need for a systems view and at the same time highlights our current limitations in applying such a view. As gains become harder to achieve through application of existing knowledge and intuition, a better understanding of systems level interactions will become increasingly necessary.

MODELS

Formal systems approaches were defined in the introduction as those which explicitly quantify dynamic interactions between system components. It was suggested that this would require the assistance of a numeric processor and would therefore almost invariably involve computer models. A very large number of models representing different farming systems and their components have been developed. A partial list of models which have application at the whole farm or whole enterprise level is provided and some general considerations in the development and use of models are discussed.

Available Models

Examples of models which might have applications at the whole farm or whole enterprise level are given in Tables 1–4. It is not the intention here to review such models in detail regarding their objectives, structure and use. Where possible, some comment taken from the original authors regarding the objectives and or special characteristics of each model is included. In some instances, models are described in a number of publications, with no single reference detailing all aspects of the model. In such cases, a primary reference has been provided which should give readers access to other publications.

TABLE 1
Sheep and Cattle Production Models

Model	*Reference*	*Remarks*
Sheep		
PROSPECT	Graham *et al.* (1976)	A model of energy and protein use in growth and production of sheep developed as an alternative to the use of feeding standards; subsequently enhanced with a rumen model (Black *et al.*, 1981).
DPI	Christian *et al.* (1978)	General pastoral sheep production model for evaluation of management strategies; includes routine for optimising management parameters.
HFRO	Sibbald *et al.* (1979)	A model for evaluating alternative management strategies for improving systems of hill sheep production.
DLRM	Arnold *et al.* (1977) and Galbraith *et al.* (1980)	Model designed to identify research opportunities for improving systems of sheep production from clover pastures in Western Australia.
GRI1	Edelston and Newton (1977)	A model to aid in development of improved systems of lowland lamb production.
GRI2	France *et al.* (1983)	A general model for evaluating feed requirements and cost and returns of alternative lamb production systems.
BREW	White *et al.* (1983)	Model of self-replacing Merino flock designed to evaluate feeding and flock management practices in Victoria, Australia.
DYNAMOF	Bowman *et al.* (1989)	A revision of BREW incorporating more flexible and comprehensive routines for pasture allocation and financial assessment.
NZAEI	Heiler *et al.* (1985)	A model to evaluate alternative farm irrigation schemes on hill and high country sheep and beef farms in New Zealand.
TAMU	Blackburn and Cartwright (1987)	A model developed for use in the USAID, Title 12, Small Ruminant CRSP, and used to investigate genotype: environment interaction in Northern Kenya.

(*continued*)

TABLE 1—*contd.*

Model	*Reference*	*Remarks*
Beef		
CONGLETN	Congleton and Goodwill (1980)	A simulation of beef herd age structure and productivity to evaluate breeding plans.
BEEF	Loewer *et al.* (1981)	A general cattle production model for use in management, planning and teaching for evaluation of the effects of alternative management and research strategies on production.
GRAZE	Loewer *et al.* (1987)	Recombination of an animal model (BEEFS156) and pasture model (GROWIT) developed independently out of BEEF.
TAMU	Sanders and Cartwright (1979)	A general cattle production model designed for evaluating genotype/environment/management interactions; used in various countries around the world.
KAHN	Kahn and Spedding (1983)	Adaptation of the TAMU model allowing representation of individual animals, stochastic treatment of discrete events, 1–30 day integration and additional management options.
SPUR	Rice *et al.* (1983)	A development of the TAMU model incorporating an enhanced herbage-animal interface.
ILCA	Konandreas and Anderson (1982)	Cattle production model for use in East Africa; includes both meat and milk production.
GUELPH	Forster *et al.* (1984)	A stochastic simulation model of growth and finishing of beef cattle on maize silage to investigate systems of lean beef production.
NATAL	Uys *et al.* (1985)	A simple model of herd structure and grazing pressure for evaluating offtake strategies in subsistence herds.
AGRI	Doyle *et al.* (1987)	A model to evaluate use of grass/clover swards in beef production systems.
IGAP	Dowle *et al.* (1988)	A model for evaluating grassland management on livestock (sheep and cattle) farms; derived from an earlier dairy model (Doyle and Edwards, 1986).

TABLE 1—*contd.*

Model	*Reference*	*Remarks*
RUA	Doyle *et al.* (1989)	A model to evaluate grassland management strategies for bull production in Waikato, New Zealand.
Dairy		
GARTNER	Gartner (1981)	A model to investigate replacement policy where heifers compete with cows for grazing.
MAINE	Congleton (1983)	A model describing herd dynamics, health and production in intensive dairy systems.
ICI	Bailie (1982)	A model to investigate the economic and management consequences of breeding efficiency.
KUIPERS	Kuipers (1982)	A model of a dairy herd for evaluation of bull and cow selection criteria.
DIJK	Dijkuisen *et al.* (1986)	A stochastic model of a dairy herd with special reference to production, reproduction and culling.
GRI	Corrall *et al.* (1981)	A model to examine grass management and ensiling method on the economics of autumn-calving cows.
AGRI1	Doyle and Edwards (1986)	A model to assess grassland management practices on dairy farms.
AGRI2	Doyle and Phipps (1987)	A model to assess the integration of maize silage with pasture on dairy farms in Southern England.
CONWAY	Conway and Killen (1987)	A linear programming model of grassland management on dairy farms.
ORACLE	Marsh *et al.* (1987)	A stochastic simulation model of a dairy herd emphasising reproduction and related issues.
SORENSEN	Sorensen (1989)	A model designed to evaluate alternative strategies of producing replacement heifers.
WADFM	Olney and Kirk (1989)	A linear programming model of a Western Australian dairy farm including herd structure, feed requirements, pasture and fodder crops, and supplementary feed.

TABLE 2
Pig and Poultry Production Models

Model	*Reference*	*Remarks*
Pigs		
BLACKIE	Blackie and Dent (1976)	A model of a pig production facility for evaluation of feeding and management strategies.
WHITT	Whittemore (1980)	A model of the nutrition and growth of the pig.
BRUCE	Bruce and Clark (1979)	A model of heat production and critical temperature in pigs for evaluating nutrition: environment interactions.
NYE	Nye *et al.* (1980)	A simple model for scheduling swine production facilities.
PURDUE	Allen and Stewart (1983)	A scheduling model for a breeding sow operation for evaluation of management parameters such as time of weaning.
SINGH	Singh (1986)	A scheduling model of a farrow to finish operation, using stochastic biological parameters, for building capacity selection.
AUSPIG	Davies *et al.* (1987)	A management package including a growth and production simulator, ration formulation LP and a production/marketing LP.
NCCI	Watt *et al.* (1987)	A comprehensive model of housing, nutrition, growth and economics of swine production based on the WHITT and BRUCE models.
Laying hens		
CHARLES	Charles (1984)	Response of laying hens in food intake and egg output to temperature, light, feeding system and feed composition.
DESHAZER	DeShazer *et al.* (1981)	A model of feed intake, environment and production of laying hens.
TIMM	Timmons and Gates (1988a)	A model to optimise laying hen performance as a function of air temperature and building thermal characteristics, including evaporative cooling.
Broilers		
TETER	Teter *et al.* (1973)	A model of broiler production dependent on environmental temperature.

TABLE 2—*contd.*

Model	*Reference*	*Remarks*
REECE	Reece and Lott (1982)	A model to interrelate house design, ventilation and heating, climate and broiler characteristics.
TIMM	Timmons and Gates (1986)	A model to optimise broiler production as affected by environment.
Turkeys		
TETER	Teter *et al.* (1976)	A model of turkey production dependent on environmental temperature.
HURWITZ	Hurwitz *et al.* (1985)	A model of nutrition and environment for evaluating the economics and management of turkey production.
TIMM	Timmons and Gates (1988b)	A generally applicable model to predict turkey growth as affected by environment and body weight.

TABLE 3
Crop Production Models

Model	*Reference*	*Remarks*
Corn		
CORNSYM	Van Ee and Kline (1979)	A model of corn planting, crop development, yield and harvest developed to provide simulated grain flow data for a corn drying and storage study.
CORNF	Stapper and Arkin (1980)	Dynamic growth and development model for maize.
CERES	Jones and Kiniry (1986)	Dynamic growth and development model for maize.
CHILDS	Childs *et al.* (1977)	A model to simulate environmental and physiological processes involved in the growth of corn.
Cotton		
GOSSYM	Baker *et al.* (1983)	Simulation of cotton crop growth and yield.
COTCROP	Brown *et al.* (1985)	Simulation of cotton crop growth and yield.
COTTAM	Jackson *et al.* (1988)	Simulation of cotton crop growth and yield.

(*continued*)

TABLE 3—*contd.*

Model	*Reference*	*Remarks*
Soybean		
GLYCIM	Acock *et al.* (1985)	Growth and development of soybean.
SOYGRO	Wilkerson *et al.* (1983)	Soybean model for crop management.
SOYMOD	Meyer *et al.* (1981)	Dynamic simulation of soybean growth development and seed yield.
REALSOY	Meyer (1985)	Redevelopment of SOYMOD.
SINCLAIR	Sinclair (1986)	A simple model to describe carbon, nitrogen and water budgets for soybeans.
TROSOY	Patron Sarti and Jones (1989)	A model of soybean growth in Mexico for assessing environmental constraints and potentials for soybeans.
Wheat		
TAMW	Maas and Arkin (1980)	A dynamic wheat model.
CERES	Ritchie and Otter (1985)	A model of the growth, development and yield of wheat for farm management and risk evaluation, yield forecasting and policy analysis.
VIC	O'Leary *et al.* (1985)	A model of the growth, development and yield of wheat
SSWC	van Keulen and Seligman	A model to provide a means for analyzing effects of soil moisture and N nutrition on growth and yield of spring wheat.
ARCWHEAT	Weir *et al.* (1984)	A winter wheat model designed to emphasise areas of needed research, interpret field experiments and estimate differences in yield due to soil and weather throughout the UK.
SIMTAG	Stapper (1984)	A wheat model accommodating different genotypes.
SIRIUS	Jamieson and Wilson (1988)	A simple model of growth, development and water use of wheat.
MACROS	Penning de Vries *et al.* (1989)	A model of growth and development of spring wheat.
Others		
SORGF	Arkin *et al.* (1976)	Sorghum growth and yield model.
SORKAM	Rosenthal *et al.* (1988)	Sorghum growth and yield model.
POTATO	Ng and Loomis (1984)	A model of the potato crop to study integrative physiology, climate–crop interaction, genotype evaluation and management strategies.

TABLE 4
Mixed Crop and Livestock Production Models

Model	*Reference*	*Remarks*
KLEIN	Klein & Sontag (1982)	A simulation model of beef, forage and grain production alternatives on farms in Western Canada.
EMBRAPA	Gutierrez-Aleman *et al.* (1986)	A linear programming model of traditional mixed farming systems in Northeast Brazil.
MIDAS	Kingwell & Pannell (1987)	A linear programming model of dryland farming in Western Australia including sheep and crops.
DAFOSYM	Rotz *et al.* (1989)	A model of forage production, harvest and feeding on dairy farms using CERESMaize and ALSIM1 (an Alfalfa model by Fick (1977)).

Models have been included which provide a vehicle for assessing a range of management considerations at the whole farm or enterprise level. Models which are at this level but which emphasise one consideration only, such as a model of lamb growth on a catch crop (Geisler *et al.* 1979), have been left out except where the production system itself is such that one or two factors dominate (e.g. feeding and environment control in poultry production). Models at less than the whole farm or enterprise level are included where they form the basis for later, whole enterprise models (e.g. the pig model of Whittemore (1980)). This distinction is less meaningful with crop models which refer either to the 'average plant' within a canopy or to the crop canopy itself as the unit of simulation.

Comparisons of model structures and performance have been made for beef models (Chudleigh and Cezar, 1982), sheep models (Elsen *et al.* 1988) and wheat models (Rimmington *et al.*, 1986). A comprehensive review of crop simulation models in general is provided by Whisler *et al.* (1986).

Use of Available Models

In the last ten years or so, production systems models have been used in a variety of circumstances to gain information and insights which would have been very difficult or very costly, if not impossible, to obtain in other ways. If modelling and model use have not come of age, then they are certainly well on their way to doing so. This is not to say that all models are well constructed and/or used appropriately and this is unlikely ever to be the case. Nevertheless, the use and usefulness of production system models are increasing. Models have been used to predict production for policy analysis; to identify production problems and guide research; to investigate the potential of different genotypes in particular environments; and to evaluate alternative production strategies and husbandry methods with respect to their management, economic and risk implications.

While the emphasis in this section is on the last of these, other uses are discussed briefly. Crop forecasting provides an example of the use of production system models in a policy context (e.g. Keener *et al.*, 1980). Use of models to guide research has been developed in a number of ways. Models have been integrated within experimental programmes designed to understand the mechanisms of system function and behaviour, although, as noted previously, this has tended to be more at a component rather than at a production system level (several examples have been cited in earlier chapters in this volume and others are given by Whisler *et al.* (1986)). Production system models have been used to evaluate the likely impact of R&D aimed at improvements in system components; for example, plant breeding research to change photosynthetic efficiency and other crop parameters (e.g. Landivar *et al.*, 1983; Elwell *et al.*, 1987). Models have also been used to identify constraints to system performance in particular circumstances which might either be investigated experimentally (e.g. causes of low growth rates in lambs; Black *et al.* (1979)), or be explained in some other way (e.g. causes of declining cotton yields; Wanjura and Barker (1988)).

Investigation of the potential of a new genotype or a variety of genotypes in different environments has been carried out with models both in the context of livestock systems (e.g. Blackburn and Cartwright, 1987*b*) and crop production systems (e.g. Aggarwal and Penning de Vries, 1989; Stapper and Harris, 1989). This is an application of models which might need to be approached with some caution since by definition the model must be operating outside the realm in which data used in its development were collected. The issue is addressed more fully in the next section on model development.

With respect to the main focus of this section, many of the livestock and crop production models listed in Tables 1–4 have been developed primarily with the objective of improving production systems on farms either through evaluation of alternative strategies of production, alternative husbandry methods or new systems of production, or through on-farm management and control of production.

There are some differences between the use of crop and livestock production models in this context simply because the dynamics of crop and livestock systems differ. Most crop models have been developed to estimate dry matter accumulation and yield in response to the major factors controlling or limiting plant growth, i.e. light, water, nutrients and pests. Within any given environment there is little that can be done under field conditions to change the amount of solar radiation received. Therefore, manipulation of yield is concerned primarily with ensuring that adequate levels of water and nutrients are available at critical times and that pests and diseases are kept under control. Once decisions have been made regarding the sequence and pattern (i.e. areas) of crops to be planted, management becomes a matter of determining input levels for these resources. These are within-season and within-enterprise resource allocation decisions and are primarily the concern of the next chapter. All that will be said here is that it is possible to address such issues in two ways. One is through evaluation of the risks and returns associated with different sets of decision rules (e.g. risk evaluation of irrigation strategies, Boggess and Ritchie (1988); and pest management strategies; Szmedra *et al.* (1988)); the other, and a primary objective in many cases, is to assist on-farm management and control directly. Development of expert systems technology has opened up new possibilities in this context and a number of decision support systems are being developed; a recent issue of the journal *Agricultural Systems* (**31**(1) (1989)) is devoted to the topic. In general, crop decision support systems seem to be better developed than livestock systems, though there are examples of the latter (e.g. Davies *et al.*, 1987).

Crop simulation models have also been used to address strategic issues such as cropping sequences and patterns, sometimes in combination with mathematical optimisation techniques (e.g. Bender *et al.*, 1984; Palmer, 1981; Tsai *et al.*, 1987) and sometimes by long run evaluation of different cropping systems (e.g. Blignaut 1986). Strategy evaluation within specific systems has also been undertaken using models of double cropping of soybeans and wheat (Chen & McClendon, 1985), and ratoon cropping of sorghum (Gerik *et al.*, 1988) for example.

As mentioned above, the dynamics of livestock systems are different from those of crop systems. Not only the status but the number of stock may change throughout a season or year and the interaction between stock and pastures or forages in pastoral systems adds an additional level of complexity to the system. Within pastoral systems models have been used to evaluate grazing management and the relative merits of rotational versus set stocking systems (e.g. Parsch and Loewer, 1987; Tharel *et al.*, 1985); the strategic use of pasture or forage resources (Doyle and Phipps, 1987; Thornton, 1988); supplementary feeding for different purposes and at different times of year (e.g. Blackburn *et al.*, 1989; Wadsworth, 1985; White and Bowman, 1987); timing of events such as lambing and shearing (Black and Bottomly, 1980; Bowman, 1989); the economics of introducing new technologies (Bowman *et al.*, 1989); and the benefits and costs of management strategies for responding to changes in market requirements (Bowman, 1989). Other areas in both pastoral and confined feeding systems which have been investigated with models include reproduction, breeding and related management considerations (Congleton and Goodwill, 1980*b*; Dijkuizen and Stelwagen, 1988; Marsh *et al.*, 1987); building capacity requirements (e.g. Singh, 1986); and environment control (e.g. Timmons and Gates, 1988*a*).

Several of these examples illustrate instances where it would be difficult if not impossible to accomplish objectives through live experimentation. Both crop and livestock production models can be used to simulate a large number of years encompassing a variability of climatic conditions that would be prohibitive to achieve in live experimentation. The analyses of White and Bowman (1987) and Bowman (1989), for example, use experimental periods of 20 to 26 years. In the case of the latter (Bowman, 1989), the analysis has been concluded before the anticipated changes in marketing (objective measurement of wool) has become widely used, which would not have been possible with live experiments. Clearly the range of factors, management interventions and environmental conditions that can be considered with a model far outweighs the scope possible with real experiments. The issue remains as to whether the results from simulation analyses can be used with the same confidence as results from experimental trials.

Some Issues in Model Development and Use

Confidence in a model and its range of application is usually built up gradually and depends on a number of considerations. It is not simply a matter of evaluating the model's behaviour relative to some real world data set; in fact, adequate test data sets have frequently not been easy to obtain. Confidence is also a function of the model's underlying assumptions and structure; whether it is a descriptive or mechanistic model (France and Thornley, 1984); and whether its structure has been derived empirically or with reference to underlying modes of action and function.

There are two respects in which production systems models may be limited in their application. The first is through cultural bias in the management systems elements of the model. An extreme example would be the difference between a drylot dairy model and a pastoral dairy model, but all models exhibit some degree of bias which reflects common production systems in the region in which they were developed, e.g. whether rotational grazing or set stocking is the norm, whether seasonal changes and conditions are consistent year to year or highly variable, whether water is abundant or deficient, etc. The objectives of the initial model development—which often reflect the discipline interests of the modeller—will also influence the balance of representation within the model. Contrast for example the grassland management orientation of models by Doyle and colleagues (Doyle and Edwards, 1986; Doyle *et al.*, 1987) with the animal genotype/management emphasis of the Texas livestock models (Sanders and Cartwright, 1979; Blackburn and Cartwright, 1987).

The second factor which may constrain model application is the generality of the biological response functions contained within it. This will depend on whether the model is descriptive or mechanistic and on whether response elements and relationships have been derived empirically or with reference to theoretical considerations. In the definition of any response element in a model, there are two steps involved; the structure (variables and functional form) of equations must be defined and then parameters must be quantified. The latter can only be done empirically in the sense that parameter values must be derived experimentally or deduced by reference to some data set. The former however may be done empirically, that is variables and equation forms may be determined on statistical grounds, as in for example step-wise regression, or it may be done on theoretical grounds based on an understanding of the relationships between, and function of, biological entities (Carson *et al.*, 1983).

Empirically defined response elements are likely to apply only where conditions are similar to those in which data from which the model was developed were collected. Where structure has been defined on theoretical grounds to accommodate specified sources of variation, the model should apply in all conditions where these are the major sources of variation, providing also that they are implicit in the data set used to parameterise the model. This highlights the necessity of clearly stating in the objectives of the model, the sources of variation to be accommodated and those to be ignored (sources of variation are accommodated when bias (a correlation between the factor concerned and residuals) has been eliminated from the model. Eliminating bias will not usually eliminate error (the existence of residuals)). This appears to be done more frequently with crop models than with livestock models, possibly because of the somewhat more limited set of sources of variation with the former.

Penning de Vries (1982) suggests that crop models go through a sequence of development stages. Preliminary models are constructed when insight at the explanatory level is vague and imprecise. Comprehensive models, which are explanatory and incorporate much of the available knowledge, can be developed when essential elements are thoroughly understood. Summary models can then be formulated in less detail, containing only elements identified as essential in comprehensive models. Reaggregation of explanatory models to develop models for extension and on-farm use was also suggested by Bywater (1984) in connection with livestock models. The contention is that summary or re-aggregated models will provide more accurate predictions over a wider range of circumstances than empirically derived models at a similar level of aggregation because they are based on identified causal factors. Evidence to support this has been developed by Oltjen (1990).

Given the above, it is almost inevitable that when a new issue is to be analyzed with a model or a model is to be used in a new environment, some modification to either the management systems elements or biological response functions will be necessary. If a model is not able to accommodate the management parameters or practices necessary for any proposed analysis, then either it has to be changed or the analysis has to be conducted some other way. In this case, the need for, and implementation of change are fairly obvious. The situation is not quite so simple with regard to the applicability of response relationships and different approaches to deal with the issue are evident in the literature. Models can be made more mechanistic or more theoretically based in order to be used in a wider range of circumstances, as described above, or they may remain at a more descriptive and empirical level and be reparameterised or recalibrated to a new environment. Examples of the former approach include the addition of a mechanistic rumen model to the PROSPECT sheep model (Black *et al.*, 1981) and the various developments and enhancements of the BREW model (see Table 1). An example of the latter are the several developments based on the TAMU cattle model (Table 1) although there are other examples of this model being used without modification. In some cases, entirely new models have been developed for each new analysis (e.g. the models AGRI, IGAP and RUA in Table 1).

The time and resource commitment necessary to restructure or reparameterise existing models may not be very different from the commitment necessary to build a new model. The choice of these approaches is likely to be more a function of the programme of which the model is a part and the personnel involved. Continued development and refinement of a model obviously requires an ongoing research or extension programme in which the model plays a central role. Many of the crop models in Table 3 and the Texas and Australian livestock models in Table 1 are part of such programmes.

A factor which may be overlooked in these circumstances is the cost of model maintenance. With a complex model which is subject to ongoing modification, there is a very real cost in simply maintaining 'clean' copies of the model and the computer programme which implements it. There are costs associated with identifying and resolving problems with both models and programmes and these tend to increase more than linearly with the complexity of each. This is rarely described in the literature and is often 'hidden' within funding requests for developments or enhancements to address new issues or new environments.

Another factor which is sometimes overlooked in the use of models, particularly at the production system level, is the need for data sets to initialise the model for analysis. Where the objective is to evaluate alternative strategies of production or production systems, data sets representing several farm situations or environments may be required. Such data sets obviously need to be compatible with the input requirements of the model; they also need to be internally consistent in the sense that the various biological components of the system are properly characterised. A simple example would be the need to ensure that in a pastoral livestock system, input data regarding stock numbers, animal genotypes, pasture growth characteristics and climatic conditions are 'in balance' and are not such as to introduce instability, or to delay attainment of stability, in model output—unless this is the purpose of the exercise. Climatic and price series are other examples of necessary data sets which may be difficult or costly to obtain. The international benchmark sites network for agrotechnology transfer (IBSNAT project; see ICRISAT (1984)) is a major international programme designed to address these sorts of issues. In a local situation it may be worth adopting a maxim similar to that often suggested to intending microcomputer purchasers regarding the necessity for a software budget; in the development of a production system model, assume it will cost at least as much to generate the data sets necessary to run the model as it will to construct it in the first place.

In summary, a number of well structured production systems models have been developed for various farming systems in the last ten to fifteen years. These models have been used to assist in policy analysis, in various research contexts, and to evaluate alternative production and management strategies for different farming systems and environments. Simulation analyses using appropriate models can accommodate a much wider range of factors, management interventions, and farm situations than can be contemplated with real experiments. They can also be completed in much less time. Confidence in a model and its range of application is usually built up gradually and depends on its underlying structure and method of derivation as much as its performance relative to test data. Introduction of mechanistic or theoretical elements into models will usually increase the range of environments to which the model is applicable, but all models have their limits in terms of either the management systems they represent or the sources of variation they accommodate. Factors which are frequently overlooked in programmes based on simulation models are the costs of model maintenance and initial data set development.

ACKNOWLEDGEMENT

The author acknowledges the support of the AGMARDT trust during preparation of this chapter.

REFERENCES

Acock, B., Reddy, V. R., Whisler, F. D., Baker, D. N., McKinion, J. M., Hodges, H. F. and Boote, K. J. (1985). *The Soybean Crop Simulator*

GLYCIM: *Model Documentation 1982.* PB85 171163/AS. Washington, DC: USDA.

Aggarwal, P. K. and Penning de Vries, F. W. T. (1989). Potential and water limited wheat yields in rice based cropping systems in Southeast Asia, *Agric. Syst.*, **30,** 49–69.

Allen, M. A. and Stewart, T. S. (1983). A simulation model for a swine breeding unit producing feeder pigs, *Agric. Syst.*, **10,** 193–211.

Arkin, G. F., Vanderlip, R. L. and Ritchie, J. T. (1976). A dynamic grain sorghum model, *Trans. of the ASAE,* **19,** 622–6, 630.

Arnold, G. W., Campbell, N. A. and Galbraith, K. A. (1977). Mathematical relationships and computer routines for a model of food intake, liveweight change and wool production in grazing sheep, *Agric. Syst.* **2,** 209–26.

Bailie, J. H. (1982). The influence of breeding management efficiency on dairy herd performance, *Anim. Prod.*, **34,** 315–23.

Baker, D. N., Lambert, J. R. and McKinion, J. M. (1983). *GOSSYM*: *A Simulator of Cotton Crop Growth and Yield.* Clemson, SC: Sth Carolina Agr. Exp. Sta. Tech. Bull. 1089.

Barlow, C., Jayasuriya, S. K., Price, E., Maranan, C. and Roxas, N. (1986). Improving the economic impact of farming systems research, *Agric. Syst.*, **22,** 109–25.

Bender, D. A., Peart, R. M., Barrett, J. R., Doster, D. H. and Baker, T. G. (1984). Optimizing cropping systems using simulation and linear programming, *ASAE Paper* 84-5017, Coll. Sta., TX: Texas A & M University.

Black, J. L. and Bottomly, G. A. (1980). Effects of shearing and lambing dates on the predicted pasture requirements of sheep in two Tasmanian locations, *Aust. J. Exp. Agric. Anim. Husb.*, **20,** 654–61.

Black, J. L., Dawe, S. T., Colebrook, W. F. and James, K. J. (1979). Protein deficiency in lambs grazing irrigated summer pasture, *Proc. Nutr. Soc. Aust.*, **4,** 126.

Black, J. L., Beever, D. E., Faichney, G. J., Howarth, B. R. and Graham, N. McC. (1981). Simulation of the effects of rumen function on the flow of nutrients from the stomach of sheep: Part 1—Description of a computer programme, *Agric. Syst.*, **6,** 221–41.

Blackburn, H. D. and Cartwright, T. C. (1987*a*). Description and validation of the Texas A & M sheep simulation model, *J. Anim. Sci.*, **65,** 373–86.

Blackburn, H. D. and Cartwright, T. C. (1987*b*). Simulated genotype, environment and interaction effects on performance characters of sheep, *J. Anim. Sci.*, **65,** 387–98.

Blackburn, H. D., Bryant, F. C., Cartwright, T. C. and Fierro, L. C. (1989). Corriedale sheep production when supplemented with ryegrass pasture in Southern Peru, *Agric. Syst.*, **30,** 101–15.

Blackie, M. J. and Dent, J. B. (1976). Analyzing hog production strategies with a simulation model, *Amer. J. Agr. Econ.*, **58,** 39–46.

Blignaut, C. S. (1986). An economic evaluation of some crop succession systems under dryland conditions in the Free State Midlands, *Agrekon,* **25,** 39–42.

Boggess, W. G. and Ritchie, J. T. (1988). Economic and risk analysis of irrigation decisions in humid regions, *J. Prod. Agric.*, **1,** 116–22.

Bowman, P. J. (1989). *Farm Management Strategies for Improving the Quality of Fine Wool,* PhD thesis, University of Canterbury, New Zealand.
Bowman, P. J., Wysel, D. A., Fowler, D. G. and White, D. H. (1989a). Evaluation of a new technology when applied to sheep production systems: Part 1—Model description, *Agric. Syst.,* **29,** 35–47.
Bowman, P. J., Fowler, D. G., Wysel, D. A. and White, D. H. (1989b). Evaluation of a new technology when applied to sheep production systems: Part II—Real-time ultrasonic scanning of ewes in mid-pregnancy, *Agric. Syst.,* **29,** 287–323.
Brown, L. G., Jones, J. W., Hesketh, J. D., Hartsog, J. D., Whisler, F. D. and Harris, F. A. (1985). *COTCROP: Computer simulation of cotton growth and yield.* MS, USA: Ag. and For. Exp. Stn., Info. Bull. 69.
Bruce, J. M. and Clark, J. J. (1979). Models of heat production and critical temperature for growing pigs, *Anim. Prod.,* **28,** 353–369.
Bryant, A. M. and Holmes, C. W. (1985). Utilisation of pasture on dairy farms, pp. 48–63 in: *The Challenge: Efficient Dairy Production,* ed. T. I. Phillips. Aust. Soc. Anim. Prod., Victoria, Australia.
Byerlee, D., Harrington, L. and Winkelmann, L. (1982). Farming systems research: Issues in research strategy and technology design, *Amer. J. Agric. Econ.,* **64,** 897–904.
Bywater, A. C. (1984). Use of models in management: Implications for development and delivery of technical information—predicting animal response, pp. 120–4 in: *Proc. 2nd Int. Workshop on Modeling Rum. Dig. and Metab.,* ed. R. L. Baldwin and A. C. Bywater. Davis, CA: University of California.
Carson, E. R., Cobelli, C. and Finkelstein, L. (1983). *The Mathematical Modeling of Metabolic and Endocrine Systems.* New York: John Wiley and Sons.
Charles, D. R. (1984). A model of egg production, *Br. Poultry Sci.,* **25,** 309–21.
Chen, L. H. and McClendon, R. W. (1985). Soybean and wheat double cropping simulation model, *Trans. of the ASAE,* **28,** 65–9.
Childs, S. W., Gilley, J. R. and Splinter, W. E. (1977). A simplified model of corn growth under moisture stress, *Trans. of the ASAE,* **20,** 858–65.
Christian, K. R., Freer, M., Donnelly, J. R., Davidson, J. L. and Armstrong, J. S. (1978). *Simulation of Grazing Systems.* Pudoc, Wageningen: Simulation Monographs.
Chudleigh, P. D. and Cezar, I. M. (1982). A review of bio-economic simulation models of beef production systems and suggestions for methodological development, *Agric. Syst.,* **8,** 273–289.
Collinson, M. P. (1980). Farming systems research in the context of an agricultural research organisation, pp. 381–9 in: *Farming Systems in the Tropics,* ed. H. Ruthenberg. Oxford: Clarendon Press.
Congleton, W. R. (1983). Dynamic model for combined simulation of dairy management strategies, *J. Dairy Sci.,* **67,** 644–60.
Congleton, W. R. and Goodwill, R. E. (1980a). Simulated comparisons of breeding plans for beef production—Part 1: A dynamic model to evaluate

the effect of mating plan on herd age structure and productivity. *Agric. Syst.*, **5,** 207–19.

Congleton, W. R. and Goodwill, R. E. (1980b). Simulated comparisons of breeding plans for beef production—Part 3: Systems for producing feeder-calves involving intensive culling and additional breeds of sire, *Agric. Syst.*, **5,** 309–18.

Conway, A. G. and Killen, L. (1987). A linear programming model of grassland management, *Agric. Syst.*, **25,** 51–71.

Corral, A. J., Neal, H. D. St. C. and Wilkinson, J. M. (1981). Economic analysis of the impact of management decisions on production and use of silage for a dairy enterprise, pp. 849–52 in: *Proc. XIV Int. Grassland Congr.*, ed. J. A. Smith and V. W. Hays. Boulder, CO: Westview Press.

Davies, G. T., Fleming, J. F. and Black, J. L. (1987). AUSPIG—A modelling package for maximising the profitability of intensive pig production, pp. 93–6 in: *Computer Assisted Management of Agricultural Production Systems,* ed. D. H. White and K. M. Weber. Victoria, Australia: Dept. Agric. and Rural Affairs.

DeShazer, J. A., Greninger, T. J. and Gleaves, E. W. (1981). Simulation model of poultry energetics for developing environmental recommendations, pp. 43–50, in: *Modelling, Design and Evaluation of Agricultural Buildings,* ed. J. A. D. MacCormack. Aberdeen: CIGR and Scot. Farm Bldg. Invest. Unit.

Dijkuizen, A. A. and Stelwagen, J. (1988). An economic comparison of four insemination and culling policies in dairy herds, by method of stochastic simulation, *Livest. Prod. Sci.*, **18,** 239–52.

Dijkuizen, A. A., Stelwagen, J. and Renkema, J. A. (1986). A stochastic model for the simulation of management decisions in dairy herds with special reference to production, reproduction, culling and income, *Prev. Vet. Med.*, **4,** 273–89.

Dowle, K., Doyle, C. J., Spedding, A. W. and Pollott, G. E. (1988). A model for evaluating grassland management decisions on beef and sheep farms in the UK, *Agric. Syst.*, **28,** 299–317.

Doyle, C. J. and Edwards, C. (1986). A model for evaluating grassland management decisions on dairy farms in the UK, *Agric. Syst.*, **21,** 243–66.

Doyle, C. J., Morrison, J. and Peel, S. (1987). Prospects for grass clover swards in beef production systems: A computer simulation of the practical and economic implications, *Agric. Syst.*, **24,** 119–48.

Doyle, C. J., Baars, J. A. and Bywater, A. C. (1989). A simulation model of bull beef production under rotational grazing in the Waikato region of New Zealand, *Agric. Syst.*, **31,** 247–78.

Doyle, C. J. and Phipps, R. H. (1987). The practical and economic consequences of integrating maize with grass production on dairy farms in Southern England: A computer simulation, *Grass and Forage Sci.*, **42,** 411–27.

Edelston, P. R. and Newton, J. E. (1977). A simulation model of a lowland sheep system, *Agric. Syst.*, **2,** 17–32.

Elsen, J. M., Wallach, D. and Charpenteau, J. L. (1988). The calculation of herbage intake of grazing sheep: A detailed comparison between models, *Agric. Syst.*, **26,** 123–60.

Elwell, D. L., Curry, R. B. and Keener, M. E. (1987). Determination of potential yield-limiting factors of soybeans using SOYMOD/OARDC *Agric. Syst.*, **24,** 221–42.

FAO (1989). *Farming Systems Development: Concept, Methods, Application.* Rome: FAO.

Fick, G. W. (1977). The mechanism of alfalfa regrowth: A computer simulation approach, *Search Agric.*, **7**(3), 1–28. NY: Cornell University.

Forster, T. G., Mowat, D. N., Jones, S. D. M., Wilton, J. W. and Stonehouse, D. P. (1984). Systems for producing leaner beef, *Agric. Syst.*, **15,** 171–88.

France, J. and Thornley, J. H. M. (1984). *Mathematical Models in Agriculture.* London: Butterworths.

France, J., Neal, H. D. StC. and Probert, D. W. (1983). A model for evaluating lamb production systems, *Agric. Syst.*, **10,** 213–44.

Galbraith, K. A., Arnold, G. W. and Carbon, B. A. (1980). Dynamics of plant and animal production of subterranean clover pasture grazed by sheep: Part 2—Structure and validation of the pasture growth model, *Agric. Syst.*, **6,** 23–43.

Garrity, D. P., Harwood, R. R., Zandstra, H. G. and Price, E. C. (1981). Determining superior cropping patterns for small farms in a dryland rice environment: Test of a methodology, *Agric. Syst.*, **6,** 269–83.

Gartner, J. A. (1981). Replacement policy in dairy herds on farms where heifers compete with the cows for grassland—Part 1: Model construction and validation, *Agric. Syst.*, **7,** 289–318.

Geisler, P. A., Newton, J. E., Sheldrick, R. D. and Mohan, A. E. (1979). A model of lamb production from an autumn catch crop, *Agric. Syst.*, **4,** 49–57.

Gerik, T. J., Rosenthal, W. D. and Duncan, R. R. (1988). Simulating grain yield and plant development of ratoon grain sorghum over diverse environments, *Field Crop Res.*, **19,** 63–74.

Graham, N. McC., Black, J. L., Faichney, G. J. and Arnold, G. W. (1976). Simulation of growth and production in sheep—Model 1. *Agric. Syst.*, **1,** 113–38.

Gutierrez-Aleman, N., De Boer, A. J. and Kehrberg, E. W. (1986). A bio-economic model of small ruminant production in the semi-arid tropics of the Northeast region of Brazil: Part 2—linear programming applications and results, *Agric. Syst.*, **19,** 159–87.

Heiler, T., King, R., Baird, J., O'Connor, K. F., Thompson, K. and Abrahamson, M. (1985). A simulation model of hill and high country pastoral systems for evaluation of irrigation investment, Proj. Rept No. 40, NZ Ag. Eng. Inst.

Hurwitz, H., Talpaz, H. and Waibel, P. E. (1985). The use of simulation in the evaluation of economics and management of turkey production: Dietary nutrient density, marketing age, and environmental temperature, *Poultry Sci.*, **64,** 1415–23.

ICRISAT (1984). *Minimum Data Sets for Agrotechnology Transfer,* Proc. of an Int'l Symp., ed. V. Krumble. Andhra Pradesh, India: ICRISAT.

Jackson, B. S., Arkin, G. F. and Hearn, A. B. (1988). The cotton simulation model 'COTTAM': Fruiting model calibration and testing, *Trans. of the ASAE,* **31,** 846–54.

Jamieson, P. D. and Wilson, D. R. (1988). Agronomic uses of a model of wheat growth, development and water use, *Proc. Agron. Soc. NZ,* **18,** 7–10.

Jones, C. A. and Kiniry, J. R. (1986). *CERES-Maize: A Simulation Model of Maize Growth and Development.* College Station, TX: Texas A & M Univ. Press.

Kahn, H. E. and Spedding, C. R. W. (1983). A dynamic model for the simulation of cattle herd production systems: Part 1—general description and the effects of simulation techniques on model results, *Agric. Syst.,* **12,** 101–11.

Keener, M. E., Runge, E. C. A. and Klugh, B. F. Jr. (1980). The testing of a limited data corn yield model for large-area yield predictions, *J. Appl. Meteor.,* **19,** 1245–1253.

Keulen, H. van and Seligman, N. G. (1987). *Simulation of Water Use, Nitrogen Nutrition and Growth of a Spring Wheat Crop.* Pudoc, Wageningen: Simulation Monographs.

Kingwell, R. S. and Pannell, D. J. (1987). *MIDAS, A Bioeconomic Model of a Dryland Farm System.* Pudoc, Wageningen: Simulation Monographs.

Klein, K. K. and Sontag, B. H. (1982). Bioeconomic firm-level model of beef, forage and grain farms in Western Canada: Structure and operation, *Agric. Syst.,* **8,** 41–53.

Konandreas, P. A. and Anderson, F. M. (1982). Cattle herd dynamics: An integer and stochastic model for evaluating production alternatives, Research Rept 2. Addis Ababa: ILCA.

Kuipers, A. (1982). Development and economic comparison of selection criteria for cows and bulls with a dairy herd simulation model. Agric. Res. Rept 913, Pudoc, Wageningen.

Lambourne, L. J. (1986). ILCA's policy towards modelling in the framework of livestock systems research, pp. 1–25 in: *Modelling of Extensive Livestock Systems,* ed. N. de Ridder, H. van Keulen, N. G. Seligman and P. J. H. Neate. Addis Ababa: ILCA.

Landivar, J. A., Baker, D. N. and Jenkins, J. N. (1983). Application of GOSSYM to genetic feasibility studies II. Analysis of increasing photosynthesis, specific leaf weight and longevity of leaves in cotton, *Crop Sci.,* **23,** 504–10.

Loewer, O. J., Smith, E. M., Benock, G., Bridges, T. C., Wells, L., Gay, N., Burgess, S., Springate, L. and Debertin, D. (1981). A simulation model for assessing alternate strategies for beef production with land, energy and economic constraints, *Trans. of the ASAE,* **24,** 164–73.

Loewer, O. J., Taul, K. L., Turner, L. W., Gay, N. and Muntifering, R. (1987). GRAZE: A model of selective grazing by beef animals, *Agric. Syst.,* **25,** 297–309.

Maas, S. J. and Arkin, G. F. (1980). pp. 80–3 in: *TAMW: A Wheat Growth and Development Simulation Model.* Texas Ag. Exp. Stn, Blackland Res. Ctr. Prog. and Model Doc.

Marsh, W. E., Dijkuisen, A. A. & Morris, R. S. (1987). An economic

comparison of four culling decision rules for reproductive failure in United States dairy herds using DairyORACLE, *J. Dairy Sci.* **70,** 1274–80.
McMeekan, C. P. and Walshe, M. J. (1963). Inter-relationships of grazing method and stocking rate in the efficiency of pasture utilisation by dairy cows, *J. Agric. Sci.,* **61,** 147–63.
Menz, K. M. (1980). Unit farms and farming systems research: The IITA experience, *Agric. Syst.,* **6,** 45–51.
Menz, K. M. and Knipscheer, H. C. (1981). The location specificity problem in farming systems research, *Agric. Syst.,* **7,** 95–103.
Meyer, G. E. (1985). Simulation of moisture stress effects on soybean yield components in Nebraska, *Trans. of the ASAE,* **28,** 118–28.
Meyer, G. E., Curry, R. B., Streeter, J. G. and Baker, C. H. (1981). Simulation of reproductive processes and senescence in indeterminate soybeans, *Trans of the ASAE,* **24,** 421–9, 435.
Ng, E. and Loomis, R. S. (1984). *Simulation of Growth and Yield of the Potato Crop.* Pudoc, Wageningen: Simulation Monographs.
Norman, D. W. (1986). Empirical results of farming systems research, pp. 128–32 in: *Perspectives on Farming Systems Research and Development,* ed. P. E. Hildebrand. Boulder, CO: Lynne Reiner.
Nye, J. C., McCarl, B. A., Nuthall, P. L., Bache, D. H. and Kadlec, J. E. (1980). Scheduling swine production facilities, *Trans. of the ASAE,* **23,** 1246–1248.
O'Leary, G. J., Connor, D. J. and White, D. H. (1985). A simulation model of the development, growth and yield of the wheat crop, *Agric. Syst.,* **17,** 1–26.
Olney, G. R. & Kirk, G. J. (1989). A management model that helps increase profit on Western Australian dairy farms. *Agric. Syst.,* **31,** 367–380.
Oltjen, J. W. (1990). Modelling growth and metabolism in cattle. In, *Proc. III Int'l Wshop on Modelling Dig. and Metab. in Farm Anims.* NZ: Lincoln Univ. (In press).
Palmer, B. C. (1981). Optimizing crop mix, planting dates and irrigation strategies. *ASAE Paper* 81–2097.
Parsch, L. D. and Loewer, O. J. (1987). Economics of simulated beef-forage rotational grazing under weather uncertainty. *Agric. Syst.,* **25,** 279–95.
Patron Sarti, R. and Jones, J. G. W. (1989). Soybean production in the tropics—A simulation case for Mexico. *Agric. Syst.,* **29,** 219–31.
Penning de Vries, F. W. T. (1982). Phases of model development, pp. 20–5 in: *Simulation of Plant Growth and Crop Production,* eds F. W. T. Penning de Vries and H. H. van Laar. Pudoc, Wageningen: Simulation Monographs.
Penning de Vries, F. W. T., Jansen, D. M., Ten Berge, H. F. M. and Bakema, A. H. (1989). *Simulation of Ecophysiological Processes in Several Annual Crops.* Pudoc, Wageningen: Simulation Monographs.
Reece, F. N. and Lott, B. D. (1982). Optimising poultry house design for broiler chickens. *Poultry Sci.,* **61,** 25–32.
Rice, R. W., MacNeil, M. D., Jenkins, T. G. and Koong, L. J. (1983). SPUR livestock component, pp. 74–87 in: *Simulation of Production and Utilization of Rangelands: A Rangeland Model for Management and Research,* ed. J. R. Wright. USDA Misc. Publ 1431.

Rimmington, G. M., McMahon, T. A. and Connor, D. J. (1986). A preliminary comparison of four wheat crop models, pp. 330–5 in: *Proc. Conf. Agric. Eng. IEAust.*

Ritchie, J. T. and Otter, S. (1985). Description and performance of CERES-Wheat: A user-oriented wheat yield model, pp. 159–75 in: *ARS Wheat yield project,* ed. W. O. Willis. USDA-ARS, ARS-38.

Rosenthal, W. D., Vanderlip, R. L., Jackson, B. S. and Arkin, G. F. (1988). *SORKAM: A Grain Sorghum Crop Growth Model.* Texas Ag. Exp. Stn, Blackland Res. Ctr. Prog. and Model Doc.

Rotz, C. A., Black, J. R., Mertens, D. R. and Buckmaster, D. R. (1989). DAFOSYM: A model of the dairy forage system. *J. Prod. Agric.,* **2,** 83–91.

Sanders, J. O. and Cartwright, T. C. (1979). A general cattle production systems model. I: Structure of the model. *Agric. Syst.,* **4,** 217–27.

Sibbald, A. R., Maxwell, T. J. and Eadie, J. (1979). A conceptual approach to the modelling of herbage intake by hill sheep. *Agric. Syst.,* **4,** 119–134.

Sinclair, T. R. (1986). Water and nitrogen limitations in soybean grain production I. model development. *Field Crop Res.,* **15,** 125–41.

Singh, D. (1986). Simulation-aided capacity selection of confinement facilities for swine production. *Trans. of the ASAE,* **29,** 807–15.

Sorensen, J. T. (1989). A model simulating the production of dual purpose heifers. *Agric. Syst.,* **30,** 15–34.

Stapper, M. (1984). *SIMTAG: A Simulation Model of Wheat Genotypes. Model Documentation,* ICARDA, Aleppo, Syria and Univ. of New England, Australia.

Stapper, M. and Arkin, G. F. (1980). *CORNF: A Dynamic Growth and Development Model for Maize (*Zea mays *L.).* Texas Ag. Exp. Stn, Blackland Res. Ctr. Prog. and Model Doc., 80–2.

Stapper, M. and Harris, H. C. (1989). Assessing the productivity of wheat genotypes in a mediterranean climate, using a crop simulation model. *Field Crop Res.,* **20,** 129–52.

Stewart, J. D. (1967). Lucerne in the farm programme, pp. 293–7 in: *The Lucerne Crop,* ed. R. H. M. Langer. Auckland, AH & AW Reed.

Szmedra, P. I., McClendon, R. W. and Wetzstein, M. E. (1988). Risk efficiency of pest management strategies: A simulation case study. *Trans. of the ASAE,* **31,** 1642–8.

Teter, N. C., DeShazer, J. A. and Thompson, T. L. (1973). Operational characteristics of meat animals part III: Broilers. *Trans. of the ASAE,* **16,** 1165–7.

Teter, N. C., DeShazer, J. A. and Thompson, T. L. (1976). Operational characteristics of meat animals part IV: Turkey, large white and bronze. *Trans. of the ASAE,* **19,** 724–7.

Tharel, L. M., Smith, E. M., Brown, M. A., Limbach, K. and Razor, B. (1985). A simulation model for managing perennial grass pastures. Part II—A simulated systems analysis of grazing management. *Agric. Syst.,* **17,** 181–96.

Thornton, P. K. (1988). An animal production model for assessing the bio-economic feasibility of various management strategies for the isohyperthermic savannas of Colombia. *Agric. Syst.,* **27,** 137–56.

Timmons, M. B. and Gates, R. S. (1986). Economic optimisation of broiler production. *Trans. of the ASAE,* **29,** 1373–8, 1384.
Timmons, M. B. and Gates, R. S. (1988a). Predictive model of laying hen performance to air temperature and evaporative cooling. *Trans of the ASAE,* **31,** 1503–9.
Timmons, M. B. and Gates, R. S. (1988b). Energetic model of production characteristics of tom turkeys. *Trans. of the ASAE,* **31,** 1544–51.
Tsai, Y. T., Jones, J. W. and Mishoe, J. W. (1987). Optimizing multiple cropping systems: A systems approach. *Trans. of the ASAE,* **30,** 1554–61.
Uys, P. W., Hearne, J. W. and Colvin, P. M. (1985). A model for estimating potential market offtake from subsistence herds. *Agric. Syst.,* **17,** 211–29.
Van Ee, G. R. and Kline, G. L. (1979). A corn production model for central Iowa. *ASAE Paper,* 79–4518.
Wadsworth, J. (1985). A model to evaluate the economic merits of dry season feeding of growing/fattening cattle in the sub-humid tropics. *Agric. Syst.,* **16,** 85–107.
Wanjura, D. F. and Barker, G. L. (1988). Simulation analyses of declining cotton yields. *Agric. Syst.,* **27,** 81–98.
Watt, D. L., DeShazer, J. A., Ewan, R. C., Harrold, R. L., Mahan, D. C. and Schwab, G. D. (1987). NCCISwine: Housing, nutrition and growth simulation model. *Appl. Agric. Res.,* **2,** 218–23.
Weir, A. H., Bragg, P. L., Porter, J. R. and Rayner, J. H. (1984). A winter wheat crop simulation model without water or nutrient limitations. *J. Agric. Sci.,* **102,** 371–82.
Whisler, F. D., Acock, B., Baker, D. N., Fye, R. E., Hodges, H. F., Lambert, J. R., Lemmon, H. E., McKinion, J. M. and Reddy, V. R. (1986). Crop simulation models in agronomic systems. *Adv. in Agron.,* **40,** 141–208.
White, D. H. and Bowman, P. J. (1987). Economics of feeding energy based supplements to grazing ewes before mating in order to increase the reproduction rate of a wool producing flock. *Aust. J. Exp. Agric.,* **27,** 11–7.
White, D. H., Bowman, P. J., Morley, F. H. W., McManus, W. R. and Filan, S. J. (1983). A simulation model of a breeding ewe flock. *Agric. Syst.,* **10,** 149–89.
Whittemore, C. T. (1980). A study of growth responses to nutrient inputs by modelling. *Proc. Nutr. Soc.,* **35,** 383.
Wilkerson, G. G., Jones, J. W., Boote, K. J., Ingram, K. I. and Mishoe, J. W. (1983). Modelling soybean growth for crop management. *Trans. of the ASAE,* **26,** 63–73.

4
Application of Systems Theory to Farm Planning and Control: Modelling Resource Allocation

C. J. DOYLE
The Scottish Agricultural College, Auchincruive, Ayr, UK

INTRODUCTION

Farm planning and control are concerned with the organisation of the farm's resources, such as land, labour, machinery, buildings and breeding livestock, and the management of the enterprises undertaken, such as milk, beef and cereal production. Thus, the production decisions facing the farmer may be broadly divided into three groups (Metcalf, 1969):

(i) resource–product problems concerned with determining the optimal amount of resources to use in the production of a commodity;
(ii) resource–resource problems concerned with the optimal combination of resources to produce a specific level of output; and
(iii) product–product problems involving the most profitable mix of crops and livestocks products to produce from the available resources.

This review is only concerned with the first two of these groups, involving decisions relating to the allocation of resources on the farm. More specifically, it focuses on the relevance and implications of systems theory to resource allocation decisions at the level of individual farm enterprises.

RELEVANCE OF SYSTEMS THEORY TO PROBLEMS OF ON-FARM RESOURCE ALLOCATION

The nub of the systems approach is a belief that the whole is more than the sum of its parts. This implies that an isolated study of the components that make up the system is inadequate to understand the complete system. This is because the separate parts are linked in an interacting manner and it is the interactions and inter-relationships between the various components which give the system its identity and organisational integrity (Dent and Anderson, 1971*a*; Rountree, 1977; Spedding, 1979). Thus, systems theory is primarily concerned with the systematic study of interactions between the different factors that make it up.

On the individual farm, many production processes can be usefully considered from a systems viewpoint (Dent and Anderson, 1971*b*; Dalton, 1982). This is because many of the processes are intrinsically linked and so have to be viewed in a holistic manner if they are to be properly understood and controlled. Most interactions in farming systems are of a complex biological nature but important economic links between processes are also involved. This is true of decisions regarding resource allocation where determination of the optimal level of resources to be used involves a knowledge not only of the biological responses but also of the price situation facing the farmer and his economic goals. Potentially the systems approach, in which the biological, economic and social aspects of a problem are examined in an integrated way, is very relevant to production and resource decisions in agriculture.

This real significance of systems theory is that its widespread application has had some profound effects on the ways in which evaluations of on-farm resource allocations are conducted. In particular, the adoption of a systems approach has presented new insights into:

(i) the concept of a resource;
(ii) the relationship between economic and technical efficiency;
(iii) the time-dependence of resource allocation decisions;
(iv) stochastic influences on resource allocation decisions;
(v) the consequences of multiple objectives in decision-making; and
(vi) the opportunities for process control.

Each of these issues are discussed in some detail.

MODELS—A WAY OF THINKING ABOUT SYSTEMS

At an early stage the development of systems theory became bound up with the use of mathematical models. The reasons for this are threefold (Wright, 1971). First, it is often impractical or impossible to study the real system. If research is concerned with designing new systems then by implication the corresponding real system does not exist. Second, even when the real system exists, experimentation may not be feasible due to factors of cost and time. For example, the national economy can be regarded as an economic system which can be observed and measured but cannot be manipulated for experimental purposes. Third, the very act of measurement may disturb the real system to such an extent that the observations relate to something that is artificial. Many of these problems can be overcome by the use of models or mathematical representations of the real world. Because the models only represent the key features of reality they are considerably easier to manipulate.

Consequently, as early as the 1950s with the advent of powerful mainframe computers which allowed more complex interactions to be studied, the rudiments of a systems approach to on-farm problems of resource allocation involving models became evident. The analysis of static, single input–output relationships (Heady and Dillon, 1961) began to be replaced by more complex mathematical models capable of determining the most profitable allocation of resources between various farm enterprises. All these early studies relied on mathematical programming techniques, notably linear programming (Swanson, 1956; Heady & Candler, 1958; Tyler, 1958) which centres on the use of formal algorithms to select an optimum solution.

However, while considerable progress in planning farming systems was made using such techniques, the rigid framework of mathematical programming made it impossible to incorporate the details of the biological and economic processes involved (Dent and Anderson, 1971). In spite of considerable ingenuity (McInerney, 1969; Rae, 1970; Agrawal and Heady, 1973), neither time considerations nor stochastic influences, which are basic to most agricultural processes, could be adequately represented by such methods. This led to the advent of simulation models (van Dyne and Abramsky, 1975), in which the emphasis on identifying optimum solutions was replaced by a concern with developing mathematical representations which could explain the detailed structure and the functioning of the processes involved. There are numerous applications of such models to problems of resource allocation on farms reported in the literature and a wide cross-section can be found in Dent and Anderson (1971*b*), Dalton (1975) and France and Thornley (1984). It is the combination of simulation modelling with systems theory which has provided the insights into on-farm resource allocation referred to earlier.

THE CHANGING CONCEPT OF A 'RESOURCE'—THE ENERGY FACTOR

The development of systems theory in respect of resource allocation problems in farming has been paralleled by conceptual changes in the way resources are envisaged and measured. This is evident in the evolution of energy as a basic resource.

The Energy Factor

Traditionally the basic agricultural resources have been taken to be land, labour and capital. However, in the early 1970s economic growth in the Western industrial nations became constrained by available supplies of fuel and a new primary resource, energy, suddenly assumed prominence in studies of on-farm resource use. The traditional triumvirate of resources was now replaced by a new triumvirate of land, labour and energy (de Wit, 1979). Inputs like machinery, fertilisers and herbicides began to be expressed in terms of the total amounts of labour and fossil energy required to produce them, including the labour and energy involved in procuring the raw materials (Leach, 1976). Some writers, such as Pimental and Pimental (1979), went further and also converted the labour into energy equivalents on the basis of presumed human energy requirements. An example of the total estimated energy inputs for maize in the United States in the mid-1970s is shown in Table 1. Such calculations permitted the energetic efficiency of different production systems to be calculated in terms of the energy input–output ratios. Other studies, for example Spedding and Walsingham (1975), even examined the efficiency with which solar radiation was utilised. As a comparatively inexhaustible source of energy, Spedding (1979) contended that the efficient capture and use of solar radiation would be a major objective of future farming systems.

TABLE 1
Typical Energy Inputs and Output for Maize Production in the US in the Mid-1970s ($GJ\ ha^{-1}$)

Inputs	
Labour	0·02
Machinery	2·34
Fuel	8·73
Fertiliser	9·44
Seeds	2·20
Irrigation	3·27
Herbicides and pesticides	1·20
Transportation	0·15
Total	27·35
Output	
Maize yield	80·16
Energy output–input ratio	2·93

A New Conceptual Approach

While criticisms can be levelled at the way in which measures of energetic efficiency have lumped together all forms of energy regardless of source or cost, the holistic way of calculating the energy inputs into agricultural enterprises is of significance. It is arguable that the conceptual approach embodied in the measures of energetic efficiency for different farm enterprises has owed much to systems theory. Conventionally, measures of land, labour and capital productivity in farming have merely expressed output as a function of the direct inputs into the enterprise. In contrast, measures of energetic efficiency have tended to consider both the direct inputs of fossil fuel into the enterprise and the quantities of fossil energy embodied in the capital and chemical inputs. This has clearly involved looking at the production of a commodity, like wheat or milk, as part of a production chain, running from the suppliers of raw materials through the manufacturers of machines and chemicals to farm output. While it must remain open to debate, it is doubtful that the study of energy use in agriculture would have progressed in the way that it did, if the idea of looking at agricultural activities from a systems perspective had not existed.

EXPLORING THE RELATIONSHIP BETWEEN ECONOMIC AND TECHNICAL EFFICIENCY—THE CASE OF CROP IRRIGATION

Considerations of Economic and Technical Efficiency

The advent of the systems approach and the use of simulation models to solve resource allocation problems has also tended to shift the emphasis away from searching for unique optimal solutions. Instead, the trend has been towards identifying sets of solutions which are efficient according to a particular technical or economic criterion. For example, in studying nitrogen use on grassland the aim might be to search out the nitrogen application rates which give (i) the highest profit, (ii) the highest yield and (iii) the lowest leaching losses. The decision-maker is then left to compare these various solutions to the problem of nitrogen use and make his own judgement about what is ideal. The value of this approach over the conventional search for a unique optimum is well illustrated by studies on irrigation.

Decisions about Water Use

Conventionally, recommended levels of irrigation for crops have been equated by engineers with the amounts of water required to maximise physical yields (for example Wiesner, 1970; Withers and Vipond, 1974). In practice, applications by farmers have often fallen significantly short of these recommendations and concern has been expressed about the deliberate underwatering of crops (for example, Falcon and Gotsch, 1971; Yates and Taylor, 1986). Carruthers and Clark (1981) have rightly observed that the so-called underwatering is an illusion stemming from the unconscious confusion of technical need with economic demand. From the viewpoint of biological efficiency it might make sense to apply water up to the point of maximum physical yield but actual applications by farmers will be governed by considerations of financial costs and returns.

It might be tempting to conclude from this that, as regards the optimal allocation of resources, considerations of biological efficiency are a total irrelevance. However, as Falcon and Gotsch (1971) observed in a study of irrigation in the Punjab, the persistent application of water below the technically optimal levels led to increasing soil salinisation. Thus, the farmers' pursuit of purely economic goals had serious long-term implications, which might well have been avoided if technical rather than economic efficiency had governed water use. Certainly, the long-term interests of the Punjabi farmers might have more closely coincided with the technically optimal solution for water use.

This example clearly illustrates that the pure pursuit of economic efficiency does not necessarily lead to what everyone recognises as an optimal allocation of farm resources. This has led Rehman (1982) to conclude that neither economists nor technologists have a uniquely valid view of what constitutes the optimum. As Spedding (1975) has observed, considerations of biological efficiency have just as much relevance in evaluating agricultural systems as ones of economic efficiency.

For this reason, simulation models, like those constructed by Flinn (1971) and Doyle (1981) to study both the physical and economic responses of individual crops to irrigation, are arguably of real value. They allow both technically and economically efficient solutions to the problems of water use to be studied. As a direct consequence, potential conflicts between technical requirements and economic efficiency can be readily identified and ways of more closely reconciling the two explored.

TIME-DEPENDENCE IN RESOURCE ALLOCATION DECISIONS—THE CASE OF WEED CONTROL

Time-Dependent Considerations in Resource Use

Most agricultural processes have important dynamic elements associated with them, so the state of the system is time-dependent. In turn, the optimal allocation of resources in such systems is also likely to be influenced by time considerations and in particular the time-horizon employed by the decision-maker. However, it is particularly difficult to study and understand time-dependent processes using conventional experimental techniques. This is because it is virtually impossible to exercise sufficient control on all the variables in the system, so as to trace the long-term consequences of individual changes. In this respect the development and use of simulation models has revolutionised the study of time-dependent processes. An example of the way in which their use has potentially extended this area of enquiry is provided by a review of weed control studies.

Studies in Weed Control

The financial implications of weed control in a crop are usually restricted to the calculation of economic thresholds on a single-year basis (Marra and Carlson, 1983). However, action taken in one year will often have repercussions in later years as result of the effect on the quantities of seed shed by the weed plants. For many weed species the seeds can remain dormant in the soil until appropriate conditions cause them to germinate. Because of the presence of these seed banks, it can take more than a one-year programme of control to produce a significant reduction in the size of the weed population. Accordingly, optimal weed control strategies need to consider both the short- and long-term consequences of control measures.

Because of constraints of cost and time, experiments are generally conducted for only one or two years and as such can only provide information on the immediate economic benefits of weed control (Doyle, 1989). For many weed species this may be a perfectly adequate basis on which to make decisions concerning control measures, but for quite a number the consequences of ignoring possible carryover effects from one year to the next are considerable. This is illustrated by a study by Doyle *et al.* (1986) which addressed the issue of what was the critical density of black grass (*Alopecurus myosuroides*) in winter wheat above which the costs of spraying with a herbicide (chlortoluron) would be less than the potential crop loss from not spraying. If only the improvement in wheat yield from spraying in the current harvest year was considered it was found that the critical black grass density was 30 plants m^{-2}. At weed densities below this the costs of the chemical sprays exceeded the value of extra crop produced. However, if the implications of current black grass populations on weed levels in subsequent years were taken into account then the economic threshold density for spraying declined to 7·5 plants m^{-2}.

For the specific conditions investigated by Doyle *et al.* (1986), the difference in projected benefits net of spray costs over a 10-year period from adopting 30 instead of 7·5 black grass plants m^{-2} as the critical density for spraying with chlortoluron are shown in Table 2. To allow for differences between the time when costs and benefits of spraying arise, both the costs and benefits were discounted at a rate of 15 per cent (for a discussion of discounting see Ritson (1978) pages 206–8; 282–92). Also shown is the difference in the frequency of spraying, expressed as the percentage of years in which chlortoluron would be

TABLE 2
Effect of Choosing Threshold Densities of 7·5 and 30 Plants m^{-2} for Applying Chlortoluron to Control Black Grass in Winter Wheat on the Net Discounted Benefits and Frequency of Spraying

Threshold density (plants m^{-2})	*7·5*	*30·0*
Net discounted benefits (£ ha^{-1})	300·0	225·0
Percentage of years in which crop sprayed	50·0	33·0

applied, assuming a dosage of 3·5 kg ha^{-1}. From this table it can be seen that the adoption of the higher threshold density for spraying could lower the requirements for spraying at the expense of a 25 per cent reduction in the projected net economic benefits.

The impact that the consideration of longer-term carryover effects can have on the optimum threshold level of weed infestation for spraying is also evident in studies by Auld *et al.* (1979), Doyle *et al.* (1984), Cousens *et al.* (1986), Moore *et al.* (1989) and Popay *et al.* (1989). Significantly, what is common to all these studies is the use of mathematical models which seek to simulate the biological cycle of weed reproduction, growth and competition. In each case, because of considerations of cost, complexity and time, it was concluded that a model of the real system would allow management decisions to be more effectively investigated than any experiment. In particular, both the long-term implications of weed control and the general instability of weed populations seem to be more easily addressed by mathematical models than field trials. Thus, the application of systems models has extended understanding of the influence of time considerations in resource allocation decisions.

STOCHASTIC INFLUENCES ON RESOURCE ALLOCATION DECISIONS—THE IMPACT ON GRASSLAND MANAGEMENT

Stochastic Variation in Agricultural Systems

A feature of most agricultural production processes is uncontrollable variation in response. Perhaps more than in industrial processes, the

TABLE 3
The Effect of Annual Rainfall Level on the Most Profitable Level of Fertiliser Nitrogen (N) to Apply to a Typical Area of Permanent Grassland in Britain

Annual rainfall (mm)	*Optimal (kg ha^{-1} year^{-1})*
250	300
350	370
450	440

response of crop yields and livestock output to inputs is subject to unplanned variation due to weather and disease, as well as unpredictable movements in market prices. One consequence of this is that there is no such thing as a unique optimum input level. This is clearly illustrated in Table 3 which shows how the most profitable level of fertiliser nitrogen to apply to a typical area of permanent grassland in Britain varies with the actual rainfall conditions during the year. The results assume a nitrogen:grass price ratio of 10:1 and are based on response curves estimated by the author from the grassland manuring trial GM20 (Morrison *et al.*, 1980). From Table 3, it is evident that the optimal level of nitrogen use can be expected to vary significantly from year to year.

Risk Analysis
Of course, if the probability distribution of rainfall is known, then it is possible to frame the optimal level of nitrogen usage in terms of the level which would maximise expected profits (Rae, 1977, pp. 367–413). However, the consequences of uncertainty in output response extend beyond the problem of defining the optimal level of a particular resource. The very fact that the response to an input is uncertain means that the decision-maker faces the risk of financial loss as a result of events not turning out as expected. As Francisco and Anderson (1972) and Bond and Wonder (1980) have observed, aversion to taking risks may significantly affect production and investment decisions by altering not only the level of input usage, but also the way inputs are combined. Thus, Doyle (1986) has shown theoretically that concern to minimise risks by grassland farmers might be responsible for fertiliser nitrogen usage being lower and concen-

trate feeding levels higher on farms than considered optimal from the viewpoint of maximising profits.

Consideration of Stochastic Effects in Grassland Farming

The existence of stochastic factors therefore greatly increases the complexity of on-farm decisions regarding resource allocation. This is particularly true for grassland management, where such variability only adds to the already difficult problem of identifying the impact of input changes on both grass production and utilisation. In this respect the development of systems models capable of simulating grass growth and the consequent livestock production has significantly increased the potential for incorporating stochastic influences in decisions regarding input levels and stocking rates. It is virtually impossible with physical experiments to make observations over a long enough sequence of years or seasons to study satisfactorily variability of grass production. On the other hand, provided suitable weather data are available, it is possible to design computer models which can simulate the impact of weather changes on both grass availability and livestock performance over long periods of time. Using such models, management strategies which maximise expected returns to livestock producers over time can be determined.

Thus, Jones and Brockington (1971), Morley and Graham (1971), Doyle (1981) and Charpenteau and Duru (1983) all used computer models to explore the implications of climatic variability on grass availability and livestock performance. Specifically, Jones and Brockington (1971) and Doyle (1981) sought to determine the physical and financial implications of installing irrigation on livestock farms in the United Kingdom. Using the models, the number of years in which irrigation could significantly boost grass production was estimated, and the average rainfall and soil conditions on the farm needed for irrigation to be economic were established. In contrast, Morley and Graham (1971) and Charpenteau and Duru (1983) used weather-driven models of grass production to establish the quantities of conserved fodder required by livestock farmers to provide security against unpredictable drought or winter conditions.

Risk Considerations in Management Decisions

In each case the models were concerned with optimising expected returns to the producer. Thus, although they allowed for output variability, they ignored the fact that the behaviour of farmers might be modified by uncertainty. In general, livestock farmers are considered to be averse to risk and to prefer a steady income to one which oscillates between extremes of high and low (Johnson and Bastiman, 1981). Thus, in deciding on how to allocate resources producers might be expected to consider both the expected income and the expected variance of that income. Theoretically, it is possible to express the way in which each individual discounts expected profits for the risks involved by means of a utility function (Officer and Halter, 1968; Dillon, 1971*a*, pages 102–48; Morris, 1974). A popular form for such a function is the quadratic (MacArthur and Dillon, 1971), namely:

$$\mathrm{U}(Z) = \mathrm{E}(Z) - b\{[\mathrm{E}(Z)]^2 + \mathrm{VAR}(Z)\} \quad (1)$$

This basically states that the utility or subjective benefit derived from a risky income, $\mathrm{U}(Z)$, is a function of the expected net income, $\mathrm{E}(Z)$ and the variance of that income, $\mathrm{VAR}(Z)$.

Using the idea of the utility function, both Anderson (1971) and MacArthur and Dillon (1971), in separate studies of Australian graziers, sought to combine stochastic variation in grass production with a model of producer behaviour which implied that farmers would reject pasture management strategies with high financial risks. In both cases the computer models indicated that, if producers were risk-averse, it would be rational for them to stock land less intensively than suggested by purely profit-maximising behaviour.

However, these studies are exceptions. Comparatively few models of grassland management address the issue of risk. The reason for this is that relatively little is known about farmers' attitudes to risk-taking, mainly because there are significant differences between individual farmers (Young, 1979). As such, it is impossible to construct generally applicable utility functions. In a study of stocking rates and fertiliser usage on UK dairy farms, Doyle and Lazenby (1984) tried to circumvent this problem by merely seeking to simulate the mean and variability of income associated with each management strategy, without attaching any weights to the outcomes. While this modelling exercise showed that there was sufficient difference between the risk-minimising and the profit-maximising stocking rates to explain the observed variance in stocking rates on farms, without some implied objective function it could not answer the question of what is the 'optimal' stocking rate.

Thus, there is no question that most models of resource allocation on farms have given insufficient consideration to the effect that imperfect knowledge and uncertainty have on the decision-maker's behaviour. Nevertheless, it has still to be said that the systems approach has the potential to offer important insights into the effect that stochastic influences have on resource allocation. As Dillon (1971*b*) observed, the essence of management is choosing between alternative risky prospects.

THE CONSEQUENCES OF MULTIPLE OBJECTIVES IN DECISION-MAKING—LIVESTOCK RATION FORMULATION

Risk analysis is merely a special case of an analysis involving multiple objectives. In practice, most decisions regarding production and resource allocation on farms are based on compromises between different objectives. As Spedding (1979) and Dalton (1982) observed, farmers rarely define their goals in terms of intangible objectives such as maximising profits. For instance, cereal farmers are more likely to express their goals in terms of target yields and prices. In this respect systems theory, which presents higher order outputs as the product of lower order systems, encourages the decomposition of objectives, such as maximising profits, into sub-objectives. It is only a short step from recognising sub-objectives to framing decisions in terms of multiple objectives.

Existence of Multiple Management Goals

Certainly, farmers' decisions tend to be based on more than just a consideration of profits. In particular, work by Gasson (1973) on the goals of farmers in Cambridgeshire, UK, revealed that factors such as liking the work, independence, job security and working close to home were all as important as maximising income in determining decisions. Similarly, research by Smith and Capstick (1976) in the United States and Cary and Holmes (1982) in Australia showed that livestock graziers had a wide variety of both economic and sociological goals. Despite this empirical evidence, most system models simplify reality by assuming that profit is the only criterion which shapes decisions.

Analysing Resource Decisions under Multiple Objectives

Nevertheless, techniques involving variants on linear programming are available for representing multiple goals and have been applied to resource allocation decisions (Romero and Rehman, 1989). The most notable of these is goal programming, which involves the simultaneous optimisation of several objectives, subject to a set of linear constraints (Romero and Rehman, 1984). There are two variants of this method; lexicographic (Charnes and Cooper, 1961) and weighted goal programming (Romero, 1986). Both involve trying to find the solution which minimises the deviations between the actual levels of attainment and the aspiration levels for the various goals. The main problem with goal programming is that the decision-maker needs to decide on the priorities for his various objectives and this becomes difficult where the number of goals is high. In addition, Zeleny and Cochrane (1973) observed that, where targets for several goals have been set at pessimistic levels, goal programming could generate an apparent optimal solution which was sub-optimal in the sense that there was another feasible solution which would lead to an increase in one of the goals without worsening the others.

In many cases the average farmer is not capable of adequately articulating his targets and setting priorities between them (Dalton, 1982). As Koopmans (1951) suggested, it may be more sensible to derive sets of feasible solutions which are Pareto optimal, in the sense that there are no other feasible solutions which can achieve the same or better performance for all the goals under consideration and for at least one of the objectives the outcome is better. For example, consider a farmer faced with three feasible solutions to the problem of irrigating grassland with the implications for profits, debt and water use presented in Table 4. Assuming that his goals are to achieve the highest profits possible, commensurate with minimising indebtedness and water use, then the second solution is inefficient. It offers the same profits as the first, but involves higher water use. On the other hand, while the third solution involves higher debt than the first, it offers

TABLE 4
Hypothetical Levels of Profit, Indebtedness and Water Use Associated With Three Feasible Irrigation Strategies

Strategy	*Profit ($\$ ha^{-1}$)*	*Indebtedness ($\$ ha^{-1}$)*	*Water use (mm)*
1	400	50 000	250
2	400	50 000	350
3	600	80 000	400

higher profits. Thus, the first and third solutions are both Pareto efficient. Which is adopted will depend on the farmer's preferences.

Multi-objective programming employs the concept of Pareto-efficient solutions in handling situations involving multiple goals. Instead of identifying an optimal solution, it generates a set of non-inferior or efficient feasible solutions and leaves the decision-maker to choose between them (Zeleny, 1982). Not surprisingly, where the number of objectives considered is large, the computational costs of defining the complete set of Pareto-efficient solutions is very high. For this reason, a variant referred to as compromise programming has been used (Zeleny, 1982; Romero and Rehman, 1989). This involves constraining the search for non-inferior, efficient and feasible solutions to a sub-set of the total. Basically, this is achieved by first defining an ideal, but non-feasible solution, and merely searching for Pareto-efficient solutions in the vicinity of the ideal one.

Multiple Criteria Decision-Making in Feed Rationing

Although the use of these techniques has found limited application in on-farm resource allocation problems, one area where it has been successfully applied is in the formulation of livestock feed rations. While conventional linear programming (LP) has been used successfully to determine least-cost rations, it does involve rather rigid formulation procedures. In the first place, the explicit view in LP that minimisation of cost is the only consideration involved in determining ration compositions is a considerable distortion of the truth. In formulating rations, a compromise is likely to be involved between considerations of cost, nutrient imbalances and food bulk. In the second place, LP is over-rigid in the way in which the animal's feed requirements are specified, with no violation of the nutritional constraints being allowed. However, in real life the constraints on the amounts of energy and protein which may be included in the animal's diet are very much more flexible. Small imbalances in nutritional supplies are unlikely to affect an animal's physical or economic performance.

Rehman and Romero (1984, 1987) have applied both goal and multi-objective programming to the problem of formulating rations for dairy cows and steers. The results showed that if, besides cost, considerations such as the levels of trace minerals and the bulk of the diet were also taken account of then quite considerable differences in the optimal ration occurred. At the same time, their studies also revealed that there was quite a range of feasible solutions for the rations, which differed very little in cost. This implied that for a comparatively small economic trade-off, the farmer could satisfy other technical goals.

Thus, the traditional framework used for analysing resource decisions, in which the choice among a range of feasible solutions is based on a single well-defined goal, is not only inadequate but potentially a serious distortion of reality. In this respect the basic principles of the systems approach, in which there is a hierarchy of relationships and linkages, have a relevance to the formulation of farmer's goals. Just as systems can be broken down into sub-systems, so conceptually objectives can be disaggregated into sub-objectives.

OPPORTUNITIES FOR PROCESS CONTROL—THE CASE OF PEST AND DISEASE CONTROL

Most models of agricultural systems remain little more than research tools which can be used to draw broad advisory guidelines on the management of crops and livestock. However, an essential task of management is one of control (Lange, 1970; Dalton, 1982) which involves keeping the production system within some preferred limits. Typical control actions by farm managers include measures to prevent crop failure through the use of herbicides and the use of medicines to combat a drop in milk production caused by mastitis. From their inception the possibility that simulation models of farm systems might be used to control processes in the real world has exercised considerable fascination (Baker and Curry, 1976; France and Thornley, 1984). However, because of the complexity of many farming systems, it has generally proved impossible to develop sufficiently accurate models of reality.

Control Models in Pest and Disease Management

In a few areas some progress has been made towards developing systems which can be used to control a biological process. A notable area is in disease and pest control. Typical of initial attempts in this field is the work of Webster (1977) which was concerned with the possibilities of devising for cereal producers decision rules capable of taking account of local conditions and circumstances when deciding on whether to spray against *Septoria* in the crop. Using subjective probabilities provided by the farmer concerning the response to spraying, Webster put forward a procedure for giving advice to individual cereal growers on when to spray.

The procedure developed by Webster (1977) simply relied on a comparison between an estimate of the likely response to spraying and the farmer's own break-even response. No attempt was made to simulate the effect of weather and disease incidence on crop losses. Later studies attempted to extend the analysis by integrating the biological and economic aspects of pest and disease control in a single model. Typical of these is the mechanistic model developed by Barlow (1985) which used data on the state of the pasture, the time of year and the pest incidence at a particular locality to assess whether a sheep farmer should spray against *Porina*. In this case the actual results from the model were used to develop a pocket slide rule for farmers which provided estimates of the threshold incidence of *Porina* above which it was economic to spray it.

However, Barlow's study, like others such as Conway *et al.* (1975), concentrated on modelling the biological aspects of the pest system in a deterministic fashion, making no allowance for uncertainty of response. As Gold and Sutton (1986) observed, most pest and disease control decisions are made under conditions of substantial uncertainty. The farmer is not generally certain about either the future extent of the pest or disease, or the crop losses caused by a given pest or disease incidence. Thus, in deciding on the best course of action he has to estimate the relative likelihood of events and the risks incurred. Accordingly, Gold and Sutton (1986), in a study of disease control in apple, sought to combine a mechanistic model, which described the dynamics of the disease progress, with a formal paradigm for making decisions under uncertainty. This basically involved calculating expected losses associated with alternative spray policies and selecting the policy with the lowest expected loss using Bayesian techniques (see de Groot, 1970; Anderson *et al.* 1977).

However, while the model developed by Gold and Sutton (1986) took account of uncertainty, it assumed that apple growers were risk-neutral in that they were only interested in the expected or average gains and losses. In practice decisions to spray often reflect an aversion to taking risks, with smaller average gains being traded-off against a reduction in the risk of occasional large losses. Thornton and Dent (1984*a*,*b*) addressed this problem when developing a farm-level computer-based management information system for advising farmers on the use of fungicide to control the leaf rust, *Puccinia hordei,* in barley. Using a model which simulated the dynamics of the disease, they derived a series of tables containing recommendations to spray and to refrain from spraying for various weather and crop conditions. Using utility functions (see section on 'Stochastic influences on resource allocation decisions'), tables of recommendations were prepared for three broad categories of farmer attitude to risk.

Notwithstanding these advances, it has to be accepted that the use of simulation models to control biological processes still has a long way to go. The very complexity of biological systems and their susceptibility to unplanned variations make it difficult to design adequate representations of the real world. Nevertheless, the systems approach to analysing processes and resource decisions on farms potentially opens the prospects of using models as aids to control of individual farm processes.

AN OVERVIEW

Although the preceding discussion represents an uneven and incomplete review of the application of systems theory to farm planning and control, it has attempted to document some of the key ways in which systems concepts have influenced the theory and practice of resource allocation on farms. In particular, the distinguishing features of a systems approach are arguably a concern with (i) holism, (ii) linkages between components, and (iii) the mechanisms responsible for changes. Each of these aspects has been echoed in the way the theory of agricultural resource allocation has developed in recent years.

The emphasis in systems theory on adopting a holistic approach to problems has apparently engendered an interest in studying both the direct and indirect consequences of resource allocation decisions on farms. As illustrated, this is especially evident in the studies made of energy efficiency in farming which have considered not only direct energy inputs into enterprises but also the energy contents of machinery and chemical inputs.

Another consequence of systems theory has been the displacement of the traditional concern for the optimum with an interest in identifying linkages and interactions between whole sets of feasible and efficient solutions. Specifically, in resource allocation studies on farms there has been an evident move away from simply searching for a unique profit-maximising outcome to finding relationships between a variety of solutions which are efficient according to different technical and economic criteria. This has partly occurred because the use of simulation models to study farm systems has increased the ability to explore large numbers of feasible solutions. At the same time, most simulation models are not goal-driven in that they search out feasible rather than optimal solutions.

While this may be considered a retrograde step, nonetheless it has had two positive effects. First, by focusing on the relationship between various feasible solutions, systems models have increased the awareness that technically and economically efficient solutions to resource problems may not only be radically different but incompatible. By encouraging both the technical and economic consequences of decisions to be explored, the dangers that an obsession with economic considerations may lead to an oversight which has unacceptable technical or social ramifications is potentially reduced. Second, following on from this concern with identifying efficient rather than optimal solutions has been a recognition that the decision-maker's ideal outcome involves some compromise between the various feasible and efficient solutions. Thus, farmers' decisions about resource allocation are unlikely to be based purely on consideration of profit, but financial risk and other technical considerations will influence them.

Lastly, the inherent stress in the systems approach on describing the mechanisms by which changes occur has been responsible for important insights in regard to resource use on farms. Because of limitations of time and cost it is difficult to employ experimental trials to explore the impact of time-dependent and stochastic influences on resource decisions. However, by using mathematical models which are capable of simulating the mechanisms responsible for change, the impact of both these influences on decisions regarding levels and combinations of farm inputs can be explored in some detail. Combined with experimentation, models of farm systems are potentially a powerful tool for analysing resource allocation problems.

However, it has to be admitted that the contribution of the systems approach has been greater to the theory than the practice of resource allocation. Disturbingly, the reasons for the failure of systems concepts and simulation models to have any practical impact on farming remain the same as those outlined by Dent (1975) some 15 years ago. In the first place, the failure of systems researchers to liaise with farm decision-makers has meant that farmers are rightly suspicious of computer-generated predictions of optimal resource use. In the second place, the preoccupation of systems researchers with model-building rather than application has greatly limited the practical use of most resource allocation models. This is reflected in the limited progress that has been made in using mathematical models to control real farm processes. Thus, if the full potential of the insights into resource allocation gained from the application of systems concepts is to be realised, then the emphasis in future will need to be on practical rather than conceptual advances.

REFERENCES

Agrawal, R. C. and Heady, E. O. (1973). *Operations Research Methods for Agricultural Decisions.* Ames: Iowa State University Press.

Anderson, J. R. (1971). Spatial diversification of high risk sheep farms, pp. 239–66 in: *Systems Analysis in Agricultural Management,* eds J. B. Dent and J. R. Anderson. Sydney: John Wiley & Sons.

Anderson, J. R., Dillon, J. L. and Hardaker, B. (1977). *Agricultural Decision Analysis.* Ames: Iowa State University Press.

Auld, B. A., Menz, K. M. and Medd, R. M. (1979). Bioeconomic model of weeds in pastures, *Agro-Ecosystems,* **5,** 69–84.

Baker, C. H. and Curry, R. B. (1976). Structure of agricultural simulation—A philosophical view, *Agricultural Systems,* **5,** 201–18.

Barlow, N. D. (1985). A model for the impact and control of *Porina* on sheep farms: development and field application, pp. 152–9 in: *Proceedings of the 4th Australasian Conference on Grassland Invertebrate Ecology,* ed. R. B. Chapman. Christchurch: Caxton Press.

Bond, G. and Wonder, B. (1980). Risk attitudes amongst Australian farmers, *Australian Journal of Agricultural Economics,* **31,** 29–44.

Carruthers, I. and Clark, C. (1981). *The Economics of Irrigation.* Liverpool: Liverpool University Press.

Cary, J. W. and Holmes, W. E. (1982). Relationships among farmers' goals and farm adjustment strategies: some empirics of a multidimensional approach, *Australian Journal of Agricultural Economics,* **26,** 114–30.

Charnes, A. and Cooper, W. W. (1961). *Management Models and Industrial Applications of Linear Programming, Vol. I.* New York: John Wiley & Sons.

Charpenteau, J-L. and Duru, M. (1983). Simulation of some strategies to reduce the effect of climatic variability on farming systems: the case of the Pyrenees Mountains, *Agricultural Systems,* **11,** 105–25.

Conway, G. R., Norton, G. A., Small, N. J. and King, J. B. (1975). A systems approach to the control of the sugar cane froghopper, pp. 193–229 in: *Study of Agricultural Systems,* ed. G. E. Dalton. London: Applied Science Publishers.

Cousens, R., Doyle, C. J., Wilson, B. J. and Cissans, G. W. (1986). Modelling the economics of controlling *Avena fatua* in winter wheat, *Pesticide Science,* **17,** 1–12.

Dalton, G. E. (ed.) (1975). *Study of Agricultural Systems.* London: Applied Science Publishers.

Dalton, G. E. (1982). *Managing Agricultural Systems.* London: Applied Science Publishers.

Dent, J. B. (1975). The application of systems theory in agriculture, pp. 107–27 in: *Study of Agricultural Systems,* ed. G. E. Dalton. London: Applied Science Publishers.

Dent, J. B. and Anderson, J. R. (1971*a*). Systems, management and agriculture, pp. 3–14 in: *Systems Analysis in Agricultural Management,* eds. J. B. Dent and J. R. Anderson. Sydney: John Wiley & Son.

Dent, J. B. & Anderson, J. R. (1971*b*). *Systems Analysis in Agricultural Management.* Sydney: John Wiley & Sons.

Dillon, J. L. (1971*a*). *The Analysis of Response in Crops and Livestock.* Oxford: Pergamon Press.

Dillon, J. L. (1971*b*). Interpreting systems simulation output for managerial decision-making, pp. 85–120 in: *Systems Analysis in Agricultural Management,* eds. J. B. Dent and J. R. Anderson. Sydney: John Wiley & Sons.

Doyle, C. J. (1981). Economics of irrigating grassland in the United Kingdom, *Grass and Forage Science,* **36,** 297–306.

Doyle, C. J. (1986). Economic considerations in the production and utilisation of herbage, pp. 217–27 in: *Managed Grasslands: Analytical Studies,* ed. R. W. Snaydon. Amsterdam: Elsevier Science Publisher.

Doyle, C. J. (1989). Modelling as an aid to weed control management, pp. 937–42 in *Proceedings of the Brighton Crop Protection Conference—Weeds—1989, Vol. 3.* Farnham: British Crop Protection Council.

Doyle, C. J. and Lazenby, A. (1984). The effect of stocking rate and fertiliser usage on income variability of dairy farms in England and Wales, *Grass and Forage Science,* **39,** 117–27.

Doyle, C. J., Oswald, A. K., Haggar, R. J. and Kirkham, F. W. (1984). A mathematical modelling approach to the study of the economics of controlling *Rumex obtusifolius* in grassland, *Weed Research,* **24,** 183–93.

Doyle, C. J., Cousens, R. and Moss, S. R. (1986). A model of the economics of controlling *Alopecurus myosuroides* Huds. in winter wheat, *Crop Protection,* **5,** 143–50.

van Dyne, G. M. and Abramsky, Z. (1975). Agricultural systems models and modelling: an overview, pp. 23–106 in *Study of Agricultural Systems,* ed. G. E. Dalton. London: Applied Science Publishers.

Falcon, W. P. and Gotsch, C. H. (1971). Relative price response, economic efficiency and technological change: A case study of Punjab agriculture, in: *Development Policy II—The Pakistan Experience,* eds. W. P. Falcon and G. F. Papanek. Cambridge: Harvard University Press.

Flinn, J. C. (1971). The simulation of crop-irrigation systems, pp. 123–51 in: *Systems Analysis in Agricultural Management,* eds. J. B. Dent and J. R. Anderson. Sydney: John Wiley & Sons.

France, J. and Thornley, J. H. M. (1984). *Mathematical Models in Agriculture.* London: Butterworth.

Francisco, E. M. and Anderson, J. R. (1972). Chance and choice west of the Darling, *Australian Journal of Agricultural Economics,* **16,** 82–93.
Gasson, R. (1973). Goals and values of farmers, *Journal of Agricultural Economics,* **24,** 521–37.
Gold, H. J. and Sutton, T. B. (1986). A decision analytic model for chemical control of sooty blotch and flyspeck diseases of apple, *Agricultural Systems,* **21,** 129–57.
De Groot, M. H. (1970). *Optimal Statistical Decisions.* New York: McGraw-Hill.
Heady, E. O. and Candler, W. V. (1958). *Linear Programming Methods.* Ames: Iowa State University Press.
Heady, E. O. and Dillon, J. L. (1961). *Agricultural Production Functions.* Ames: Iowa State University Press.
Johnson, J. and Bastiman, B. (1981). Social and human factors in grassland farming, pp. 268–88 in: *Grassland in the British Economy,* ed. J. L. Jollans. Reading: Centre for Agricultural Strategy.
Jones, J. G. W. and Brockington, N. R. (1971). Intensive grazing systems, pp. 188–211 in: *Systems Analysis in Agricultural Management,* eds J. B. Dent and J. R. Anderson. Sydney: John Wiley & Sons.
Koopmans, T. C. (1951). Analysis of production as an efficient combination of activities, pp. 33–97 in: *Activity Analysis of Production and Allocation,* ed. T. C. Koopmans. New York: John Wiley & Sons.
Lange, O. (1970). *Introduction to Economic Cybernetics.* Oxford: Pergamon Press.
Leach, G. (1976). *Energy and Food Production.* Guildford: IPC Science and Technology Press.
MacArthur, I. D. and Dillon, J. L. (1971). Risk, utility and the stocking rate, *Australian Journal of Agricultural Economics,* **15,** 20–35.
McInerney, J. P. (1969). Linear programming and game theory models—some extensions, *Journal of Agricultural Economics,* **20,** 269–78.
Marra, M. C. and Carlson, G. A. (1983). An economic threshold model of weeds in soyabeans (*Glycine max.*), *Weed Science,* **31,** 604–39.
Metcalf, D. (1969). *The Economics of Agriculture.* Harmondsworth: Penguin Books.
Moore, W. B., Doyle, C. J. and Rahman, A. (1989). The economics of controlling *Carduus nutans* on grazed pasture in New Zealand, *Crop Protection,* **8,** 16–24.
Morley, F. H. W. and Graham, G. T. (1971). Fodder conservation for drought, pp. 212–36 in: *Systems Analysis in Agricultural Management,* eds J. B. Dent and J. R. Anderson. Sydney: John Wiley & Sons.
Morris, J. (1974). The utility approach to making decisions under uncertainty, *Oxford Agrarian Affairs,* **3,** 15–28.
Morrison, J., Jackson, M. V. and Sparrow, P. E. (1980). The response of perennial ryegrass to fertiliser nitrogen in relation to climate and soil. *Report of the Joint ADAS/GRI Grassland Manuring Trial—GM20.* Grassland Research Institute Technical Report no. 27. Hurley: The Grassland Research Institute.
Officer, R. R. and Halter, A. N. (1968). Utility analysis in a practical setting, *American Journal of Agricultural Economics,* **50,** 257–77.

Pimental, D. and Pimental, A. (1979). *Food, Energy and Society*. London: Edward Arnold.

Popay, A. J., Barlow, N. D. and Bourdot, G. W. (1989). Economics of controlling *Ranunculus acris* in New Zealand Dairy Pastures, pp. 943–8 in: *Proceedings of the Brighton Crop Protection Conference—Weeds—1989, Vol. 3.* Farnham: British Crop Protection Council.

Rae, A. N. (1970). Profit maximisation and imperfect competition: an application of quadratic programming to horticulture, *Journal of Agricultural Economics,* **21,** 133–46.

Rae, A. N. (1977). *Crop Management Economics*. London: Granada.

Rehman, T. (1982). Technical and economic criteria in agricultural production: a case for a systems approach to farm decision-making in the Pakistan Punjab, *Agricultural Systems,* **9,** 43–55.

Rehman, T. and Romero, C. (1984). Multiple-criteria decision-making techniques and their role in livestock ration formulation, *Agricultural Systems,* **15,** 23–49.

Rehman, T. and Romero, C. (1987). Goal programming with penalty functions and livestock ration formulation, *Agricultural Systems,* **23,** 117–32.

Ritson, C. (1978). *Agricultural Economics: Principles and Policy*. London: Crosby Lockwood Staples.

Romero, C. (1986). A survey of generalised goal programming (1970–82), *European Journal of Operational Research,* **25,** 183–91.

Romero, C. and Rehman, T. (1984). Goal programming and multiple criteria decision-making in farm planning: an expository analysis, *Journal of Agricultural Economics,* **35,** 177–90.

Romero, C. and Rehman, T. (1989). *Multiple Criteria Analysis for Agricultural Decisions*. Amsterdam: Elsevier.

Rountree, J. H. (1977). Systems thinking—Some fundamental aspects, *Agricultural Systems,* **2**, 247–54.

Smith, D. and Capstick, D. (1976). Establishing priorities among multiple management goals, *Southern Journal of Agricultural Economics,* **2,** 37–43.

Spedding, C. R. W. (1975). The study of agricultural systems, pp. 3–19 in: *Study of Agricultural Systems,* ed. G. E. Dalton. London: Applied Science Publishers.

Spedding, C. R. W. (1979). *An Introduction to Agricultural Systems*. London: Elsevier Applied Science Publishers.

Spedding, C. R. W. and Walsingham, J. M. (1975). The production and use of energy in agriculture, *Journal of Agricultural Economics,* **27,** 19–30.

Swanson, E. R. (1956). Application of programming analysis to corn belt farms, *Journal of Farm Economics,* **38**.

Thornton, P. K. and Dent, J. B. (1984*a*). An information system for the control of *Puccinia hordei*: I—Design and operation, *Agricultural Systems,* **15,** 209–24.

Thornton, P. K. and Dent, J. B. (1984*b*). An information system for the control of *Puccinia hordei*: II—Implementation, *Agricultural Systems,* **15,** 224–43.

Tyler, G. T. (1958). An application of linear programming, *Journal of Agricultural Economics,* **13,** 473–85.

Webster, J. P. G. (1977). The analysis of risky farm management decisions:

advising farmers about the use of pesticides, *Journal of Agricultural Economics,* **28,** 243–53.
Wiesner, C. J. (1970). *Climate, Irrigation and Agriculture.* Sydney: Angus and Robertson.
Withers, B. and Vipond, S. (1974). *Irrigation: Design and Practice.* London: B. T. Batsford.
de Wit, C. T. (1979). The efficient use of labour, land and energy in agriculture, *Agricultural Systems,* **4,** 279–87.
Wright, A. (1971). Farming systems, models and simulation, pp. 17–33 in: *Systems Analysis in Agricultural Management,* eds J. B. Dent and J. R. Anderson. Sydney: John Wiley & Sons.
Yates, R. A. and Taylor, R. D. (1986). Water use efficiencies in relation to sugar cane yields, *Soil Use and Management,* **2,** 70–6.
Young, D. L. (1979). Risk preferences of agricultural producers: their use in extension and research, *American Journal of Agricultural Economics,* **61,** 1063–70.
Zeleny, M. (1982). *Multiple Criteria Decision Making.* New York: McGraw-Hill.
Zeleny, M. and Cochrane, J. L. (1973). A priori and a posteriori goals in macro-economic policy making, pp. 373–91 in: *Multiple Criteria Decision Making,* eds J. L. Cochrane and M. Zeleny. South Carolina: University of South Carolina Press.

5
Optimising the Mixture of Enterprises in a Farming System

J. B. Dent
Division of Rural Resource Management, The Scottish Agricultural College, Edinburgh, UK

INTRODUCTION

It is a basic tenet of systems theory that systems comprise subsystems that interact. For systems directed by man, control is attempted by managing these sub-systems and the interactions between them in an unpredictable environment.

> *Whole agricultural systems deserve to be studied in their own right.*
>
> (Spedding, 1976)

To this extent the 'exploitation' of the systems approach in designing agricultural enterprises (on the one hand) and farming systems (on the other) are based on the same principles. A paper concerned with optimising the mixture of enterprises then essentially should be redundant following one concerned with enterprise management. Furthermore, farming in much of the developed world has become an issue of managing a set of individual enterprises. Individual farm enterprises driven by advancing technology have developed almost in isolation. Industrial inputs into farming have almost broken the sub-system (enterprise) interaction in farming systems. Certain dependencies between enterprises, of course, remain; these are related to the need to distribute scarce resources within the farm business. The management of a farm as a system has been relegated to resource acquisition and an allocation problem between (almost) independent enterprises.

It is not surprising that this situation has arisen: the whole infrastructure for agriculture is set within a commodity framework. This is true in all parts of the world. Almost all research is organised and funded on a commodity basis. Most farming policy is independently determined for separate commodities. The majority of advice to farmers is provided at enterprise or sub-enterprise level. Economies of scale and the feasibility of technical advancement in Western agriculture has been in sympathy with this fragmented approach to farming.

While infrastructure has encouraged a similar trend in Third World agriculture, lower levels of industrial inputs have maintained the interdependence between sub-systems in the whole farm system: balance between restorative crops and exploitative crops is required as is the interaction between crop enterprises (in, for example, intercropping systems) and livestock and crop enterprises. Some agricultures in the developed world have maintained the same characteristics but these relate to more extensive and/or integrated production patterns, e.g. sheep and wheat systems in parts of Australia and dairying in New Zealand.

Extension and Farming Systems Adjustment

In the developed world, economic changes have meant that there is a reducing capacity within the agricultural infrastructure for translating research into practice. Free advice from government agencies is becoming more restricted in many countries and state-owned demonstration farms are less numerous than a decade ago. Major facilities to demonstrate whole-farm systems as representative of a farm type have not been a favoured mechanism for extension. Those facilities that do exist often have the whole-farm message to farmers and advisers clouded by an overlay of field-scale enterprise research. A serious and successful attempt at whole-farm system demonstration was instituted in New Zealand by the establishment of dairy farmlets (Hutton, 1968). In the main, however, extension effort has been directed towards enterprise management or to sub-systems of enterprises (for example, winter feeding systems for cattle). Operations thus tend to be demonstrated out of context of the totality of the whole farm.

As agricultural policy in Europe moves slowly towards structural reform within a 'stabilizer'-type price support system for commodities and as encouragement is given towards more environmentally sensitive farming practices, the structure of some farming systems in Europe is likely to change to incorporate more rotational interdependence.

During this change, planning new systems will require a more obvious application of biological knowledge and an understanding of the biological interaction between the enterprises involved. It will not be a trivial adjustment and appropriate modelling procedures will undoubtedly assist the process: such models will need to recognise the socio-economic sub-systems as well as the biological (Anderson *et al.*, 1985).

> *The subject of agricultural systems is envisaged as one that combines the relevant economics and social science with the essential underlying biology.*
>
> (Spedding, 1976)

Farming Systems Research

In the Third World, Farming Systems Research (FSR) has become a major force in extension. At least FSR purports to be a holistic concept. Four sequential phases to FSR have been identified (Dent and Thornton, 1988) where knowledge of the district is limited:

(a) the examination of existing production systems with respect to constraints (appraisal);
(b) the identification of potential improvements (experimentation);
(c) the evaluation of promising production possibilities under local farmers' conditions (design); and
(d) extension to more farmers' fields (implementation).

The identification of potential improvements may relate to modifications in the enterprise sub-systems or a change of balance in enterprises or, indeed, the introduction of a new enterprise may be at the expense of an existing one. Experimentation in the district may involve field trials of new cultivars of existing crops, new crops or new technology packages on existing cultivars. In any case, the experimental phase must obviously be a long one, perhaps over several seasons, but never long enough to expose the true variability of the field trial to external variables such as rainfall and temperature. The results of experiments are heavily dependent on the climate sequence of the particular season, the specific soil type and the numerous management factors under the control of the investigator—for example, the established plant population, the timing of cultivations and sowing and pest control procedures. The critical conceptual issue is whether the results from such field trials can bear much relationship to the way 'technology' developed from them will function in the fields of smallholder farmers in a different place, in a different year, on a different soil type and where farming is constrained by a range of socio-economic conditions neither experienced nor sometimes even perceived at the research station.

The holistic nature of FSR is evident in that any potential change is assessed during demonstration in a whole-farm context. In the fields of collaborating farmers, interaction between the various sub-systems during the course of a season are implicitly involved in the evaluation. Further, an appreciation of the resource demands of the farm is possible and the social and economic implications for the farmers and their families may be judged. However, it is clear that the FSR sequence (involving experimentation and demonstration) should normally be worked out over many years, a time delay which is rarely acceptable.

MODELLING IN THE STUDY OF FARM SYSTEMS

Mathematical Programming Models

Study of agricultural systems is usually assisted by the creation of abstract models.

> *Mathematical model-building offers new dimensions in aids to decision-making and comprehension.*
>
> (Spedding, 1976)

It is interesting to note that the model type most frequently used to consider the enterprise mix problem is of a mathematical programming format. This is to be expected since this type of model is well suited to resource allocation issues: it is not so well suited, for example, to the dynamic expression of the biology of farming activities. Linear programming is the simplest and probably the most frequently applied model to explore alternative enterprise combinations within a stated resource framework of the farm. Other mathematical programming methods allow the modeller to specify the system under study in alternative (perhaps more realistic) ways. It is not the purpose of this paper to review the characteristics nor applications of such methods: this has been done on many previous occasions (e.g. Hazell, 1971; Smith, 1973; McCarl *et al.*, 1977; Hardaker, 1979; Hazell and Norton, 1986; McGregor, 1986).

There are few examples where mathematical programming methods have been used in practice (but, for example, see Thompson, 1976 and Pannell and Falconer, 1988) to assist individual farmers make decisions about which enterprises to run and how large they should be (enterprise mix optimisation). In so far as farmers are concerned with the enterprise mix problem, their planning would seem to be intuitive with hardly any formal appreciation of the underlying biology and with emphasis on general input–output estimates and financial outcomes. Simple financial models of whole farms are used by some advisers avoiding mathematical programming techniques. These models take a standard form such as a forward annual budget or a projected monthly cashflow statement. While such models fulfil a specific purpose, any systems view which they may represent is subjugated to a broad financial assessment of a pre-determined enterprise mix.

Simulation Models

Formal whole-farm systems analysis and subsequent modelling to display the complex dynamics and uncertainty of the biology, sociology and economics of farm systems has rarely been attempted. A few research papers can be cited (e.g. Baum and Schertz, 1983; Beck and Dent, 1987) but it is possible that data difficulties and even conceptual uncertainty have dampened enthusiasm for simulation modelling at whole-farm level except with relatively simple systems. Such studies as have been completed are simplistic and appear to have concentrated on more extensive grazing livestock farm systems with beef and/or sheep products being the sole output from the farm (Monteiro *et al.*, 1980; Beck *et al.*, 1982). Work in this area has tended to be orientated to the normative study of farmer behaviour and therefore to farm systems change in response to change in the (economic) environment. Further, no great attention has been paid to the variability inherent in enterprise systems and any models developed are unlikely to be of immediate value in the study of alternative options for enterprise adjustment within real farm systems. Nor is it likely in the foreseeable future that whole-farm systems simulation models will be of value to individual farmers as a basis for operational decision support.

Nevertheless, simulation modelling of farming systems has the potential to overcome some of the difficulties related to FSR mentioned previously and to speed the transition from experimentation to adoption. The advantage of farm system modelling is that a specific technology can be assessed *ex ante* at enterprise level and also within a whole-farm context.

While insignificant research has been carried out in the area of whole-farm simulation modelling, a great deal of international research effort has been directed towards crop simulation (e.g. Ritchie, 1986; Godwin *et al.*, 1989; Penning de Vries *et al.*, 1989) and to a slightly lesser extent livestock simulation (e.g. Di Marco *et al.*, 1989). For simplicity, the following discussion relates to crop systems and is restricted to that type of simulation model that takes account of the dynamics of biological processes such as light interception and water uptake. Such models have been termed mechanistic (Thornley and France, 1984).

Enterprise Models: Strategy Comparison

A suitable crop model can simulate daily crop growth, development and final yield, taking account of soil, climate, cultivar types and managerially controlled inputs. Assuming that confidence has been established that the crop model can produce similar outcomes to those experienced from field trials, it is possible to take an alternative approach to the experimental stage of FSR in identifying potential improvements. Field trials would not be the source of primary data, rather they would be designed to validate and calibrate the model for a particular location. The model itself would then be used to generate data for extension purposes much more quickly and comprehensively than from real field trials. In particular, running the model for a specific site (agro-climatic zone) would involve the use of simulated or historical weather time series. Repeats of an 'experiment' can be carried out over many years of representative simulated climate to produce a frequency distribution for yield for each treatment. Production stability in relation to weather over time can then be investigated explicitly. The frequency distribution for output may be expressed in cumulative form to facilitate analysis (Anderson, 1976). In this way, models can be used to explore any question that involves factors to which the model is sensitive. Figure 1, for example, illustrates the simulated results of five wheat varieties grown at a locality in Australia in rainfed conditions at the same point in the rotation. Yield variability in this environment is considerable but the analysis shows that the cultivar Titan stochastically dominates all others in grain yield. What can be achieved for simple 'treatments' such as cultivar types equally can be done for more complex, strategy-type 'combination treatments', e.g. a specified date of sowing, seed rate, fertiliser rate, cultivar combination.

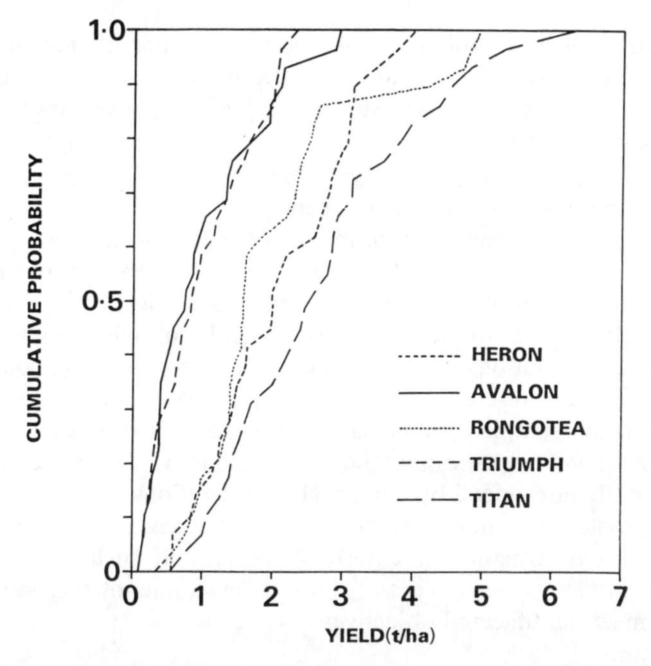

Fig. 1. Yields for five varieties of wheat simulated over 30 years in Australian rainfed conditions (adapted from McGregor and Thornton, in press).

Strategy comparison using first-degree stochastic dominance analysis is not always as unequivocal as indicated in Fig. 1 where Titan is the preferred option for all seasons that might be experienced in the locality. An alternative 'traditional' strategy, for example, may never yield below a certain amount without fail, and provide a stable though low average output and in addition may outyield an alternative strategy in, say, 3 years out of 10. Furthermore, cash or resource carryover from one season to another may be of paramount importance (in broad terms, the farming system's economic sustainability); low but dependable performance may allow the household to subsist from one year to another while high average performance but poor performance under difficult climatic conditions may spell disaster if poor years are experienced in succession. The decision to adopt a new strategy for the crop under consideration will thus be complex, but at least with yield frequency distributions generated from a model there is more information on which to formulate a sensible extension package. Such information is superior by an order of magnitude to the data from a replicated field trial which indicates at some degree of statistical confidence that one strategy is preferable over another in a given year.

Whole-Farm Systems Models: Concepts

The decision to be made by a farmer is, of course, further complicated by the fact that he normally does not rely on a single enterprise for the survival of his family. In spite of remarks earlier, many farmers, particularly in the tropics, still do depend on integrated farming systems using rotations for the maintenance of soil fertility, control of weeds, pests and diseases, and inter-cropping systems of various types to exploit the complementary nature of crops and ostensibly to limit risk. These influences cannot be determined by a single enterprise system study nor indeed by biological considerations alone.

The problem for individual farmers then becomes one of designing the scale and structure of enterprise systems in such a way as to operate within the resource and cultural constraints of the household and to meet multifaceted objectives.

> *Agricultural activities are carried out for many different purposes, from making money or saving imports to providing products and employment.*
>
> (Spedding, 1976)

The concept of optimisation in some economic sense, as implied in the title of this chapter, is excessively simple in this context, where trade-offs between a large number of farming, personal, business, family and group objectives will always be the norm. For those interested in studying such whole-farm systems by way of models, social, cultural and economic sub-systems must be added to the biological components of crop or livestock models in order to prescribe integrated enterprise strategies that provide more successfully for the well-being of the farm household.

There is an obvious need to have appropriate biological models for major farm enterprises for any whole-farm representation. Some reorganisation of those models currently available will be necessary not least to permit individual farmers' fields to be simulated in parallel. Such transformation is a relatively simple task (Thornton and Blair-Fish, 1990). Because of the technical nature of many of the alternative strategies that may be suggested it is doubtful that much reduction in biological detail of the present range of crop models be feasible. This may be considered as the technological base fo whole-farm model and will provide data to a larger structure which w permit the examination of alternative technologies for enterpris sub-systems within the total farm–household framework. In thi context, alternative strategies can be assessed in terms of the potential overall impact on production output and on household values. Clearer notions also may be gained about likely adoption. Equally important, estimates can be made about the way adoption likelihood may be improved by minor adjustments in technology or by creating a more favourable environment for adoption (e.g. credit provision).

> *There is great need to try to understand the influence of the socio-economic environment on the choice of systems.*
>
> (Spedding, 1976)

Farm household modelling studies have all stressed that account be taken of consumption behaviour of the farm family, attitudes to risk, borrowing and investment and of seasonal labour and other resource availability (Kingma and Kerridge, 1977). Marginal propensity to consume has been determined in a number of situations to give meaning to a regression model linking household income to family consumption from survey data (Barnum and Squire, 1979). The simplicity of this model resembles the plant growth models of 50 years ago (for example, Crowther and Yates, 1942). Asymptotic yield response curves represented the state of biological understanding at that time. Research has probed the processes of growth and development in plants to the point where these can be largely represented in terms of a mechanistic model. Comparatively, the resources deployed towards revealing the mechanisms determining the behaviour of the farm family in respect to consumption (and investment and risk and so on) have been trivial. Scant attention has been paid to the equally important social parameters including attitudes, values, traditions, peer group pressure and culture (Gasson, 1973; Symes and Appleton, 1985). Consequently, levels of understanding in these areas seem to be based on limited empirical observation. That so much is appreciated and quantified about the physiology of a crop plant and so little about the social system in which it is but one of many sub-systems seems to be contrary to systems thinking. Certainly, it reflects the 'reductionist' approach to research direction and expenditure in agriculture which

...undly criticised when he called for research to be ...anised within a holistic framework.

... recalled that whole-farm systems themselves are part ...of systems and that the typical assessment procedures ... or Farming Systems Research development externalises ... interactions with, for example, environmental systems, ...n systems and regional economic systems. Such exter... ...e not necessarily neglected but rarely do they impact on the ...-making at farm level and to this extent the value systems of ... are ignored.

> *By considering the whole system it is possible to suggest which changes in what parts might have a particular effect on other parts in which the investigator is interested.*
>
> (Spedding, 1976)

There is little doubt that in the short term neither experimentation nor observational studies in social anthropology will yield data that will permit the same mechanistic approach to modelling of whole-farm systems as has been used in modelling crop sub-systems. This has been a severe limitation on modelling whole-farm systems and probably much of the reason why unsatisfactory models have emerged incorporating untenable behavioural assumptions about household aspirations and objectives. Recent advances in the discipline of artificial intelligence has suggested ways in which these elements can be encapsulated in qualitative terms. Rule-based algorithms familiar to the expert systems researchers (Pederson, 1989), have recently been used, for example, to model smallholder agro-forestry systems in India (Sharma *et al.*, 1990).

Whole-Farm Systems Models: Potential New Structure

An expert system has been defined as a computer structure incorporating expert human reasoning which will reach the same conclusions as a human expert would when working with identical data (Weis and Kulikowski, 1984). Just as human reasoning can make use of databases and mechanistic models so too can expert systems. When faced with data inadequacy or deficiency or with a new situation a human expert will apply past experience in a questioning sequence and using background knowledge will reach a conclusion. Expert systems attempt to formulate the rules by which experts proceed in such situations and are therefore ideal constructs for dealing with qualitative (soft) data.

Such structures can be set up to represent the whole-farm system and to call enterprise sub-system mechanistic models and other databases as appropriate. The approach may be best appreciated by way of a simple example related to the decision about whether a smallholder farmer should introduce a new crop cultivar—in this case a new variety of *Phaseolus* bean—into the farming systems of farmers near Jutiapa in Guatemala. The following factors will have a bearing on the decision:

(a) required seasonal labour changes for jobs such as weeding;
(b) required cash input additions for, say, phosphatic fertiliser;
(c) varietal preference of the household if the beans are to be consumed in the family, in particular whether the new beans are an acceptable colour and shape;
(d) the general innovativeness of the particular household; whether the resource base is adequate to take on a new technology such as this and whether family motivation is sufficient;
(e) information availability (advice) about the new cultivar, particularly if unfamiliar husbandry techniques are involved;
(f) estimated production and variability of production of both current and new cultivars in the locality;
(g) rotational considerations about whether the new cultivar will adequately replace the old in the cropping sequence;
(h) marketability of the new cultivar if the beans are to be sold off the farm: particularly related to the palatability of the bean for the market and whether the new cultivar can be stored under farm conditions without deterioration.

Each factor is influenced by a tapestry of technical, economic and social conditions, exemplified in the following discussion relating to (d) above, the general innovativeness of the household. Farmers in the area may be obviously classified into tenants and those who farm their own land (Table 1). At a second level, the classification distinguishes between households with either low or high cash reserves. For each of these four typified farm households an ordered ranking of attributes is hypothesised in terms of the perceived relative importance in influencing household innovation. Thus, for a tenant farming household with low cash reserves it is suggested that risk aversion will be the major motivating force in relation to innovation, followed by household size and the level of discretionary consumption possible indicating the base production for subsistence of the family. For such farmers these

TABLE 1
The Household's Potential for Innovation

Land owner		*Tenant*	
Cash reserves		*Cash reserves*	
Low	*High*	*Low*	*High*
Risk aversion	Income level	Risk aversion	Social aspirations
Social aspirations	Age	Household size	Income level
Family size	Education	Discretionary consumption	Age
	Acess to information		
Income level	Risk aversion		Risk attitude
Age	Social aspirations		Household size
Cash growth			
Off-farm earning activity			

factors would appear to be so overwhelming in motivating the household that other issues such as social aspirations or recent income levels would not be relevant.

The potential model structure for this particular classification to determine likely innovation in an expert systems framework is shown in Fig. 2. At each step, questions preserving the hierarchy established in Table 1 are arranged in descending order of importance. Sequential examination is carried out until an outcome is achieved as a potential for household innovativeness (high, medium or low). In the scheme of Fig. 2 a tenant farmer with high cash reserves and with high social aspirations might be expected to have a high potential for innovation. A farm family with not such a high social aspiration but with household incomes reasonably high in recent years might have a medium status for innovation. A structure such as this involves trade-offs between ranked attributes; high income levels in recent years partly compensate for lack of social aspiration. At the other end of the scale, a tenant farmer with low cash reserves, lowly social aspirations, who has achieved low recent income levels, who is no longer young and is risk averse will have a low potential for innovation.

It is important to understand that the classification of farm types and

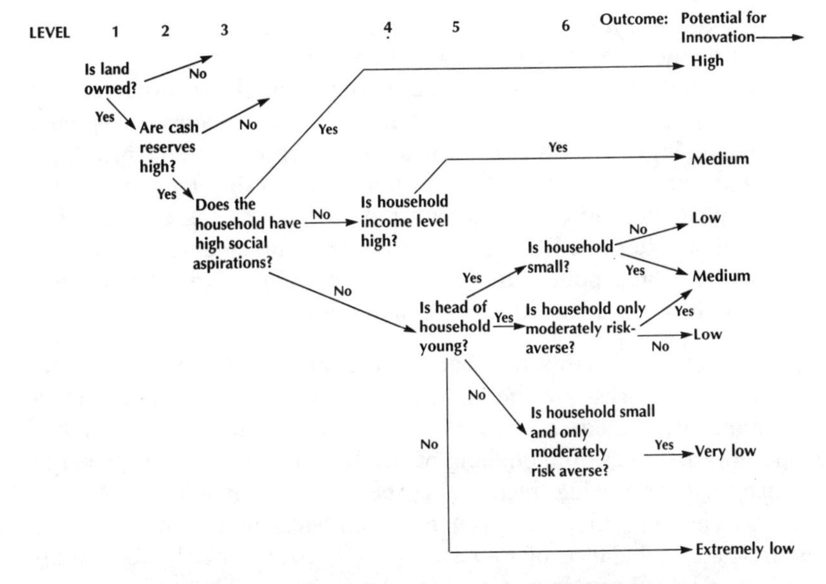

Fig. 2. A ranking of factors that may influence household innovation.

the ranking of attributes occur as a result of socio-economic study, data collection and analysis. With suitable methods large study areas can be covered reasonably quickly (Chambers, 1983; Carter, 1990). Such activity is of crucial importance to the study of whole-farm systems, particularly in the Third World, and can be seen as akin to biological exploration in the study of plant or livestock sub-systems.

Similar structures can be established essentially as lower level systems to formulate a complete model of a whole-farm system. Some of these may rely totally on soft data, while others may require access to the results of crop models operated in the agro-climatic zone under study. Clearly, access to climatic and economic databases will be necessary to provide representative samples of, for example, rainfall and input costs.

A whole-farm systems model constructed around these notions could be applied in two major ways (Dent and Thornton, 1989).

(1) It could be used to estimate the timing and extent of the likely social, cultural and economic impact on farm households of alternative farm enterprise strategies, perhaps based on new technology for an area. It should be anticipated that such

estimates will provide technologists, sociologists and economists with guidance on research priorities.

(2) It could be used to determine the policy infrastructure for agriculture and for the region necessary to establish a specific technology in a district, such as price support, credit provision and extension input. It is irrational for design (research) and implementation of technology not to be determined alongside and simultaneously with the design and implementation of farm and regional policy. In the nature of things, farm systems are simply sub-systems in regional and national systems.

A number of attempts to create whole-farm models within expert systems frameworks are being attempted at present, albeit in a methodological research context. Nevertheless, these new studies would appear to permit modelling of whole-farm systems to proceed in a conceptual sense while recognising the extent of data limitations. In this way, impetus may be given to sociological and anthropological research and a balance of understanding between social sub-systems and biological sub-systems may at last be achieved. This will require a major reorientation of agricultural research funding.

CONCLUSIONS

The foregoing discussion has attempted to demonstrate that effective study of whole-farm systems has progressed hardly at all since conceptual statements were made 20 years ago (Dent and Anderson, 1971; Spedding, 1971). Systems analysis and associated modelling procedures are well recognised and recently have been integrated with research into plant and animal systems. Whole-farm systems research represents the route by which the biological understanding of crop and livestock systems is formally translated into practice. In so far as whole-farm systems research has been inadequate, research into agricultural sub-systems, by definition, must have been undirected and has resulted in less than achievable welfare for farm families. The whole process of what has been termed 'adaptive research', whereby developed and tested technology goes through a process of adaptation, evaluation and adjustment in real farming situations before the adoption process becomes effective, appears to be without the accepted professionalism of agricultural research. Linkages between extension agents and scientists are informal and the experiences of advisers are not reported in the research literature (A. McRae, personal communication). In a holistic sense, however, technology should be developed only in relation to the farming systems in which it will play a part: knowledge of the potential impact on enterprise systems, farm household systems and therefore potential adoption rates should be assessed in advance.

> *This . . . is . . . to emphasise the role of agricultural systems studies in establishing the relevance of the developments in the supporting sciences.*
>
> (Spedding, 1976)

Ideally, adaptive research should be concerned only with technology which *ex ante* is known to be relevant to the resource base of farmers and which is likely to impact in a favourable way on the value sets of the farm households.

It is sobering to reflect that even the concept of 'optimising' whole-farm organisation (as presented within the title of this paper) is not understood in the context of a farm household. Rather, trivial surrogates have been used for the complex value sets involved (for example, risk aversion) when it is known that these are either insufficient or misleading.

Perhaps the only way in which improvement can be achieved is by the construction and application of suitable whole-farm models. The lack of progress in this is indicative of the limited understanding of the social, cultural and economic sub-systems involved. A sharp contrast with the knowledge of the biological sub-systems is inescapable and of concern. There appears to be an acceptance that agricultural research should be involved only or mainly with systems at the biological level. Hence, the major aspects of farming systems are ignored in research and development funding. Recent computer software development in expert systems may provide the basis for a start in modelling of whole-farm systems even with incomplete conceptual understanding and data sets. In turn, such work should lead to improved ability to assist farmers in both the Western world and the Third World to adjust to rapidly changing circumstances.

At the same time, research directions will need to be reconsidered, not in a trivial way within biological disciplines but in the sense of a major root and branch overhaul. There is a view that national agricultural research systems are characterised by fragmentation of effort, with a good deal of duplication and with lack of success in bringing about coordination. The undoubted success of agricultural research in improving output, it has been suggested, has been in spite of, rather than because of, national organisational structures (Arnon, 1989). There can be little prospect of much fundamental adjustment while the agricultural research structures of national and international organisations are driven by bureaucracies under political direction. Recent experience in the UK has shown that political will can bring about significant change within national agricultural research organisations. As far as research is concerned, political will is usually driven by financial considerations and the end result of change may simply be a less satisfactory version of a previous pattern without any real change of direction.

REFERENCES

Anderson, J. R. (1976). Essential probabilistics in modelling, *Agricultural Systems,* **1,** 173–84.

Anderson, J. R., Dillon, J. L. and Hardaker, J. B. (1985). Socio-economic modelling of farming systems, in: *Agricultural Systems Research for Developing Countries,* ed. J. V. Remenyi. Proc. Int. Workshop, Hawkesbury Agricultural College, Richmond, Australia, ACIAR Proc. no. 11.

Arnon, I. (1989). *Agricultural Research and Technology Transfer.* London: Elsevier Applied Science.

Barnum, H. N. and Squire, L. (1979). *A Model of an Agricultural Household: Theory and Evidence.* World Bank Staff Occasional Paper no 27, Washington, DC: World Bank.

Baum, K. H. and Schertz, L. P. (Editors) (1983). *Modelling Farm Decisions for Policy Analysis.* Boulder, Colorado: Westview Press.

Beck, A. C. and Dent, J. B. (1987). A farm growth model for policy analysis in an extenive pastoral production system, *Australian Journal of Agricultural Economics,* **31,** 29–44.

Beck, A. C., Harrison, I. and Johnston, J. H. (1982). Using simulation to assess the risks and returns from pasture improvement for beef production, *Agricultural Systems,* **8,** 55–71.

Carter, S. E. (1990). A survey method to characterise spatial variation for rural development projects, *Agricultural Systems,* **34**(3), in press.

Chambers, R. (1983). *Rural Development: Putting the Last First.* Harlow: Longman.

Crowther, E. M. and Yates, F. (1942). Fertilizer policy in wartime. The fertilizer requirements of arable crops, *Empire Journal of Experimental Agriculture,* **9,** 77–98.

Dent, J. B. and Anderson, J. R. (Editors) (1971). *Systems Analysis in Agricultural Management.* Sydney: Wiley International.

Dent, J. B. and Thornton, P. K. (1988). The role of biological simulation models in farming systems research, *Agricultural Administration and Extension,* **29,** 111–22.

Dent, J. B. and Thornton, P. K. (1989). Whole-farm systems simulation using biological crop models. *Proceedings of the American Society of Agronomy Meetings,* Las Vegas.

Di Marco, O. N., Baldwin, R. L. and Calvert, C. C. (1989). Simulation of DNA, protein and fat accretion in growing steers, *Agricultural Systems,* **29,** 21–34.

Dillon, J. L. (1976). Economics of systems research, *Agricultural Systems,* **1,** 5–22.

Gasson, R. (1973). Goals and values of farmers, *Journal of Agricultural Economics* **24,** 521–38.

Godwin, D. C., Ritchie, J. T., Singh, U. and Hunt, L. A. (1989). *A Users Guide to Ceres-Wheat, V.2.10.* Muscle Shoals, Alabama: IFDC.

Hardaker, J. B. (1979). A review of some farm management research methods for small farm development in LDCs, *Journal of Agricultural Economics,* **30,** 315–27.

Hazell, P. B. R. (1971). A linear alternative to quadratic and semivariance programming for farm planning under uncertainty, *American Journal of Agricultural Economics,* **53,** 53–62.

Hazell, P. B. R. and Norton, R. D. (1986). *Mathematical Programming for Economic Analysis in Agriculture.* New York: Macmillan Publishing Company.

Hutton, J. B. (1968). At Ruakura, three season intensive dairy production, *Proceedings of the Ruakura Farmers Conference,* 25–9.

Jones, C. A. and Kiniry, J. R. (1986). *Ceres-Maize: A Simulation Model of Maize Growth and Development.* College Station, Texas A & M University Press.

Kingma, O. T. and Kerridge, K. W. (1977). Towards an analytical base for studying farm adjustment problems—A recursive stochastic model of the farm firm, *Quarterly Review of Agricultural Economics,* **30,** 91–116.

McCarl, B., Candler, W. V., Doster, D. H. and Robbins, P. R. (1977). Experiences with farmer-oriented linear programming for crop planning, *Canadian Journal of Agricultural Economics,* **25,** 17–30.

McGregor, M. J. (1986). A Multiple Objective Planning Framework for the Analysis of Water and Soil Resource Conflict in New Zealand. Unpublished PhD Thesis, Lincoln College, University of Canterbury, New Zealand.

McGregor, M. J. and Thornton, P. K. (in press). Information systems for crop management: prospects and problems, *Journal of Agricultural Economics.*

Monteiro, L. A., Gardner, A. L. and Chudleigh, P. D. (1980). Bio-economic analysis of ranch improvement schemes and management strategies for beef production in the Cerrado Region, *World Animal Review,* **37,** 37–44.

Ockwell, A. P. (1979). Lender Policy and Farm-Firm Growth. Unpublished PhD Thesis, Department of Agricultural Economics, University of Sydney, Australia.

Pannell, J. D. and Falconer, D. A. (1988). The relative contributions to profit of fixed and applied nitrogen in a crop-livestock system, *Agricultural Systems*, **26**, 1–17.

Pederson, K. (1989). *Expert Systems Programming: Practical Techniques for Rule-Based Systems*. Chichester: John Wiley & Sons.

Penning de Vries, F. W. T., Jansen, D. M., ten Berge, H. F. M. and Bakema, A. (1989). *Simulation of Ecophysiological Processes of Growth in Several Annual Crops*. Wageningen: PUDOC.

Ritchie, J. T. (1986). Using computerized crop models for management decisions, pp. 27–41 in: *Proc. Int. DLG–Congress for Computer Technology*, Hanover, FRG.

Sharma, R. A., McGregor, M. J. and Blyth, J. F. (1990). The socio-economic evaluation of social forestry in Orissa (India), *International Tree Crops Journal*, in press.

Spedding, C. R. W. (1971). *Grassland Ecology*. Oxford: Oxford University Press.

Spedding, C. R. W. (1976). Editorial, *Agricultural Systems*, **1**, 1–3.

Smith, B. J. (1973). Dynamic programming of the dairy cow replacement problem, *American Journal of Agricultural Economics*, **55**, 100–4.

Symes, D. and Appleton, J. (1985). Family goals and survival strategies. The role of kinship in an English upland farming community, *Sociologia Ruralis*, **27**, 21–37.

Thompson, S. C. (1976). Canfarm: A farm management information system, *Agricultural Administration*, **3**, 181–92.

Thornley, J. H. M. and France, J. (1984). Role of modelling in animal production research and extension work. pp 4–9 in: *Modelling Ruminant Digestion and Metabolism*, eds R. L. Baldwin and A. C. Bywater. Davis: University of California.

Thornton, P. K. and Blair-Fish, J. A. (1990) Crop simulation model using a transputer-based parallel computer, *Agricultural Systems*, in press.

Weis, S. M. and Kulikowski, C. A. (1984). *A Practical Guide to Designing Expert Systems*. NJ: Rowman and Allan.

6
Farming Systems Research–Extension

P. E. HILDEBRAND
Food and Resource Economics Department, University of Florida, USA

INTRODUCTION

As associated today with agricultural research and extension, the term 'farming systems' began to be applied in the mid-1970s to technology development activities oriented to small-scale, limited-resource farmers. These activities were under way in several countries around the world (Colombia, Guatemala, Kenya, Nigeria, Philippines and Thailand to name a few) and in both national and international agricultural research (and to a lesser extent extension) organizations. The use of the term 'farming systems' as in farming systems research/extension (FSR/E) is different from its use by Ruthenberg (1971) which is descriptive of farming systems in the tropics. The francophone farming systems concept (Fresco, 1984), which began in West Africa at about the time of the Ruthenberg book, could be described as lying between the descriptive use of Ruthenberg and the more pragmatic approach to technology development associated with anglophone proponents. The anglophone influence on farming systems research and extension has its roots more in cropping systems concepts such as those which Bradfield (1966) pioneered at the International Rice Research Institute (IRRI) in the Philippines in the late 1960s and early 1970s and the ecological crop systems work in the early 1970s at the Tropical Agricultural Research and Training Center in Costa Rica (Hart, 1982). Some practitioners use the term in a way more akin to economists' use of the term 'farm management' (Marz, 1990), or have evolved from a farm management perspective (Ruthenberg, 1971; Norman *et al.*,

1982; Collinson, 1983). Of course, there is also the perspective of systems analysis and development (or replication and innovation as in Spedding and Brockington; 1976) which involves formal modelling of agricultural systems.

Obviously, the term 'farming systems' is not applied to a single approach, notwithstanding that it generally refers to methodologies associated with the diagnosis of farm problems, the design of alternative solutions or technologies, the evaluation of these solutions on farms and the diffusion of the technologies to identified target farmers. In this characteristic, FSR/E differs from farm management, which deals with the reallocation of resources with a constant technology base. Schultz (1964), in arguing that traditional farmers are efficient allocators of available resources given their knowledge of technologies, set the stage for FSR/E which creates disequilibrium within the systems with new technology, thus allowing for reallocation to a higher level of productivity.

Most farming systems activities are associated with the use of teams comprised of biophysical and socio-economic scientists or technicians working with farmers in a collaborative and integrated effort. This last characteristic of 'farming systems', along with the fact that it is still an evolving methodology and is being practised widely and on all continents, is undoubtedly responsible for the multitude of interests, points of emphasis and emerging themes associated with the term 'farming systems' and with 'farming systems research/extension' methodology. In this paper, some of this rich diversity is discussed from a methodological perspective.

UNDERSTANDING THE SYSTEM

One source of difference in emphasis is associated with the completeness of 'understanding' of the system felt necessary before suggesting interventions. Spedding and Brockington (1976) recognize three kinds of experimentation in the study of agricultural systems, each with an increasing requirement for understanding: (a) practical operation; (b) repair and modification to components; and (c) replication and innovation. Of these, the second, for which Spedding and Brockington recognize the necessity of multidisciplinary teams, is most closely related to FSR/E efforts today:

> *Both repair and some modification most frequently relate to sections of a system and these sections have to be understood in detail . . . it is essential that any change wrought in a part of the system should be studied in relation to its effect on the output of the whole system.*

Rapid rural appraisal (Chambers, 1985; Khon Kaen University, 1987) or *sondeos* (Hildebrand, 1981), sometimes called informal surveys, are one of the best known features of farming systems methodology. These provide a rapid means of assessing problems, needs and constraints in an area prior to initiating intervention. They depart drastically from the more traditional questionnaire survey procedure which attempts to obtain quantitative information from a sample of respondents which hopefully represents a larger population. Rapid reconnaissance tends to use a more conversational approach with informants, where the direction of the interchange can be controlled as much by the informant as the interviewer. Proponents of rapid reconnaissance argue that sufficient information on the component which may be the subject for intervention (the cropping sub-system), and changes in this system on the whole system (the farm) can be obtained in a few days and that intervention can be initiated while additional studies continue:

> *Overall, the data in this paper support the hypothesis that the informal survey is an effective and sufficient method for developing an understanding of farming systems and planning experimental programs for farmers. It also suggests that a formal survey may be replaced by (1) a slightly longer and more carefully managed informal survey than would otherwise be conducted, or (2) two or more informal surveys.*
>
> (Franzel, 1986)

Opponents feel that intervention should be initiated only when a complete knowledge base has been formed. The difference is in part related to the magnitude of the intervention proposed. Those who favor rapid reconnaissance are usually inclined toward intervention in components of sub-systems (e.g. varieties, fertilizers, or Spedding and Brockington's B level). Those who insist on a complete knowledge base often have in mind major changes in the system (new crops or cropping systems, or Spedding and Brockington's C level). Systems modellers appear to fall into an intermediate group. They require quantitative data, but are willing to use estimates where no real data exist because their interventions tend to affect only computer solutions or units isolated on experiment stations and not, directly, farmers themselves.

SOURCE OF KNOWLEDGE

Another area of divergence in emphasis radiates from concern with the source of the knowledge base. Some people argue for an increase in farmer participation in not only diagnosis, but also in the design, evaluation and diffusion steps in farming systems methodology (Richards, 1979; Rhoades and Booth, 1982; Chambers and Ghildyal, 1984; Lightfoot *et al.*, 1988, to name a few). This concern stems largely from the socio-economic influence on multi-disciplinary FSR/E teams, but it is also recognized by such pre-eminent biological scientists as Bunting (1979):

> *The knowledge system should involve the producers—not merely as the targets of advisory exhortation, as pupils at farming training centers, or as the passive victims of development done to them by remote government from afar: they have much to tell about soils, weather, crops, animals, diseases and pests, as well as about their own purposes and difficulties . . . . How many of us, who are so wise in international gatherings about what other people should do, could emulate them in winning subsistence, survival, dignity, and fortitude in the face of calamity from the meager resources of traditional rural society in tropical environments?*

DESIGNING INTERVENTIONS

The source of knowledge aside, the nature of the interventions proposed to 'improve' the systems under consideration is influenced by (1) the degree to which the users or clients are allowed to participate and have a voice, (2) the flexibility of the 'interveners', partly related to (3) the mandate of the organization or institution from which the interveners originate, and (4) the magnitude of the proposed change. Rural residents, whether or not they are farmers, are often more interested in improved sources of water than in a new variety of one of their staple crops. However, persons involved in development, and in particular FSR/E, more often than not come from national or international organizations involved in the development of agricultural technology and often with a specific crop mandate. These factors have an obvious bias on the nature of interventions designed.

The CIMMYT (International Center for the Improvement of Maize and Wheat, headquartered in Mexico) farming systems program in East Africa has been active for many years but mostly in farm systems in which maize is a component. Their influence on national programs with which they collaborate is also, of course, related to maize. Whether or not farmers would put top priority on maize technology, this is the kind of intervention they will be exposed to.

The bias introduced by institutional mandate is powerful, yet attempts are made in most farming systems efforts to incorporate the farmers—men and women—into the technology development process. Some form of rapid reconnaissance is nearly always included. However, there are those who argue that this process does not go far enough: 'Rapid rural appraisal has been developed to understand the circumstances of resource-poor farmers but does not help farmers to identify issues for experiment.' (Lightfoot *et al.*, 1988).

An interesting and not atypical experience is related by Maurya (1989):

> *For the moment I wish to stress that even in the absence of on-farm research involving outsiders, farmers regularly innovate and make their own selection of appropriate technologies. Sometimes, indeed, they select technologies which have been rejected by official research. The most striking example is the paddy variety Mahsuri which was introduced into India from Malaysia for tests during 1967–68. After two years of work, this variety was rejected by rice breeders on account of its lodging behaviour. But somehow the seed reached some villages through a farm labourer in Andhra Pradesh. Farmers who tried it found its performance excellent. As a result, it spread from Andhra to Orissa, and then to West Bengal, Bihar, Uttar Pradesh and part of Madhya Pradesh. As a result of this 'farmer-to-farmer' extension, Mahsuri is now the third most popular variety among Indian farmers, after IR8 and Jaya Dwarf rice.*

EVALUATION AND/OR EXPERIMENTATION

Most farmers are experimenters and the degree of their participation in on-farm research has been a concern for a number of years with social scientists urging more and the biological scientists resisting the loss of control associated with too much farmer participation. An often used model has several stages of on-farm research with decreasing researcher control and increasing farmer participation (Hildebrand and Poey, 1985). Others argue that this process requires too much time before farmer participation is sufficient to have an impact, or that it ignores the farmers' natural ability to experiment (Lightfoot, 1987). The difference affects both the nature of the intervention and the evaluation process.

Few farmers have the capability to analyze complex experimental designs such as those used by plant breeders selecting from a large number of genetic lines. At initial stages in the selection process, it is simply not practical for farmers to participate. However, farmers' effects on the nature of the technology selected is important and too often is ignored until late in the selection process. A danger of not creating a mechanism for farmer participation in evaluation at an early stage is that some materials which farmers might have selected for further evaluation may already be rejected by the time they are able to evaluate what is left. As stated by Maurya *et al.* (1988):

> *As a result of the divergences in conditions and evaluation criteria between farms and research stations, the selections resulting from breeding programmes have commonly exhibited two types of inadequacy, even in relatively homogenous irrigated areas. First, much of the material officially released has been of limited relevance to farmers; second, breeders have rejected material which has subsequently found wide acceptance among farmers.*

Before farmers can effectively evaluate more complex technology such as the use of a herbicide, it is necessary for them to learn to use it. The more simple the technology, the more rapidly farmers can become proficient with it (Wake *et al.*, 1988). Simple changes are therefore more readily evaluated and adopted by farmers. This fact guides many farming systems practitioners in the direction of simple methods as well as simple technology.

In an excellent review of farmer participation in agricultural research (FPR), Farrington and Martin (1988) state:

> *Considerable confusion has arisen over the relationship between FSR and FPR . . . some proponents of FPR seek to distance themselves from conventional agricultural research institutes which are seen as defending the status quo in relations between researcher and farmer, and, ultimately, in the imbalance between rich and poor farmers in ldcs. On the other hand, even in its earliest formulations, FSR stressed the need to involve and learn from the farmer in research, and where departures from this principle occur they are generally attributable to poor interpretation of FSR's objectives or to funding constraints.*

They conclude that farmer participation is not, as some might claim, a substitute for FSR but a complement to client-oriented ('problem-focussed') research and development which, in turn, is one component of the agenda for research in ldcs.

DIFFUSION

On-farm technology evaluation in which researchers, extensionists and farmers work in close collaboration provides for the natural merging of technology development, evaluation and diffusion (Hildebrand, 1988). Farmers who are actively participating in evaluation by hosting on-farm trials are moving along the learning curve through first-hand experience. Neighbors and others such as extension agents who have not had previous knowledge about the technology can learn through observation. Researchers or technology developers can help adapt the technology to local conditions through the familiarity gained by their participation in this process.

INSTITUTIONAL CONSTRAINTS

Though this process seems logical and efficient, it is surprising that institutional inertia in many countries both developing and developed, and in the international sphere can suppress it. One of the most frequently levelled criticisms of farming systems is that there is little or no effective collaboration between research and extension organizations even though some of their functions may be blended by those working in the field (McDermott and Bathrick, 1982). The institutional separation of functions is even perpetuated by such international organizations as ISNAR (International Service for National Agricultural Research) which was established by the CGIAR 'for the purpose of assisting governments of developing countries to strengthen their agricultural research'. This institutional mandate even has an impact on the terminology used. For example, ISNAR uses 'the generic term "on-farm, client-oriented, research" (OFCOR) as distinct from "farming systems research" (FSR) because the latter has come to have too many different and confusing meanings' (Merrill-Sands, 1989). Because of its limited commodity focus, CIMMYT has coined the term 'on-farm research with a farming systems perspective'. Both of these international institutions emphasize the research but not the extension of FSR/E because of mandate.

One can argue successfully that FSR/E is as old as the Land Grant University system in the United States. Farmers were the focus and technology development was the activity. Most researchers were farm-raised and had a farming perspective and capability to communicate with farmers. Later an extension service was created to facilitate transfer of technology to wider audiences. As technology became more sophisticated, researchers and extension workers became increasingly specialized. Increasing specialization created communication problems as each speciality developed its own vocabulary. This narrowing of focus away from the whole farm and towards its components fomented a drift away from farmers as the primary clientele group. Specialized agriculturalists were less able to communicate with farmers who were unable to understand the specialized vocabulary. As a result, the new clientele group for those in the Land Grant system became their own peers—the only persons with whom there was easy communication. This institutional drift has made it difficult to generate an FSR/E program in the Land Grant University system of the US today.

ENTERPRISE FOCUS

The first two decades of FSR/E activity has seen a heavy emphasis toward intervention in annual crop systems. This in part reflects its cropping systems antecedents which concentrated on annual crops. It also reflects the relative ease of on-farm research with annual crops as distinct from perennials or animals. Though some argue against the use of the term 'farming systems' when it applies to the process of technology development for annual crops only, it is merited because attempts are made to understand the impact of the annual crop sub-system, or technological modifications in this sub-system, on the farm as a whole.

Livestock were not being ignored. As early as 1978 a Bellagio conference was organized to study the integration of crop and livestock production in developing countries (McDowell and Hildebrand, 1980). A conclusion was 'that animals form an integral and essential part of small-farm systems in most of the developing countries and that efforts should be made to create awareness of the importance of this integration among training institutions and government agencies.' In 1985 a workshop was held at ICARDA (International Center for Agricultural Research in the Dry Areas) in Aleppo, Syria, on research methodology for livestock on-farm trials (Nordblom *et al.*, 1985). More recently Winrock International and the International Development Research Centre (Canada) have published a book on conducting on-farm animal research (Amir and Knipscheer, 1989). A number of workshops have been held in Africa on livestock and on-farm research with emphasis on animal traction. This aspect has become an important component of the West African Farming Systems Research Network (WAFSRN) now headquartered in Ouagadougou, Burkina Faso.

GENDER ISSUES

In retrospect it is hard to believe that through many early years, women on farms and women farmers were ignored by farming systems practitioners who were espousing a 'holistic' systems methodology. Nor were many women included in the research and extension groups who were working with farmers. This latter short sight began to be rectified with the creation of the efforts variously described as Women in Development (WID), Women in Agriculture (WIA), and Women in Agricultural Development (WIAD):

> *The WID field, similar to FSR/E, began with a concern for the distribution of development benefits. Like farming systems, women and development is far from a unified field of knowledge. Not only does it include many strands of*

> *research and practice, but the field has evolved rapidly over the approximately 20 years of its existence, since economist Ester Boserup published her ground-breaking work* Women's Role in Economic Development *in 1970.*
>
> (Poats *et al.*, 1988)

Early contact between those concerned with the inclusion of women in farming systems was sometimes confrontational. However, through the efforts of the Farming Systems Support Project (FSSP) with its worldwide networking activities and other groups, an historic conference was held on the campus of the University of Florida in 1986. Confrontation was absent as the participants concentrated on methods for incorporating gender issues into farming systems at its various stages. Case studies, developed with the collaboration of the Population Council, were also evaluated at the conference. 'The Conference offered an excellent opportunity for exploring and testing new ideas and successful approaches for incorporating gender sensitivity in agricultural research and development' (Poats *et al.*, 1988). Improved means of gender analysis are now an accepted procedure in virtually all farming systems projects.

SUMMARY

Farming systems research and extension methods evolved from a concern with the distribution of benefits from the national and international investment in agricultural research and extension efforts. It has required a blending of the disciplinary methods used by plant breeders, agronomists, anthropologists, economists, animal scientists, geographers and others. In the diverse areas of the world where it is being used, local conditions, varying severity of budget restrictions, institutional base, and the individuals involved have all placed their impact on it. In a very real sense FSR/E is not a method, but rather a philosophy with a number of common methods and one common goal. Perhaps this view can best be expressed by the recently organized Association for Farming Systems Research–Extension (AFSRE), itself a very diverse group. The AFSRE is:

> *An international society organized to promote the development and dissemination of methods and results of participatory on-farm systems research and extension. The objective of such research is the development and adoption*

through the participation by farm household members—male and female—of improved and appropriate technologies to meet the socioeconomic needs of farm families; adequately supply global food, feed and fiber requirements; and utilize resources in a sustainable and efficient manner.

(IFAS, 1989)

BIBLIOGRAPHY

Amir, P. and Knipscheer, H. C. (1989). *Conducting On-farm Animal Research: Procedures & Economic Analysis.* Morrilton, AR: Winrock International Institute for Agricultural Development and International Development Research Centre.

Avila, M., Whingwiri, E. E. and Mombeshora, B. G. (1989). *Zimbabwe: Organization and Management of On-farm Research in the Department of Research and Specialist Services, Ministry of Lands, Agriculture and Rural Resettlement.* OFCOR Case Study no. 5. The Hague: ISNAR.

Bradfield, R. (1966). *Toward More and Better Food for the Filipino People and More Income for her Farmers.* New York: The Agricultural Development Council.

Bunting, A. H. (1979). *Science and Technology for Human Needs, Rural Developments, and the Relief of Poverty.* New York: IADS Occasional Paper.

Chambers, R. (1985). Shortcut methods of gathering social information for rural development projects, in: *Putting People First: Sociological Variables in Rural Development,* ed. M. M. Cernea. New York: Oxford University Press.

Chambers, R. and Ghildyal, B. P. (1984). Agricultural research for resource-poor farmers: the 'farmer-first-and-last' model. Paper prepared for the National Agricultural Research Project Workshop on National Agricultural Research Management, National Academy of Agricultural Research Management, Rajendranagar, Hyderabad, India.

Chambers, R., Pacey, A. and Thrupp, L. A. (Editors). (1989). *Farmer First: Farmer Innovation and Agricultural Research.* London: Intermediate Technology Publications.

Collinson, M. (1983). *Farm Management in Peasant Agriculture.* Boulder, CO: Westview Press.

Farrington, J. and Martin, A. (1988). *Farmer Participation in Agricultural Research: A Review of Concepts and Practices.* London: Overseas Development Institute.

Franzel, S. C. (1986). Comparing informal and formal surveys, pp. 98–102 in: *Perspectives on Farming Systems Research and Extension,* ed. P. E. Hildebrand. Boulder, CO: Lynne Rienner.

Fresco, L. (1984). *Comparing Anglophone and Francophone Approaches to*

Farming Systems Research and Extension. Working Paper no. 1. Gainesville: Farming Systems Support Project, University of Florida.

Gilbert, E. H., Norman, D. W. and Winch, F. E. (1980). *Farming Systems Research: A Critical Appraisal.* MSU Rural Development Paper no. 6. East Lansing: Department of Agricultural Economics, Michigan State University.

Hart, R. D. (1982). An ecological systems conceptual framework for agricultural research and development, pp. 44–58 in: *Readings in Farming Systems Research and Development,* eds W. W. Sharer, P. F. Philipp and W. R. Schmehl. Boulder, CO: Westview Press.

Hildebrand, P. E. (1981). Combining disciplines in rapid appraisal: the sondeo approach, *Agricultural Administration* **8,** 423–32.

Hildebrand, P. E. (1988). Technology diffusion in farming systems research and extension, *Horticultural Science,* **23,** 488–90.

Hildebrand, P. E. and Poey, F. (1985). *On-farm Agronomic Trials in Farming Systems Research and Extension.* Boulder, CO: Lynne Rienner Publishers.

IFAS (1989). p. 2 in: *Farming Systems Research Newsletter, No. 3.* University of Florida, Gainesville: Institute of Food and Agricultural Sciences.

Khon Kaen University (1987). *Proceedings of the 1985 International Conference on Rapid Appraisal.* Khon Kaen, Thailand: Rural Systems Research and Farming Systems Research Projects.

Lightfoot, C. (1987). Indigenous research and on-farm trials, *Agricultural Administration & Extension,* **24,** 79–89.

Lightfoot, C., de Guia, Jr, O. and Ocado, F. (1988). A participatory method for systems-problem research: Rehabilitating marginal uplands in the Philippines, *Experimental Agriculture,* **24,** 301–9.

Marz, U. (1990). *Farm Classification and Impact Analysis of Mixed Farming Systems in Northern Syria.* (*Farming Systems and Resource Economics in the Tropics,* vol. 7). Kiel, FRG: Wissenschaftsverlag Vauk.

Maurya, D. M. (1989). The innovative approach of Indian farmers, p. 10 in: *Farmer First: Farmer Innovation and Agricultural Research,* eds R. Chambers, A. Pacey and L. A. Thrupp. London: Intermediate Technology Publications.

Maurya, D. M., Bottrall, A. and Farrington, J. (1988). Improved livelihoods, genetic diversity and farmer participation: A strategy for rice breeding in rainfed areas of India, *Experimental Agriculture,* **24,** 311–20.

McDermott, J. K. and Bathrick, D. (1982). *Guatemala: Development of the Institute of Agricultural Science and Technology (ICTA) and its Impact on Agricultural Research and Farm Productivity.* Project Impact Evaluation no. 30. Washington, DC: USAID.

McDowell, R. E. and Hildebrand, P. E. (1980). Integrated crop and animal production: making the most of resources available to small farms in developing countries. Working Papers. New York: The Rockefeller Foundation.

Merrill-Sands, D. (1989). Introduction to the ISNAR study on organization and management of on-farm client-oriented research, p. iii in: *Zimbabwe: Organization and Management of On-Farm Research in the Department of Research and Specialist Services, Ministry of Lands, Agriculture and Rural Resettlement,* eds M. Avila, E. E. Whingwiri and B. G. Mombeshora. OFCOR Case Study no 5. The Hague: ISNAR.

Nordblom, T. L., Ahmed, A. K. H. and Potts, G. R. (eds) (1985). *Research Methodology for Livestock On-Farm Trials*. Proceedings of a workshop held at Aleppo, Syria, 25–28 March, 1985. ICARDA and IDRC.

Norman, D. W., Simmons, E. B. and Hays, H. M. (1982). Farming Systems in the Nigerian Savanna: Research and Strategies for Development. Boulder, CO: Westview Press.

Poats, S. V., Schmink, M. and Spring, A. (1988). *Gender Issues in Farming Systems Research and Extension*. Boulder, CO: Westview Press.

Rhoades, R. E. and Booth, R. H. (1982). Farmer-back-to-farmer: a model for generating acceptable agricultural technology, *Agricultural Administration*, **11,** 127–37.

Richards, P. (1979). Community environmental knowledge in African rural development, pp. 28–36 in: *Rural Development*: *Whose Knowledge Counts*? ed. R. Chambers. IDS Bulletin 10 (2).

Ruthenberg, H. (1971). *Farming Systems in the Tropics*. Oxford: Clarendon Press.

Schultz, T. W. (1964). *Transforming Traditional Agriculture*. New Haven and London: Yale University Press.

Simmonds, N. W. (1985). *Farming Systems Research*: *A Review*. Technical Paper no. 43. Washington, DC: World Bank.

Spedding, C. R. W. and Brockington, N. R. (1978). Experimentation in agricultural systems, *Agricultural Systems,* **1,** 47–56.

Wake, J. L., Kiker, C. F. and Hildebrand, P. E. (1988). Systematic learning of agricultural technologies, *Agricultural Systems,* **27,** 179–93.

Whyte, W. F. and Boynton, D. (Editors) (1983). *Higher Yielding Human Systems for Agriculture*. Ithaca and London: Cornell University Press.

7

Food Policy and Food Security Planning: Institutional Approaches to Modelling Grain Markets and Food Security in Sub-Saharan Africa

J. G. GRAY

Food Studies Group, International Development Centre, University of Oxford, UK

Many developing countries experience difficulty in the formulation of appropriate food policies. It is difficult to design policies where powerful vested interests exist; there are usually sharp trade-offs between conflicting objectives such as support to urban and rural groups, marketing structures are often complex and vary between commodities, and it is difficult to identify politically acceptable policies which are financially feasible. As a result, shortcomings of national food policies have been identified as constituting a major structural obstacle to accelerated growth (World Bank, 1981). As such the field of food policy lends itself to a systems approach which allows the impact of a range of types of environmental, behavioural and policy factors to be taken into account.

Food policy has in fact been an area of active application of systems methods, primarily though not exclusively through the construction of econometric simulation models. This paper presents a rather eclectic review of some approaches which have been taken to the modelling of food policy issues. This is not a comprehensive review but, rather, concentrates on a particular group of exercises. The focus for this paper is on the approaches to the modelling of food policy issues which have been taken by institutions involved in the development process with particular reference to sub-Saharan Africa (SSA). The review accordingly bypasses a wide range of modelling exercises undertaken by, for example, individual academics and institutions, such as OECD and the EEC, not directly involved in sub-Saharan Africa. A further restriction is that the review is confined to models which address domestic food policy issues and especially staple food marketing problems.

The main aspects to be considered are:

(i) Has adequate attention been paid to modelling food marketing issues?
(ii) Have technically adequate modelling structures been developed?
(iii) What contribution have these models made to resolving the massive food policy problems of the region?
(iv) Have the modelling initiatives been effective vehicles for the transfer of technical analytical capability to institutions in the region?

CHARACTERISTICS OF THE REGIONAL STAPLE FOOD MARKETS

A key aspect of the debate on food policy in sub-Saharan Africa in the past decade has been the performance of the foodgrain economy. While governments in the region have emphasised a foodgrain production-based self-sufficiency orientation to food security, issues relating to the management of domestic grain markets and the effectiveness of the marketing boards as an instrument of policy have increasingly come to the fore. Almost all countries of the region are characterised by heavy dependence on grains both as the major crop and as the dominant food staple. Policy debate on the management of the cereal economy has encompassed a number of major areas: the need to identify and remove major distortions and obstacles to economic efficiency in production; the role of grain reserve stocks as opposed to external trade in stabilising aggregate food availability; whether and how to stabilise domestic cereal markets; the problems and distortions associated with the emergence of strongly dualistic grain marketing structures often accompanied by supply rationing in parts of the domestic market; the need to reduce the financial losses incurred through marketing parastatals on their grain account and to reduce the contribution of internal inefficiencies in those losses; how to promote the development of less accessible regions within countries without incurring an excessive cost in economic efficiency.

In many SSA countries most or even all of these issues have needed to be confronted simultaneously in the process of trying to identify economically and financially feasible and politically acceptable policy directions. Since the set of issues are closely inter-related and policies set in one area have implications in all other areas there is a strong *prima facie* case for adopting a systems approach to policy analysis. The remainder of this section reviews the approaches taken by institutions to modelling this inter-related grain system as an aid to policy formulation.

WORLD BANK MULTI-MARKET MODELLING

As might be expected, given the importance attached to food policy and marketing issues in adjustment debates, the World Bank has been the source of important approaches to modelling SSA's food policy problems. We concentrate here on the multi-market modelling approach developed by Braverman, Hammer and others. World Bank work in this area (Singh *et al.*, 1985; Braverman and Hammer, 1985) started from recognition of the limitations of some standard procedures applied in policy assessment arising from the need to take account of interactions between markets and general equilibrium effects. Examples of inadequacies cited are (Braverman *et al.*, 1983): the use of the domestic resource cost (DRC) as a measure of comparative advantage; the DRC does not provide a quantitative indication of the adjustment to be expected from policy change; the use of single-market partial equilibrium analysis can be misleading where markets are closely inter-related; nominal and effective exchange rates, used as indices of trade distortions, are problematic in policy assessment where other distortions exist in the economy or where the move to a fully liberal regime is not politically feasible.

The multi-market approach sought to address these methodological inadequacies by creating a modelling framework which would allow the key market interactions to be captured without the cost, delay and complexity of producing a full-blown general equilibrium model. In particular the approach sought a modelling framework whereby the impact of key general equilibrium effects could be identified and quantified without losing touch with the policy-maker or his adviser in the context of adjustment policy dialogue.

One of the early applications of this approach was a model which focused on pricing policy issues in sub-Saharan Africa (Singh *et al.*, 1985). The price policy model was designed to quantify the trade-offs between objectives under alternative settings of policy instruments. The objectives selected were foodgrain self-sufficiency, economic efficiency, income distribution, producer incomes and the external balance of payments. The instruments examined in the model included parastatal trading monopolies, commodity and export taxes, wholesale and retail prices, import/export controls and the exchange rate. A feature of the model is the care taken to capture the essential institutional features which affect market clearing procedures. In particular the interactions between official and unofficial domestic markets were explicitly modelled, with the direction and quantities of 'leakage' between market sectors dependent primarily on the official:unofficial market price relativities. This feature has been incorporated in most subsequent modelling of SSA grain markets.

A basic finding of Singh's pricing model was that partial equilibrium models of single markets can indeed be expected to give results which are misleading and sometimes wrong in direction compared to results obtained using the multi-market model. This result is not surprising in a theoretical sense, but the early applications of the multi-market model showed that such results were likely and also quantitatively highly significant in the typical conditions prevailing in SSA food markets.

Accessibility of the model framework has been achieved by deriving a linear formulation (from non-linear underlying structural equations) which allows the model to be run and adapted simply on a microcomputer without complex and expensive software. 'The method proceeds by assembling what is known about supplies and demands for the important commodities, the institutional structures of government policies and the mechanisms for market clearing. This information is arranged in a set of equations which is totally differentiated so that changes in the outcomes of interest can be solved in terms of changes in the available policy options . . . the resulting model is linear.' (Braverman *et al.*, 1985).

Since 1985 the multi-market approach has been pursued in an increasing number of country cases, often in association with World Bank agricultural policy-lending missions, and has become a major mechanism for upgrading the quality of analysis underlying policy-lending programmes. It has also been used as a training tool in the context of short-term courses provided by the Economic Development Institute of the World Bank.

The main limitation of the multi-market approach as it has been applied stems from the fact that it is comparatively static or single-period, that is to say it does not track the economy over time. The quantitative policy responses relate to a single time period which must, of course, be defined in establishing elasticities and other behavioural parameter values. Hence the multi-market approach does not lend itself to issues arising from instability such as stocking survey.

THE INTERNATIONAL FOOD POLICY RESEARCH INSTITUTE

The International Food Policy Research Institute (IFPRI) has not developed a generic model of the sub-Saharan food sector. However, as an institution it has used a characteristic approach to addressing policy issues through simulation modelling which has proved very effective in highlighting for public debate specific issues in food policy through the application of relatively simple modelled structures.

Examples of this type of IFPRI models are the analysis of stocking and trade policy options for food security in the Sahel (McIntyre, 1981) and Pinkney's analysis of grain stocking, pricing and trade policies in Kenya (Pinkney and Gotsch, 1987).

IFPRI's work has been primarily aimed at highlighting issues through the use of selective country-specific analyses. The results of its research have been particularly important in changing policy perceptions in the academic community and amongst the main lending institutions. In the author's experience the results of IFPRI's research are, however, not well known amongst policy-makers in public sector institutions in SSA, which is regrettable in view of the relevance of that research for the policy problems of the region. It is perhaps worth noting that until the last few years the bulk of IFPRI's work has lain outside sub-Saharan Africa. This has changed markedly in recent years and a number of studies are in the pipeline.

THE EUROPEAN COMMUNITY TRAINING MODEL—HARAMBEE

The European Community (EC), through its Statistical Office in Luxembourg, has developed a model of the food sector under sub-Saharan African conditions. The model, known as 'Harambee', was designed specifically as a training tool in the use of statistical information in food policy formulation (Corbett and Röder, 1987; Greener and Eele, 1989). The underlying model is a simplified dynamic general equilibrium model concentrating on the agricultural sector. It aims to capture characteristic conditions in the food sector in countries in the region (grain dependence, dual marketing structures, problems in assuring national and household-level food security, foreign exchange and budgetary constraints, controlled external trade regimes). The model contains modules covering agricultural (crop) production, grain marketing, consumer demand, government finance, the balance of payments and nutrition. This model is used as a basis for a role-playing simulation exercise, the prime purpose of which is to train participants in the choice and application of statistical information as an aid to policy formulation.

Participants in the exercise are given roles in the policy formulation process (minister of finance, manager of grain parastatal, minister of health and so on). Basic information is made available in the form of historical time series and other information. Further information must be purchased by use of the participant's statistics budget. One round of the exercise involves the collective setting of policy instruments (official prices, exchange rate, grain external trade targets and others) and the allocation of statistical budgets. The model is then run to obtain the outturn for the period. The model generates the period solution on the basis of parameter specification, policy instrument setting and random components in, for example, food production and external market prices. Output indicators are made available for the participants to assess the effectiveness of their previous policy decisions and to prepare the next round of policy settings.

In allocating their statistical budget, participants can choose from a menu of characteristic sources of information including routine administrative records and surveys. More accurate surveys are naturally more expensive and are subject to longer delays before the data become available. Participants are informed of the (correct) standard error of estimates produced by surveys they may purchase. Output from the model as reported to participants are subject to stochastic error terms which capture the degree of accuracy of the surveys purchased. True output variables from the model, i.e. excluding statistical error terms, are made available at the end of the session. In this way participants are encouraged to assess the cost-effectiveness of incremental information in terms of improved policy formulation.

The main strength of the Harambee model is that it provides an imaginative training device. Participants come to recognise the interrelatedness of apparently distinct aspects of food policy. The model has to date been used in a series of seminars (in Mali, Tanzania and Zimbabwe) to allow testing. Clearly if the EC model is to have a significant impact then it must be incorporated into the routine training programme of some dedicated training institution to reach the appropriate target trainees. The next stage in use of the model will be distribution to training institutions and training instructors in its use with the aim of the exercise becoming a standard feature of food policy training with an emphasis on statistics. It is hoped that the Harambee exercise will assist in addressing problems on the demand side of statistical development; too often statistics bureaux are producing routine statistics which have little or no impact on the policy process at least in part because demands for appropriate statistics are not well articulated.

FOOD STUDIES GROUP/OVERSEAS DEVELOPMENT NATURAL RESOURCES INSTITUTE

The Cereal Market Policy Model developed by the Food Studies Group (FSG) in association with the Overseas Development Natural Resources Institute (ODNRI) of the UK was designed as a modelling structure which could be customised and applied within food policy institutions in sub-Saharan African countries (Van der Geest, 1989). The stimulus to its development lay in previous work on modelling the rice market in Indonesia (Mitchell *et al.*, 1986; Farrington, 1986). The model consists of a small, computable general equilibrium agricultural sector model which gives prominence to grain markets. The main compoments of the sector model are a set of supply functions, a set of demand functions and market clearing equations for a dual-market grain economy (Rogers *et al.*, 1988).

The main innovative feature of the FSG/ODNRI model is the extent to which institutional detail can be incorporated. To achieve this the model operates over two distinct time periods. The underlying equilibration process is annual and related to the cropping year. Within this, however, modules relating to the operations of the marketing board provide monthly output variables which permit the peak seasonal logistical, market price and financial implications of policies to be explored. These features are designed to help bridge the gap between policy analysis and the practical aspects of policy implementation which are the prime concern of most food sector managers.

One of the main areas which the FSG/ODNRI model is designed to address is food reserve stocking policy and the expected costs under alternative specifications of stocking and external trade rules. The model allows the user to explore the trade-off between financial cost and food security (measured by the probability of food reserve stock-out) using stochastic simulation over a 10-year-run period. The simulation processes employed allow the user to define the parameters of non-normal disturbance terms. This is an important aspect as there is now considerable evidence of non-normality in the distribution of production and market supply disturbances (Rogers and Van der Geest, 1988).

As with the FAO stock policy simulation model and World Bank multi-market modelling exercises, the FSG/ODNRI model is a positive rather than an optimising model; that is to say it permits the user to explore the implications of alternative sets of policy instruments on a number of policy outcomes without attempting to find an optimising result in terms of an objective function which aggregates over different policy objectives.

It is still too early to judge the effectiveness of the FSG/ODNRI model in terms of the aims behind its development. The model is to be implemented in a series of country applications firmly based in food policy and management institutions in the region, starting with Zimbabwe (Ministry of Agriculture) and Ethiopia (Office of National Committee for Central Planning). These applications are likely to involve significant modifications of the prototype version of the model.

FOOD AND AGRICULTURE ORGANISATION STOCK SIMULATION MODEL

The Food and Agricultural Organisation (FAO) Stock Simulation Model (Abbott, 1988; Prantilla, 1989) has been developed as a 3-year project which commenced in July 1986. Under the project a generic model has been developed with a view to examination of stocking and related food policy options in five Asian countries—Bangladesh, Bhutan, Nepal, Pakistan and Sri Lanka. Although not designed with a view to SSA conditions the model is in fact highly apposite to the marketing structures and problems encountered in SSA and is accordingly mentioned here. The model has been designed to facilitate the assessment of the impact of a range of types of policy (trade policy, stocking strategy and consumption policy) on a set of indicators representing government objectives—budgetary cost, an index of food security and price stability.

The model permits dynamic multi-period simulations of up to three commodity markets over a 5-year simulation period. Urban and rural sectors are distinguished with three categories of income class in each sector. The model distinguishes two domestic trading and stockholding sectors, one private and one official/public. The private market meets effective demand in excess of public distribution.

Once developed, the model has provided a framework for application in a number of Asian countries. This has been achieved by country-specific projects which have been undertaken collaboratively by staff from FAO with local institutions. A significant role for the model has accordingly been the training function.

LIMITATIONS OF THE MODELLING APPROACH

As this brief review has clearly illustrated, the modelling approach can be adopted for a number of reasons: the World Bank models have been designed to support policy dialogue, the EEC model is a training tool, IFPRI and FAO models have typically been designed to shed light on specific generic policy issues and the FSG/ODNRI model is designed as a device for accelerating institutional development. In view of these differing objectives it is not appropriate to ask which is the best. They are different because they serve different functions and must be judged according to their specific objectives.

Some of the limitations to the modelling approach to food policy in the SSA region have been as follows:

(1) The difficulty of capturing some of the key features of the systems of markets which make up the food system in African countries. Some of the areas which have posed particular problems for modellers have been: (a) smallholder food stock retention and sales behaviour which is motivated as much by poorly understood household food security criteria as conventional economic criteria; coupled with the absence of statistics

on farm level stocks this has constituted a significant weakness; (b) the role of expectations and information in price formation in food markets (especially under food crisis conditions) which has been an active area of empirical research (e.g. Ravallion, 1988) and the conditions under which trader expectations may tend to destabilise markets; the models discussed tend to assume simple market-clearing functions which ignore many of the real life problems associated with oligopsonistic food markets in rural areas; (c) difficulty in modelling external trade flows in foodgrains, which in the sub-Saharan African region are usually unrecorded and illegal.

(2) The inability to address policy options involving really major structural change. Econometric models are designed and estimated around a structure and cannot in general be used for forecasting or extrapolating outside that structure. This is a major weakness in the field of marketing policy, where much of the debate has centered on radical options for market liberalisation. Most models can cope with modest experimentation with parameter change but not with this magnitude of structural change

(3) The poor availability of statistical data has been a key constraining factor in model design although some models, such as the World Bank multi-market models, have been designed precisely to be useful in data-sparse environments.

(4) The poor understanding of some aspects of the functioning of grain markets, as, for example, the stockholding and release behaviour of households and traders. Coupled with lack of data on private sector transactions these render most of the attempts to model dual-market structures somewhat arbitrary.

(5) The dearth of trained local staff in the relevant policy institutions in the sub-Saharan region has limited the effectiveness of the modelling approach in terms of its impact on both the local perception of specific policy issues and in terms of the possibilities for internalising the model development capacity.

(6) While these models have proved effective in addressing some issues (notably pricing, tax, subsidy and stocking questions), a range of issues are not amenable to a modelling approach without adding considerably to complexity. Examples are the spatial and temporal aspects of food pricing structures which are at the centre of the policy debate but which cannot be modelled

without adding spatial and temporal (within year) disaggregation to model structures respectively.

It should also be recognised that there has been a reluctance on the part of some donors to support food policy-modelling initiatives. In some cases this represents an understandable caution in the presence of staffing and data constraints in countries in the region. In other cases, however, it has arisen from a reluctance to transfer modelling and analytical capability to countries in the throes of structural adjustment negotiations with the International Monetary Fund (IMF) and the World Bank.

HOW USEFUL ARE THESE MODELLING APPROACHES?

The models briefly reviewed above have a number of features in common. Although designed for different purposes they all represent attempts to capture for the policy analyst the essential interactions in larger complex general equilibrium systems without the costs required to build full system models. The generally poor experience with the larger multi-sectoral modelling projects, at least in sub-Saharan African conditions, suggest that this has been a sensible approach.

The great majority of modelling exercises in food policy in sub-Saharan Africa have been designed and to a large extent developed by external agencies. In view of the wide range of mechanisms for modelling initiatives to influence development this is not necessarily problematic. The multi-market modelling approach has contributed significantly, in the author's opinion, to sharpening the policy debate on food sector issues. The principal benefit of model development, however, often does not lie in the quantitative results of the model in terms of output. Few who have been closely involved in developing policy models defend the precision of output variables and most emphasise that their models are not to be conceived as forecasting tools. The main benefit accrues rather to the modeller himself who sharpens his analytical perception of the functioning of the system modelled. This implies that for the sub-Saharan region to benefit from the systems modelling approach to food policy, the design and development of models must be transferred to institutions and individuals within the region.

DIRECTIONS FOR MODELLING WORK IN THE 1990s

As argued above, policy modelling has played a significant role in shaping the food marketing policy debate in sub-Saharan Africa in the past decade, especially through the multi-market model applications used in the context of World Bank policy on lending. The need for such approaches will continue for some years. However, it can also be argued that, with the accelerated adoption of marketing policy reform packages in the region, the characteristics of models and the institutional approach to their application will need to be adapted. During the 1980s policy modelling for sub-Saharan Africa has been essentially an external activity financed, implemented and utilised by First-World institutions to prepare their positions for policy discussions with African governments. This will need to change primarily because, with the removal of the grosser policy distortions in these countries, the issues to be addressed will typically be more complex, arising from the second round of the policy adjustment process. The need for carefully designed and calibrated models which can capture the interactions between sub-sectors in agriculture is certainly going to increase. It will become increasingly implausible for external modellers to patch together a credible policy framework in a month or two as has been the common practice during the 1980s.

These considerations point to the need for food policy institutions in the region to develop their own full agricultural sector planning models to form the basis for routine year-to-year policy adjustment. For some years to come staffing constraints will effectively restrict the extent of the use of models in domestic policy formulation. Attention needs, accordingly, to be given to assuring that modelling projects are designed so as to maximise their institutional capacity-enhancing effect in this respect.

We conclude from this brief review that food policy modelling for SSA has been an active area on behalf of a number of development institutions. The models developed have been able to cope with many key institutional features of grain markets in the region while some important aspects of policy have not been addressed. In relation to the magnitude and urgency of SSA's food problems, however, it must be argued that the scope of modelling activities has been rather modest. The most disappointing feature has been the limited role played by national research institutions in the SSA region and the very limited modelling capacity in the countries in the region with the most pressing food sector problems. It is to be hoped that the increased attention apparently being afforded by donors to issues of institutional development will be reflected in expanded initiatives in the field of food market modelling.

REFERENCES

Abbott, P. C. (1988). *A Methodology for Evaluating National Grain Stocking Strategies* (revised edition). Rome: Food and Agriculture Organisation.

Braverman, A. and Hammer, J. S. (1985). Multi-market analysis of agricultural pricing policies in Senegal, in *Agricultural Household Models: Extensions, Applications and Policy,* eds I. Singh *et al.* Washington, DC: World Bank.

Braverman, A., Hammer, J. S. and Ahn, C. Y. (1983). *Alternative Agricultural Pricing Policies in the Republic of Korea.* Staff Working Paper no. 62. Washington, DC: World Bank.

Corbett, J. and Röder, K. (1987). *Harambee—An Exercise for Food Policy Statisticians.* The Courier, European Commission, Brussels.

Farrington, J. (1986). Modelling staple food stocks management: can methodologies developed in Asia be applied to Africa? in *Proceedings of Workshop on Statistics in Support of Africa Food Strategies and Policies,* Brussels, May 1986.

Greener, R. and Eele, G. (1989). *Harambee Simulation Exercise Participant's Manual.* Munich Centre for Advanced Training.

Kirchner, J., Singh, I. and Squire, L. (1984). *Agricultural Pricing and Marketing Policies in Malawi.* Country Policy Department Discussion Paper. Washington, DC: World Bank.

McIntyre, J. (1981). *Food Security in the Sahel: Variable Import Levy, Grain Reserves, and Foreign Exchange Assistance.* Washington, DC: International Food Policy Research Institute Research Report 26.

Mitchell, M., Gray, J. G. and Street, P. R. (1986). *Report of a Study of Minimum Grain Reserves.* London: Tropical Development and Research Institute.

Pinkney, T. C. and Gotsch, C. H. (1987). *Storage, Trade and Price Policy Under Production Instability: Maize in Kenya.* Washington, DC: International Food Policy Research Institute Research Report 71.

Prantilla, E. B. (1989). Assessment of alternative stocking policies using the FAO stock simulation model. In *Proceedings of Workshop on Inventory Problems in Developing Countries.* Hungary: International Society for Inventory Research.

Ravallion, M. (1988). *Markets and Famines.* Oxford: Clarendon Press.

Rogers, J. B. and Van der Geest, W. (1988). *Empirical Approaches to Public Foodgrain Marketing and Food Security in East and South Africa.* Draft Working Paper. Oxford: Food Studies Group.

Rogers, J. B., Van der Geest, W. and Greener, R. (1988). *The ODNRI/FSG Cereal Market Model: A Diagrammatic Presentation*. Oxford: Food Studies Group.

Singh, I., Squire, L. and Kirchner, J. (1985). *Agricultural Pricing and Marketing Policies in an African Context. A Framework for Analysis*. Staff Working Paper no. 743. Washington, DC: World Bank.

Van der Geest, W. (1989). *The Analysis of Foodgrain Stock Requirements and National Food Security, Vols 1 and 2.* Oxford: Food Studies Group.

Van der Geest, W. and Greener, R. (1988). *A Model for Food and Agricultural Policy Analysis*. Oxford: Food Studies Group/Overseas Development and Natural Resources Institue (draft).

World Bank (1981). *Accelerated Development in Sub-Saharan Africa.* Washington, DC: World Bank.

8
A Systems View of Commercial Supply and Marketing Links

P. R. STREET
Department of Agriculture, University of Reading UK

INTRODUCTION

For the purposes of this chapter the food chain is the physical and organisational system linking producers and consumers (Fig. 1). The basic supply of raw materials entering the food chain emanates from the agricultural ecosystem.

> *Every agricultural ecosystem operates within a context that includes all those parts of the world outside that affect or are affected by the system and its activities.*
>
> (Spedding, 1975)

This chapter paints a picture of the agricultural ecosystem as part of a wider food chain system which is constantly adapting in response to market opportunity, and institutional and technical pressures. A systems view can improve the understanding of the constant restructuring taking place and of the associated formation of new supply and marketing links. The concepts and principles which underlie these changes are discussed and considered from the commercial viewpoint and in terms of the longer term and broader interests of society.

OVERVIEW

Supply Issues

In Europe and the Western world only a small proportion of the products of agriculture are consumed in their natural state. Increas-

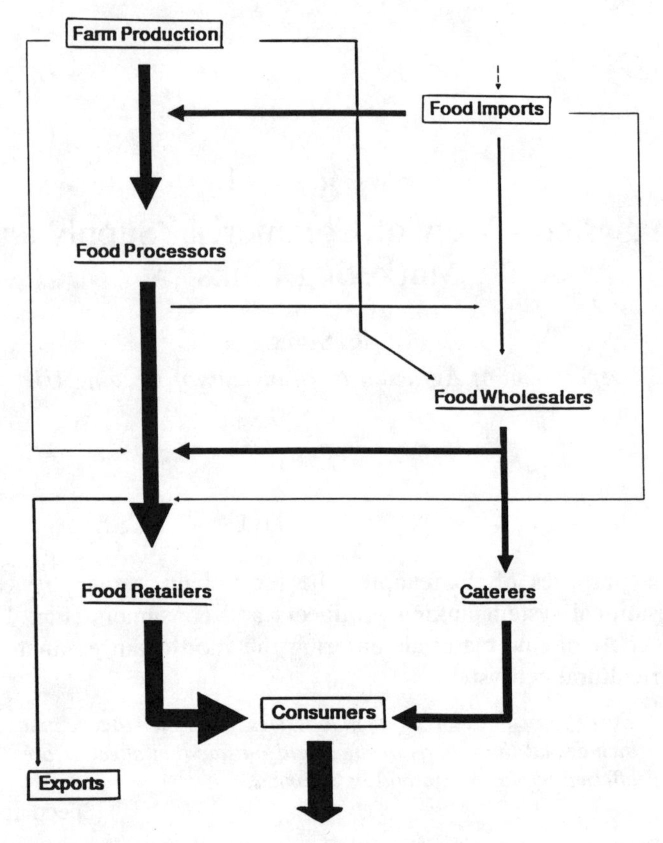

Fig. 1. The food chain system—approximate relative values of product flow.

ingly they provide the raw materials and value-adding opportunities for the food-processing, retailing and catering sectors. Strong price and supply competition provide for import as well as product substitution; resultant price pressure is significant.

In addition, as the support costs of agriculture have proved too high, price protection has had to be progressively removed. Whilst this has caused severe reduction in producer incomes, which continues, it also results in greater market transparency making growers more aware of commercial market messages and requirements. Until recently the main criterion for success was the ability to increase output. There was invariably an institutional market which absorbed the product, guaranteed a fair price and was not too discerning about quality or specification. The producer was rarely paid a premium for quality or penalised for lack of it, and in many cases, e.g. grains and meat, the incentive was for increased volume and not quality. This in turn created an artificial stability in the market which stifled the development of product variants and technical differentiation between producers, which in a less protected situation are the competitive mechanisms and the means by which markets are developed and growth occurs.

Demand Issues

Downstream from the farm the market for food is rapidly changing both in terms of the characteristics of demand as well as the organisation by which it is satisfied. The characteristics of the demand for food have moved progressively, driven by changing life styles, levels of income and its distribution as well as age and social structure of the community. However, in no small part this change has been driven by the success of agriculture itself. Agriculture has succeeded in providing more than adequate supplies of relatively cheap food. This in turn has permitted the consumer the luxury and the food industry the opportunity to move attention from the satisfaction of basic 'needs' for food to the development of 'wants'. In other words, to set more precise conditions regarding the quality, presentation and degree of processing of goods on offer, and increasingly to determine the very technology by which it is produced and processed.

The rapidly changing sophistication of the demand for food has become the means of growth in the sector and so the specification of the raw materials required has had to change too. Raw material product specifications are more exacting than ever before and supplies which do not meet requirements are frequently very highly discounted or may even be difficult to market at any price. Gone, too, is the stability of the specification; this now changes as the market develops increasingly towards higher quality. The market is taking a serious interest in how the farmer husbands his crop, i.e. what variety he plants, what pesticides and chemicals he uses, and all activities from field to retail (Street, 1988*a*). This is not limited by national boundaries. Increasingly, interest and influence are creeping back up the supply chain to assure quality and reduce supply risk. Production is increasingly programmed through formal and informal links between producers and users.

The processing and retailing sectors have both responded to and led this development and, through the resultant change in demand, have grown and prospered by the provision of more sophisticated and highly processed food products. This is progressively changing the balance of power towards major retail and catering groups, with resultant repercussions along the whole food chain. The retail sector in particular has become adept at passing their problems back upstream to maintain their margins, i.e. to the packer/processor/manufacturer and, in turn, on to the weakest seller, the producer. The effect of the price pressures discussed above end up at the farm gate. In an oversupplied market there is generally somebody who will supply, however eroded the price. If a business fails in the process there is always another which acquires its assets cheaply and takes its market share.

Increasingly, high-technology investment in production, assembly, grading, packing, processing and distribution facilities is needed to service an increasingly sophisticated market. The retailers have shown little propensity to share in the upstream investment requirements or risks. They have no need to if others will take the risk to assure a market. Thus processors, packers and grower groups have had to make these investments. Increasingly, the grower has become the supplier of raw material, which is a weak sector compared with the food processors and manufacturers. The latter increasingly enjoy the value-adding opportunities, not the former. This is all a natural development as the sophistication of consumer requirements is developed and as free market forces operate in an oversupplied market.

Market-Led Change

Thus, the farmer is facing declining real prices and an increasing debt burden against a background of declining wealth and borrowing power (Harrison and Tranter, 1989). He is now fighting to arrest the decline in his standard of living. He is therefore understandably much more critical of his husbandry and farming system and is seeking new opportunities and new strategies for survival. Farming systems are in a highly disturbed and adaptive state. Increasingly, farmers are having to seek new formal and informal linkages in the market to minimise input costs and to assure markets for oversupplied products. This too brings with it adoption of new technical innovation as well as structural change in the sector and socio-economic effects. There is also overcapacity in processing and retailing facilities and hence a highly competitive and adaptive sector downstream from the farm which is developing the market and seeking adjustments to supply. Under these pressures the food chain system is constantly evolving and restructuring.

It is to the understanding of the commercial actions and reactions involved in this change that the remainder of this chapter is addressed. Understanding this behaviour is essential for the identity of opportunity and the development of strategies for improvement and can hardly be achieved without resort to systems thinking.

DETERMINANTS OF CHANGE

The food chain system is changing under the influence of economic, market, institutional and technical pressures which are inter-related; each has implications for all others and a systems approach is essential to understanding the resultant behaviour of the players in the system and for managing change in the sector. These pressures and their influence are described below.

Economic and Regulatory Policy

There are major institutional changes taking place in relation to the development of firms in the supply chain. To understand these changes it is necessary briefly to discuss the economic policy framework which provides the environment for business growth and development. The current economic policy framework results from the Keynesian, neo-classical synthesis of full macroeconomic employment and optimal microeconomic allocation of resources. Policy instruments are designed against the major objective of maximisation of the 'value' of economic activity, i.e. gross domestic product (GDP). To a large degree this approach ignores psychic satisfaction, distribution, resource depletion or pollution.

The target is a bigger economic pie, and any social dissatisfaction with that is largely averted by growth in each slice rather than in change in relative sizes. Full employment and efficient allocation serve to increase the growth of the pie, i.e. GDP, and so by fiscal policies growth through investment is encouraged. If investment is not forthcoming today then unemployment will result, but if enough investment is made today then even more will be required tomorrow (Domar, 1947). This results in a 'growth, more growth' treadmill both nationally and for individual firms. Thus economic policy is expressly designed to promote the required growth through taxation incentives and investment grants. Politicians, and hence governments, are also increasingly concerned with issues of food safety, health and environmental degradation. This leads to legislative action and regulation of business activity.

The above simplistic analysis is introduced purely to demonstrate that, in the short term and commercially, current economic policy provides every encouragement for growth. The commercial operator has little option but to respond. He is under very significant pressure to expand in order to safeguard the interest of shareholders through improved profit and stock market ratings as well as to protect his personal position and progress his career.

The Market and Marketing

Within this treadmill the market for the products of agriculture is constantly changing. In this sense the market is defined as the opportunity for buying or selling. The market results from an intensity and variety of forms of demand which determine the size, composition and flow of goods appearing in the market chain (Quilkey, 1987). These demand forms are determined by key driving factors such as consumer numbers, wealth, income, change in life style, relative prices of products, knowledge of quality and range of goods and services available, as well as perceptions of health, nutrition and pollutant effects, and not least in developed economies by product development and promotion effort. In the Western world at least, consumers are becoming more affluent, the average age of the community is increasing, a greater proportion of meals are eaten out and convenience foods used at home, and, not least, there has been a major change in shopping habits towards one-stop shopping. Consumers are becoming more sophisticated and health-conscious and are demanding ever improving range, quality, presentation and novelty in their products. Above all, they are increasingly demanding knowledge of the full history of what they purchase.

In the UK and in Europe in general there has been little recent growth in real terms in spending on food. However, since the value-added component of this expenditure is increasing, this exerts extreme downward price pressure on raw materials. This is illustrated by the major changes in the composition of expenditure (PSL, 1987). Expenditure on meals eaten outside the home, convenience foods and snacks is taking an ever-increasing share. Since 1980 total convenience food sales are estimated to have increased by 43 per cent and frozen food sales by over 60 per cent. Major developments in the catering industry through concentration and fast-food retailing add to this momentum. Spending on meals outside the home has shown spectacular growth (135 per cent between 1975 and 1982) and has risen by 25 per cent in the last 5 years. With the decline of the formal shared family meal there has been a spectacular growth in snack food sales which is now a very dynamic sector and estimated to account for between 6 and 22 per cent of consumer expenditure on food in different EEC member states (private communication, PROMAR, Produce Studies Ltd, Northcroft House, West Street, Newbury, Berks, RG13 1HD, UK). For every product group the market is constantly changing through a balance of positive and negative influences. This is illustrated in Fig. 2 in relation to the development of the $40 billion European snack food market.

A range of players including importers, packers, processors and distributors play an increasingly important role in transforming the raw materials of agriculture into forms and service arrangements suitable to retailers and caterers who are the direct interface between the producing sectors and consumers. These players are provided with opportunities by the changing characteristics of demand but are constrained in their ability to meet the demand challenges by such factors as the availability of capital, competition, location, research and development. Marketing is the reconciliation of these opportunities and constraints and is the process of getting the product from the producer to the consumer. This starts with the farmer's decision to produce a saleable farm commodity and involves all subsequent aspects of the marketing structure or system, both functional and institutional, with technical and economic considerations. At the end of the day it is a demand-driven exercise and as Adam Smith (1776) writing over 200 years ago in his *Wealth of Nations* stated:

> *Consumption is the sole and end purpose of all production; and the interests of the producer ought to be attended to, only so far as it may be necessary for promoting that of the consumer.*

In other words 'the customer is king' and in its strict commercial sense marketing has the objectives of survival, efficiency and profit. Some firms, increasingly the less successful, treat the market merely as a sink for their products. The more progressive treat the market as a

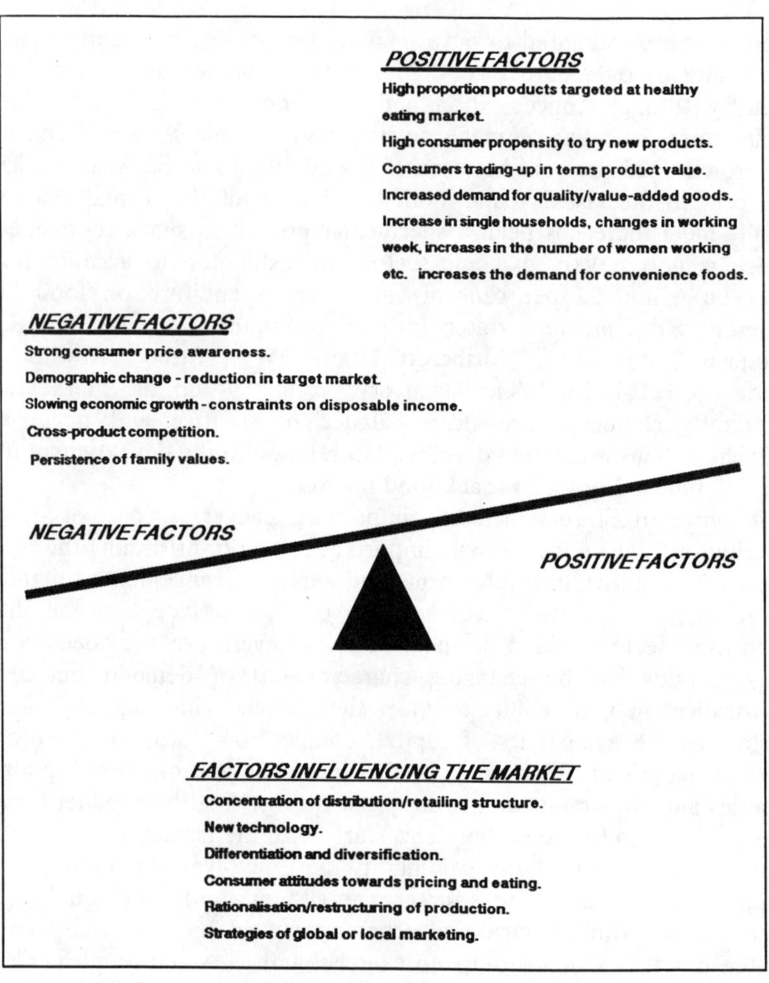

Fig. 2. The snack food market—positive and negative influences.

continuous and changing source of 'needs' and 'wants' to be identified, created and satisfied. In this sense needs are the essentials or staples of life, i.e. food, clothing and shelter, whilst wants are goods and services for which demand can be created (brown eggs, square crisps, the uniformly red apple, multiple layer packaging) and on which in developed economies business success increasingly depends. It is the development of wants in particular that gives rise to much of the dynamics of the food chain. To survive, firms and advertisers have had to create a desire and appetite where nature did not provide for it. They have reinforced and exploited the consumer's desire for novelty and to be different. Increasingly, products are built for short life, obsolescence and replacement by new ones and a less stable market results (Weisskopf, 1973).

Increasingly, in the food sector new products have to be developed or introduced and a whole series of services are required to get goods from the point of production to consumption. It is these services which increasingly make up the process of marketing. In this sense, a service is any function which alters the commodity in form, place, time or possession and usually also increases the value of the product. Effective exploitation of the market potential requires a systems view of the factors involved (Fig. 3).

The total service system comprises a range of stages and players as shown in Fig. 4, which transfer and transform raw materials from production to consumer products at retail. It is a complex of market channels, involves the processes of concentration, equalisation and dispersion and is unique for each product and buyer. Formal and informal arrangements in the marketing system link these processes into a system. *Concentration* is the process of assembly or bulking-up involving group activity, agents and wholesale markets, and takes place at various levels in the food chain system. *Equalisation* includes the processes of storage, grading, packing and processing and activities which render raw materials more uniform and satisfactory to the consumer. *Dispersion* involves the final steps in the food chain system by which bulk is broken and products are distributed through marketing channels to consumers, increasingly through third party distribution systems which break bulk for individual items and reassemble the required product range into in-store loads.

A major component of marketing is planning for profit and survival. These plans encompass the whole service system and include product range, price and delivery terms, and organisational linkages. Without such planning, businesses would fail. The farmer has to plan what to produce and is largely concerned with market opportunities and prices. Wholesalers have to plan procurement requirements, attempting to match them to those of processor and retail customers, and the latter have to plan their supplies to match consumer requirements. At any point within the food chain system, backward and forward planning and linkages are required. Thus, market planning involves considera-

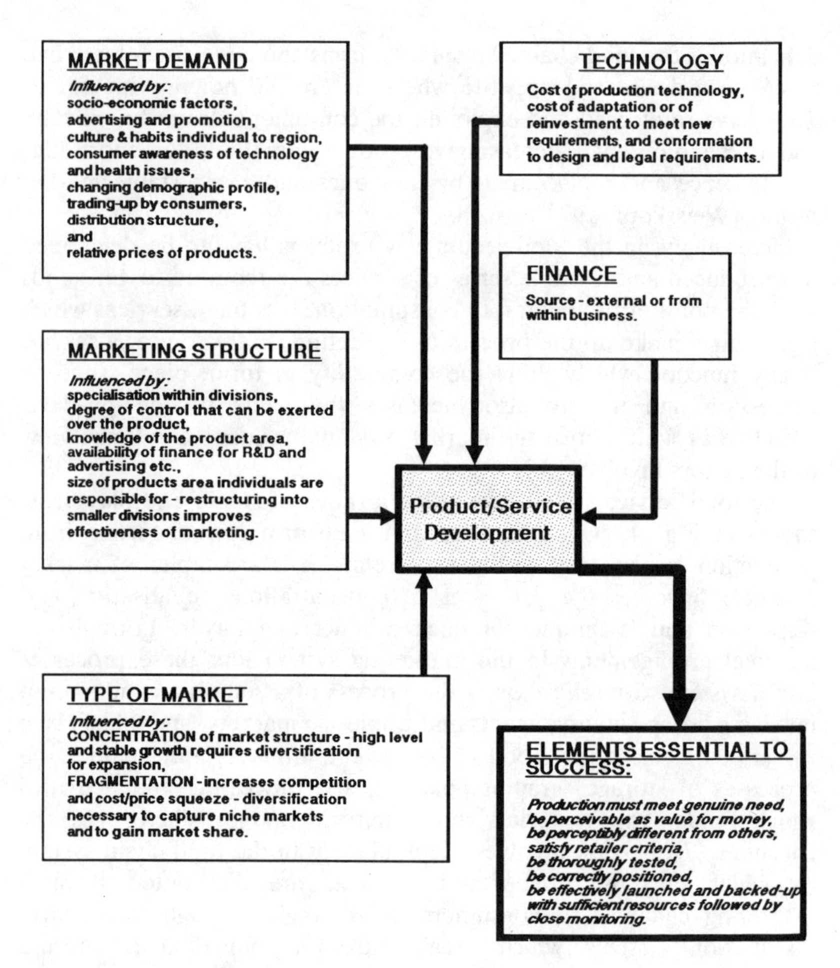

Fig. 3. Product and service development.

tion of options for improved efficiency at each level in the service system and more frequently by reorganisation of vertical links within it so that progressively the user (Fig. 5) can specify what is produced, how it is produced and dictate price and delivery terms. In an aggregate sense the market is the result of arbitrage between the many and different self-interests at various levels in the food chain which are often conflicting.

However, beyond this strict commercial definition the market must

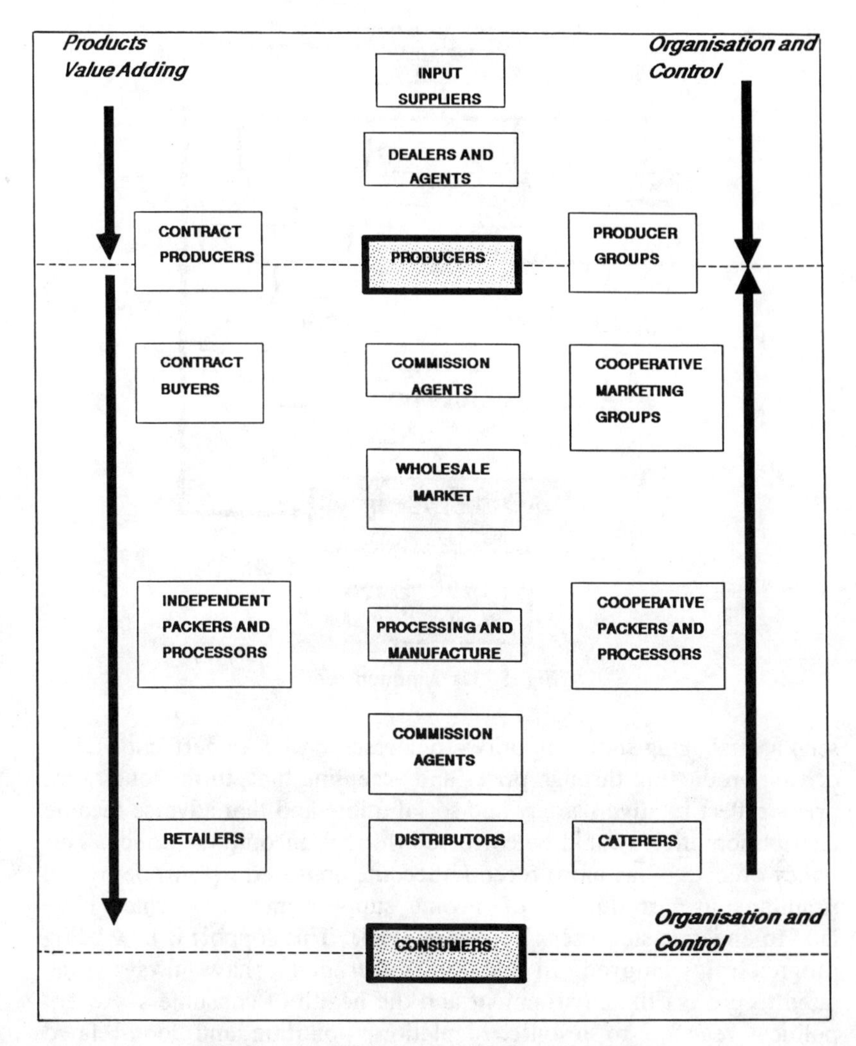

Fig. 4. The food chain—stages and players.

be seen as part of the social system and commercial activity must be regulated by governments and directed through policy instruments (fiscal and other) to meet social, safety and environmental objectives. Historically and simplistically, the main role of governments in agricultural markets has been through managing prices to equate long-run supply with long-run demand. In this respect the goal was

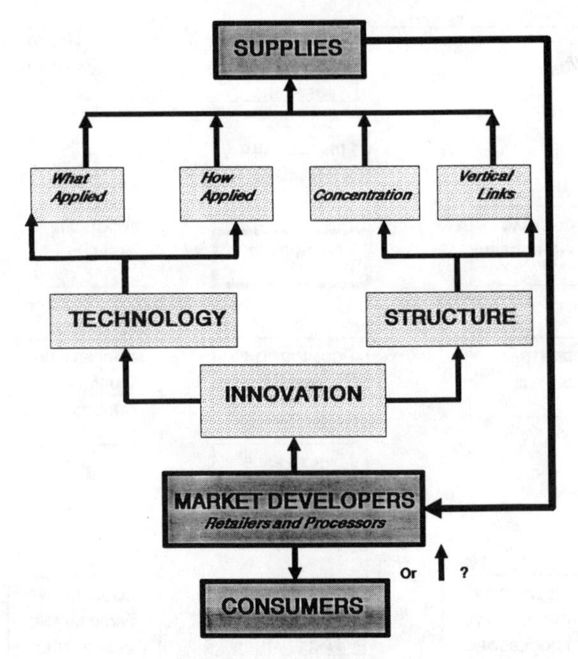

Fig. 5. User influence.

seen as managing short-run prices to correct market imperfections, i.e. driving production through prices and accepting that, in the long term, prices reflect relative scarcity and social utility and that adverse income distribution effects could be corrected through income tax policy. This policy objective has more recently become confused in Europe by the simultaneous introduction of income support measures which have lead to endemic surpluses (Boussard, 1988). This support is now being progressively removed. In addition, governments have always legislated to protect the environment and the health of consumers. Recent political reaction to resource depletion, pollution and food-related health risks has raised the profile of mechanisms to regulate agriculture and food industry activity.

The market is therefore an ever-changing consensus reached between the different interests of the range of players involved, from producers through consumers to governments. There are contradictions at any level in the chain and conflicting interests between levels. This is best illustrated with an example. Figure 6 shows some of the issues involved in increasing the market penetration of 'organic' foods.

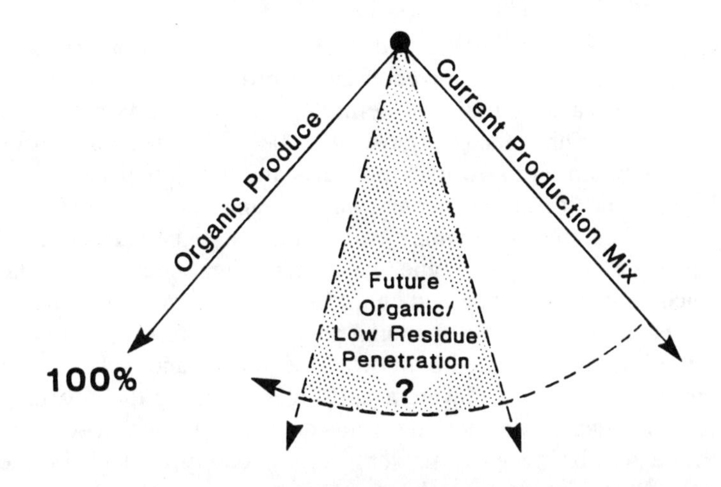

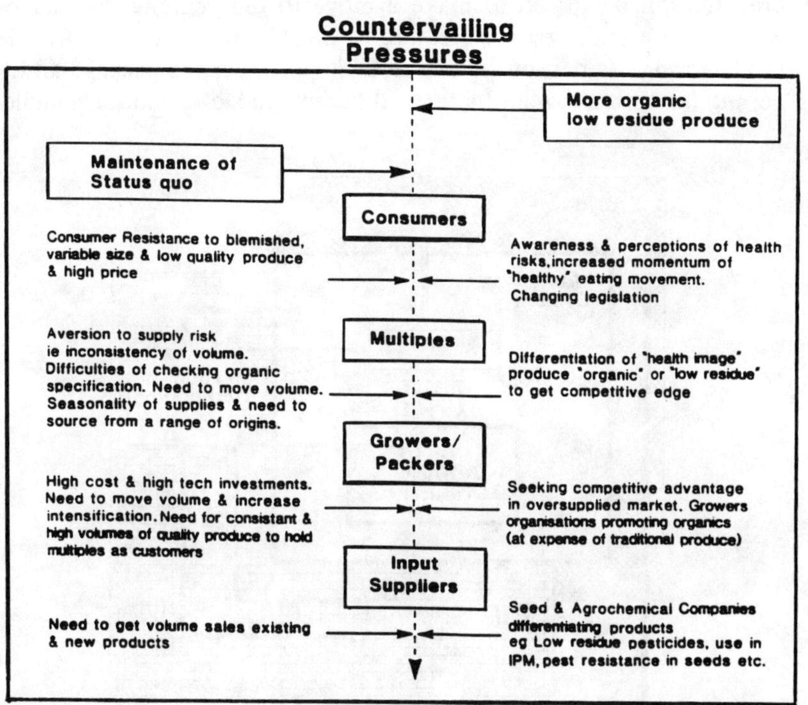

Fig. 6. Organic produce. Market penetration—the countervailing pressures.

In this case the 'greening' of the consumer creates both risks and opportunities. Risks are involved because promotion of the 'organic' product range leads naturally to demotion of the 'traditional' around which businesses and markets have been built. Against this must be weighed the potential to diversify products and develop niche markets. In this case, and for all food commodity groups where formerly there was one market, 'the traditional', two additional niche markets can be developed, i.e. the 'organic' and the 'low/minimal chemical residue'. Each niche has its own demand characteristics determined by perceptions of 'healthy eating' and spending power, and offers a potential for differentiation, for trading-up between segments and hence for increasing the total spend. Players at all stages weigh up the advantages and disadvantages of development; however high the risks, sooner or later there will be an early mover, i.e. somebody through business pressure will be forced to make a move to differentiate themselves. Retailers are under strong pressure to do this; they have to offer the product range as part of the one-stop shopping service package and to permit business growth. In turn, different and often more complex

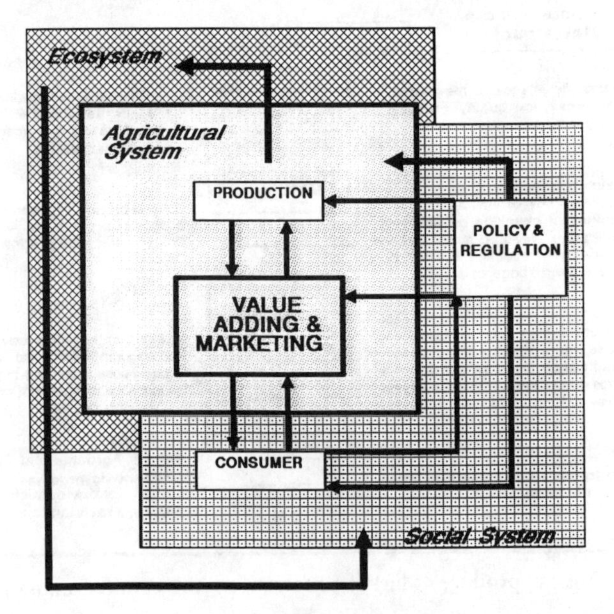

Fig. 7. The marketing system. An interface between agriculture, the ecosystem and the social system.

supply mechanisms and linkages have to be formed to satisfy this product diversification; each provides conflict, risk and opportunity at other levels in the chain and so the total system adjusts. In a wider sense still (Fig. 7), the marketing system must be seen as the interface of the agricultural, social and ecosystems (Spedding, 1975) and results from the feedbacks and control mechanisms linking these systems; rightly or wrongly, it is the manifestation of the balance of interests. In the example above, the 'greening' of the consumer results initially at least from perceptions of the adverse affect of agricultural practices on the environment, the safety of food and the welfare of animals. Consumer pressure for change builds and impacts on suppliers and on governments, the former offering new products and using the change in consumer perceptions to build new market segments and the latter reacting with new legislation to regulate practices. Thus, in turn, the effect of agriculture on the ecosystem changes and so the total system adapts.

Institutional Pressures

Economic policy and the changing demand for food result in significant institutional pressures.

Growth, Scale, Concentration and Market Power

The pressure to increase scale is intense. In the economic policy environment described above, and in view of the intense competition, successful firms must continue to grow either by organic means or by merger. In a relatively static market successful growth is through acquisition and invariably results in increased concentration at various levels in the supply chain. A change in scale and/or concentration of businesses alters the balance of production and consumption and *inter alia* of buying and selling power (Burns and Stalker, 1961). Mass markets have created and in turn been created by techniques of mass production and sophisticated retailing. Economies have derived from the size of plant or retail operation. The agriculture and food sectors are not immune, particularly processing and retailing operations, and we have seen recent and major concentration in the fruit and vegetable sectors, the meat and cereal-processing sectors and, perhaps most importantly, in the retail sector. The essential features of the growth and concentration cycle are described (private communication, PROMAR, *vide supra*) in relation to the development of the snack food market in Europe and summarised in Fig. 8a. Demand changes

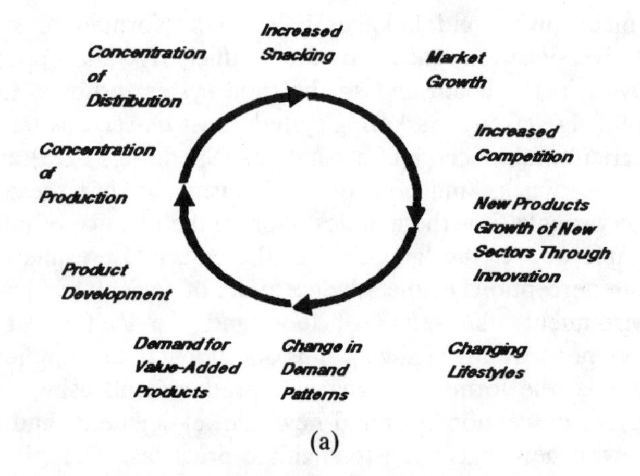

(a)

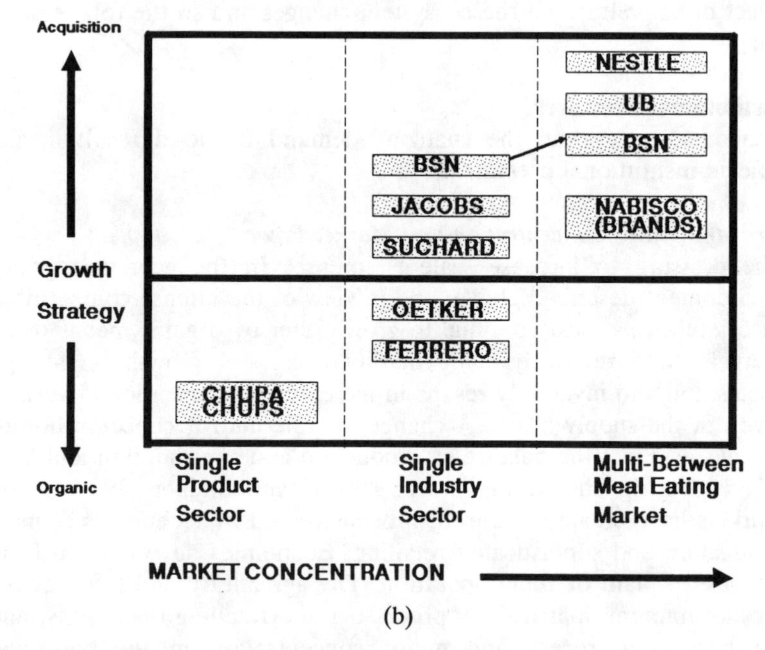

(b)

Fig. 8. Growth and concentration (a) The concentration cycle. (b) Growth—the snack food market.

and is developed; this provides opportunity which is satisfied through product development, business growth and concentration. In turn, these create the pressure for further demand development and growth and so the cycle continues. The corporate position of firms and their relative attitudes towards growth is shown in Fig. 8b in relation to the development of the European snack food market. A number of the market leaders have grown, concentrated and diversified their product ranges. The most successful have achieved this through acquisition and significant research expenditure on product development.

The effects of scale and concentration on market power in the food sector have been well described by McDonald *et al.* (1987, 1989) and Howe (1983). These authors describe an industry with stable demand and lack of growth where competition for market share is intense. They review a process of progressive rationalisation amongst the strongest players, the retailers, and upstream of them due to their powerful influence. These retailers are highly concentrated with about 70 per cent of packaged grocery sales through the multiples. The top six own over 80 per cent of the shops, and the top two groups account for 54 per cent. Five of the groups account for about 55 per cent of grocery sales with one group having a 30 per cent share. This power has placed the initiative for food chain system development into the hands of a few major groups who as a result enjoy 'supernormal' profits at the expense of those upstream (McDonald *et al.*, 1989). Own brand shares range between about 30 per cent and 60 per cent depending on product, and total market share varies widely between product groups. As indicated above, the multiples have been most successful with groceries but now have shares exceeding 40 per cent for produce, bacon and poultry with significant scope for extra growth in red meat and pork where in all cases shares are below 30 per cent.

Further upstream are the food manufacturers who have also grown and concentrated in response to market pressures. The five-firm concentration is in excess of 60 per cent of all sales for a number of commodity groups and over 80 per cent for biscuits. Two companies dominate the biscuit sector; Nabisco and United Biscuits having about 63 per cent of the savoury biscuit and 45 per cent of the sweet biscuit markets (private communication, PROMAR, *vide supra*). Despite this concentration, profit on sales in the food manufacturing sector is low, generally less than 5 per cent, and the author's own research amongst several firms in the fruit and vegetable sectors indicate profits as low as 3·8 per cent expressed as a percentage of fixed assets. This results from overcapacity in the sector, lower than required throughputs and hence high unit costs as well as from price pressure from retailers. As a result manufacturers and processors exert strong downward price pressure back onto their suppliers.

Producers have reacted to this market power downstream through group activity, attempting to increase scale through horizontal integration to gain buying and selling power and cost efficiency and through vertical integration to share in value-adding opportunities. In Britain, cooperation has not been an overriding success. Requisites, marketing and service cooperatives have grown in number from 192 in 1965 to 636 in 1986 (Plunkett Foundation, 1987). Marketing cooperatives are the most numerous group (72 per cent) and account for about 53 per cent of turnover whilst requisites cooperatives are only 17 per cent numerically but account for 46 per cent of turnover. It remains overwhelmingly a small business sector, having grown by only seven points ahead of the retail price index in 5 years (Curnock Cook, 1989).

Ownership and Control

As part of this growth process and again through the influence of economic policy in the processing and retailing sectors, there has been encouragement for a separation of ownership and control from the traditional entrepreneurs to public companies and within the latter from the holding of shares and control of the policy and activities of the company by management itself to holders of minority shares. This is inherent in capitalist enterprise and is the tendency to monopoly in the economy arising out of the division between ownership and the use of property (Burns and Stalker, 1961). Very recently there has been reorganisation of ownership of slaughter and meat-processing facilities and of produce grading and packing from private to public ownership and even the conversion of cooperatives to public companies. Once down this road the effect is to stimulate further scale and growth.

In terms of the primary processing of food, i.e. grading, storing, and pre-preparation, there was a move towards cooperative ownership frequently encouraged by generous investment grants. This form of ownership presents a number of problems. Many stem from the competence of the management and the forms of decision-making, frequently by committee (Plunkett Foundation, 1982). However, there are other issues, not least being the high level of overhead attendant upon processing requirements. Plants are financially efficient only with sufficient volume throughput. This requires dedication to plant management including the acquisition and marketing of third party produce from other domestic and, increasingly, import sources to achieve required throughput targets and acceptable unit costs. Handling third party produce is often more profitable and this confuses the corporate objective as there is a dichotomy between efficiency in plant operation and maximisation of grower returns. Finally, and not least, is the problem related to the financing of expensive developments. Funding is frequently derived from members by withholding profits from distribution or by capital injection as well as from bank loans and grants. However, this form of ownership brings inflexibility in capital mobilisation and conflict between members and their societies. Mechanisms such as revolving funds, whereby bonuses and interest on shares are allocated to members but cannot be withdrawn until the end of the period, allocation of asset ownership by usage and closed membership all make for inflexibility and difficulties of exit and entry.

Linkages

An increase in scale brings with it the requirement for investment to provide the increasingly sophisticated technologies for the manufacture of consumer products. With this comes an increase in financial exposure and risk. To offset this, the availability of continuous supplies of adequate volumes of raw material of constant specification becomes a prerequisite for success and it is essential to spread overheads, reduce unit costs and to provide quality products which are cost-competitive. Increasingly, manufacturers and retailers resist procurement through commodity markets and turn away from unreliable spot and wholesale markets as sources of supply. To assure greater process efficiency, security and reliability they set their own specifications and enter into supply contracts of formal and informal types and even in some cases into vertical integration, even with associated horizontal integration at various levels.

In the fresh and lightly processed food sectors, many of the supply arrangements are with cooperatives. Ignoring milk, which is a special case, the proportion of agricultural products sold through cooperatives in the UK between 1976 and 1986 grew from 16 per cent to 33 per cent for fruit and from 8 per cent to 17 per cent for vegetables but is fairly static for other products of which only two have significant shares i.e. eggs 28 per cent and cereals 20 per cent (Eurostats, 1977, 1982, 1987). Compared with France these shares are low. However, in important commodity areas where retail and processing power is strong, e.g.

produce, poultry meat and cereals, then group organisation and/or other means of vertical and horizontal integration have grown and flourished.

Intensive livestock is the area where vertical integration (i.e. where a firm controls two or more stages of the production and distribution process) has flourished. Intensive livestock breeding, production, fattening and processing lends itself readily to this type of activity. In other areas forward production to formal contract or programme agreements has been the method of integrating suppliers and users. This is a favoured method of production of vegetables, e.g. peas and carrots for processing, and between retailers and producers for fresh produce. These contracts take many forms ranging through statements of product specification and delivery terms to management transfers where buyers provide inputs, working capital and technical control, and take ownership of the product. Full management transfer is typical in the broiler industry, whilst it is more commonly found in diluted form in the pig industry where inputs are supplied but ownership of products does not occur until they leave the farm. More typical of vegetables is production to informally agreed programmes with strict specifications on quality, cool chain conditions and volume but with no formal agreements on price or guarantees to purchase.

Implications

Thus, the business treadmill described so graphically by England and Bluestone (1973) results from, and progressive impacts on, scale, concentration and ownership, and reforms the linkages and relationships in the food chain. The above analysis says much about the strategies firms adopt including growth by amalgamation, takeover and investment but also the way they have to differentiate their products and grow their market share to cover the ever-increasing need to deploy yet more productive resources to survive. They are on a treadmill perpetuated at the national level through the preoccupation with GDP growth, a system described by Johnson (1973) as a train going faster and faster, increasingly difficult to get on and impossible to get off, consuming more and more finite resources, which is unsustainable in a world where resources are finite and the ability to handle pollution and energy waste limited. Nevertheless, this is the situation which currently prevails in the environment where commercial decisions are made. Survival strategies provide for a strong interest in obsolescence and short-life products and, in the longer term, adverse reaction from the social system and ecosystem.

Elasticity of Resources

Concentration and public ownership in the manufacturing and retail sectors in particular, and through cooperative group formation at the producer level means that the resources deployed in the food chain have become more inelastic. High levels of capital and social commitment are involved which are hard to unravel, and the birth and death cycle and relocation of assets are less likely than was the case when the sector was largely in the hands of small, privately owned businesses. In these new corporate circumstances the shareholders' interests are paramount and survival of the business the crucial issue. Survival means adaptation, but from change within the business and not through extinction, mobilisation of pecuniary assets and replacement. An entrepreneur can maximise profit for any period of time since profit-taking today benefits survival by offering a wide choice of opportunities for reinvestment and consumption. How different for the corporation, where time is an important function of profit-taking, since today even profit is subordinate to survival. In this latter case survival takes place through amalgamation and takeover. Unsuccessful businesses continue through absorption when others acquire them, rationalise their assets, take on their market share and benefit from improved economies of scale, and so the birth and death cycle continues but the basic business remains, albeit under different management.

Technical Development Pressures

Technical development and progress is the very essence of the evolutionary processes described above and business development depends upon it. Change in institutional structure cannot be viewed without recourse to discussion of the impact of technical developments. In Burns and Stalker's (1961) review of the management of innovation two opposing views are voiced regarding technology and institutional development. Marx (1846) views technical change as underlying every kind of change in the social order whilst Durkheim (1893) considers that technical progress results from changes in the institutions of society. The truth is that both views are correct and that the two go hand in hand. Technology both pushes change in the institutional framework and is pulled by it (Fig. 9).

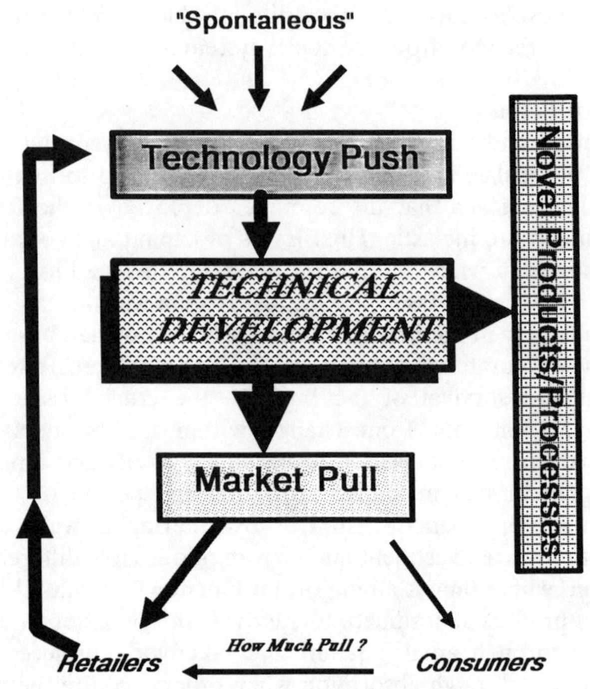

Fig. 9. Retailers as agents for change.

Commercially, companies respond by the adoption and adaptation of spontaneous technology development but are increasingly responsible for its creation through research and development funding in an effort to differentiate, and to develop and satisfy wants. Whilst at the farm level much technical development has resulted from basic scientific research both institutionally and commercially funded, the nature of competition and growth in the food industry means that further downstream much of the research expenditure is of a highly applied nature and concerned with product and process development. There is far less emphasis on basic scientific research than in, say, the chemical and pharmaceutical sectors. The objective of research is generally short-term pay-off with less support for longer-term science-based fundamental work (Cottrell, 1989; Gurr, 1989).

Many technologies once introduced have the potential quite naturally to lead to reorganisation of the stratification, organisational and locational characteristics of the supply and marketing system. In the agricultural and food chain system, technical developments have in many cases been the key to industry restructuring. Large processing, retailing and catering operations are difficult to manage efficiently without reliable supplies of raw materials continuously available to a tight and preset specification. Repeatability is the key; only then can processes be controlled, scale efficiencies realised and branded products result. When technical breakthroughs permit such production, e.g. hybridisation of vegetables, factory methods for white meat production, new animal breeding technology and developments in food packaging and preservation, then the downstream sector will have reorganised vertically and horizontally and created the means to pull the required raw materials through the marketing chain. They will have done this to increase the sophistication of their products, to brand them, to round their product range, to differentiate themselves from their competitors and concomitantly to grow their businesses. This is a non-stop process of continuous adjustment and restructuring and is clearly seen in the history of the broiler industry, the pig industry and more recently the fruit and vegetable sector.

The soft drinks industry provides a classic example of the influence of technical developments. The recent introduction of new packaging technology in the form of PET bottles for carbonated drinks, and Tetrabriks and Combibloc carton systems for the packaging of fruit juices provided for flexibility of size, ease of handling, lower cost and greater convenience and transportability for the consumer. This encouraged the mass movement of soft drink sales into the multiple grocers and led to a stimulation of demand. As a result of these technical developments the multiples increased their share dramatically in this important market segment. In France over 70 per cent of soft drink consumption is in the home, and the supermarket and hypermarket share of this has risen from 68 per cent in 1983 to 84 per cent in 1988. In the UK price cutting is a major feature of the market, about 75 per cent of product is consumed in the home and the grocers' share of the total trade has grown from 48 per cent in 1983 to 56 per currently. Today around 45 per cent of fruit juice sales in Europe are in cartons and as much as 75 per cent of the major carbonated drink, cola, is in PET. In the UK the proportion of all carbonated soft drinks packed in PET increased from 19 per cent in 1982 to 34 per cent in 1987 and approaching 90 per cent of fruit juice is now sold in cartons. In France 73 per cent of colas and 86 per cent of lemon/limes are now sold in PET bottles and 77 per cent of fruit juices (private communication, PROMAR, *vide supra*). This development took place at the same time as a price squeeze and the traditional supply from numerous regional bottling plants was inefficient and did not meet the requirements of the multiples who demanded low cost production, central buying and bulk deliveries to distribution depots. The whole production, packaging and distribution system was rationalised and now supplies are derived from a few large central production units which provide economies of scale.

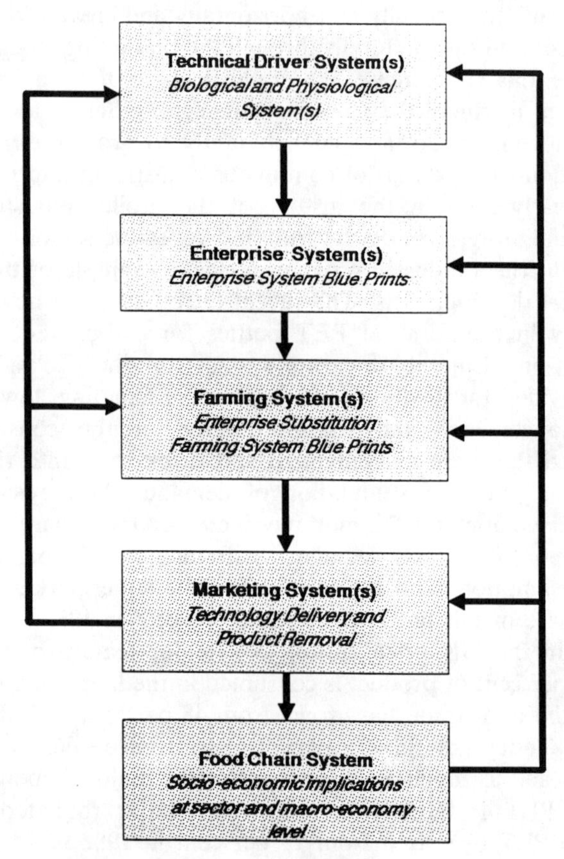

Fig. 10. The technology transfer system. Effective technology requires system redesign at all levels.

Implications

Thus, technical developments in agriculture and the food chain can in some cases be the causal factor of food chain reorganisation and often naturally lead to specialisation, concentration and geographical relocation at all levels in the food chain both within countries and across national boundaries. The introduction of new technology at any point in the food chain can effectively disturb the whole food chain system (Fig. 10) since it will have forward and backward implications from the point of introduction. Effective technology transfer will require a systems understanding of these implications and effects.

Under the circumstances described above the development of the food chain is a complex and dynamic process which only makes sense in a systems context. Market development, organisational restructuring and technology development and transfer go hand in hand and it is difficult to separate cause and effect. Change results from a consensus of conflicting interests within and between each level in the food chain. Ripples move backwards and forwards along the chain setting up action, reaction and feedback. These ripples move beyond the food chain into the environment and the broader economy, and reaction there in turn results in regulation and control through legislation and change in consumer attitudes. Successful business development is generally only effective if management takes a systems view of the development of the market and considers the system not just horizontally but several stages forward and backward from its position to assess opportunities. Successful entrepreneurs are, therefore, systems thinkers by design albeit in many cases subconsciously.

THE DEVELOPMENT OF BUSINESS STRATEGY

Insight into the considerations involved in determining strategy assist greatly in understanding how the pressures described above impact and why the food chain system evolves. A successful firm is involved in a continuous matching process between its output and the requirements of its customers. The effect is seen in the constant evolution of the total service package it provides. Commercial success depends on how well a firm identifies and anticipates needs and wants and on how well it matches its output to them and anticipates the response of competitors. As the supplier progressively creates and/or satisfies wants rather than needs then the economist's traditional measures of the market in terms of the econometric links between the duality of supply and demand and prices become less useful for planning since the structure of the market upon which they are based is changing and basic shifts in demand are being engineered. Wants are developed and markets built and the strategist has to resort to market research and product testing to assess positive product and advertising attributes, to design and test products and to determine price position and volume prospects (Jones, 1985; Quilkey, 1987). Market research applies a systems approach where the changing relationships of technology, preferences, social institutions and profitability are all taken into account (Manilay, 1987).

The formulation of effective and competitive business strategy also requires an understanding of the shifts of power which accompany the rationalisation in the sector including identification of the strengths and weaknesses at any level of suppliers, customers and competitors. In view of the intense competition in the food industry discussed previously, each player is attempting to increase or at least hold market share and to achieve above-average performance in the long run to satisfy shareholders. He has therefore to achieve sustainable competitive advantage and develop his business strategy to this end. The soft drinks industry provides a good illustration of these shifts in power. Until recently the industry was characterised by a multitude of loose brands but has recently been rationalised. Resulting from the volume demands of the multiples and the intense price competition in the market the major players sought to increase scale and hence cost efficiency and market power through acquisition. Starting in 1986 a series of mergers and reorganisations took place between the major players. Today the industry is dichotomised into a group of low-cost producers providing own-label products for the multiples and two large high-cost producers of branded products, e.g. Coca Cola and Schweppes Beverages Ltd, and Britvic Corona. In between the two exists a spectrum of small, independent companies decreasing in number and market share and driven back into niche markets. As a result of these changes the production base is now highly concentrated and Coca Cola and Schweppes Beverages Ltd now account for about 23 per cent of the soft drinks market some way ahead of Britvic Corona with 17 per cent (private communication, PROMAR, *vide supra*).

Strategy Objectives

To achieve sustainable competitive advantage, business strategy is formulated around three competitive planks (Fig. 11): (1) *Cost advantage,* which is achieved if the cumulative cost of performing all value activities is lower than the competitor's costs; (2) *Cost leadership,* which is the provision of superior value by offering lower prices than competitors for similar benefits; and (3) *Differentiation,* which is the provision of unique benefits that more than offset a higher price.

Cost Competitiveness

Cost competitiveness includes advantage and leadership and is achieved from sound analysis of the business as a system, i.e. through

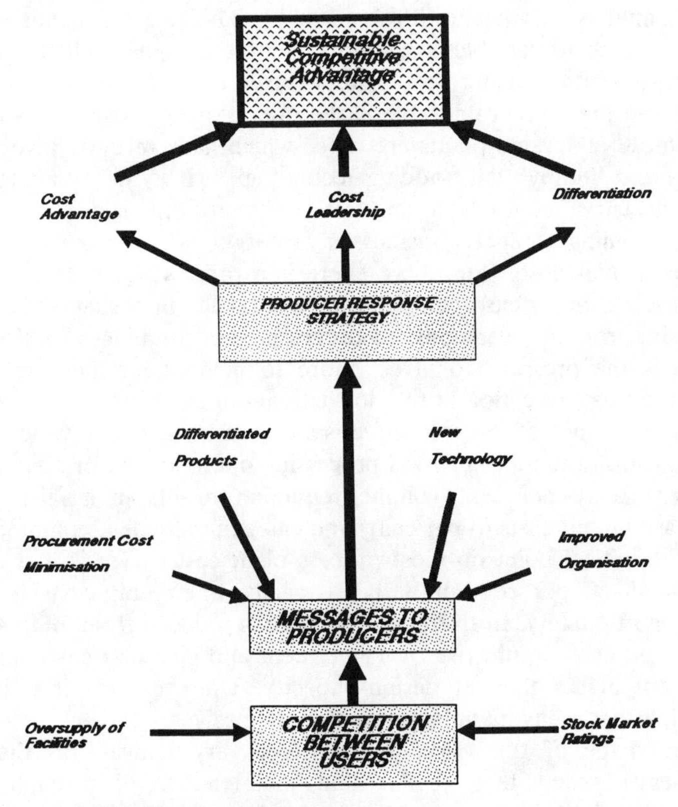

Fig. 11. The development of business strategy.

the identification of the value chain, the cost drivers of each value activity and how they interact. These are compared with competitor value chains, alternative methods and technologies, and strategies are developed to lower the relative cost position without erosion of differentiation. These main cost drivers include:

(i) *Economies of Scale and Capacity Utilisation.* Scale economies involve the reduction of unit costs through overhead spreading. Firms deliberately set policies to reinforce scale economies in scale-sensitive areas. The benefits of scale are mainly reduced unit costs, improved security against the adverse cost effects of future volume erosion and market power in procurement or disposal.

As shown in Fig. 12a, which is derived from research with a series of produce packing companies, the relationship between unit cost and throughput is curvilinear (Street, 1988*b*). The shape depends on the ratio of fixed to variable costs and the curve is higher and steeper for businesses with a higher proportion of fixed costs. Clearly, the steeper the curve the more vulnerable is the operation to erosion of volume. The most vulnerable plants are those which have recently invested in costly and 'lumpy' value-adding technologies. It is important to note that this curve moves each time fixed costs are increased, i.e. upwards and becoming steeper. Businesses face two major pressures which influence unit costs (Fig. 12a): there is a pressure to reduce volume because of the intense competition that results in a stagnant market suffering from an oversupply of processing and retailing capacity; and there is the pressure to invest more to meet the requirements for increased sophistication in the downstream operations.

The relevance of the cost curve is clear: in the example which is not untypical of a number of food processing operations at or near design capacity a 10 per cent volume reduction results in a fairly small increase in unit costs (3 per cent) and can generally be accommodated; but the critical point on most process plant cost curves is if they run below 45–50 per cent of design capacity when unit costs increase disproportionately. In this case if volume is reduced from 50 to 40 per cent, unit costs would rise by 11 per cent and give unit costs some 34 per cent higher than at design capacity. Unfortunately it is in this position that many manufacturers find themselves.

The shape of this curve is therefore very relevant to the way businesses react: they realise that maintenance of throughput is fundamental to cost containment; they know that any measure that

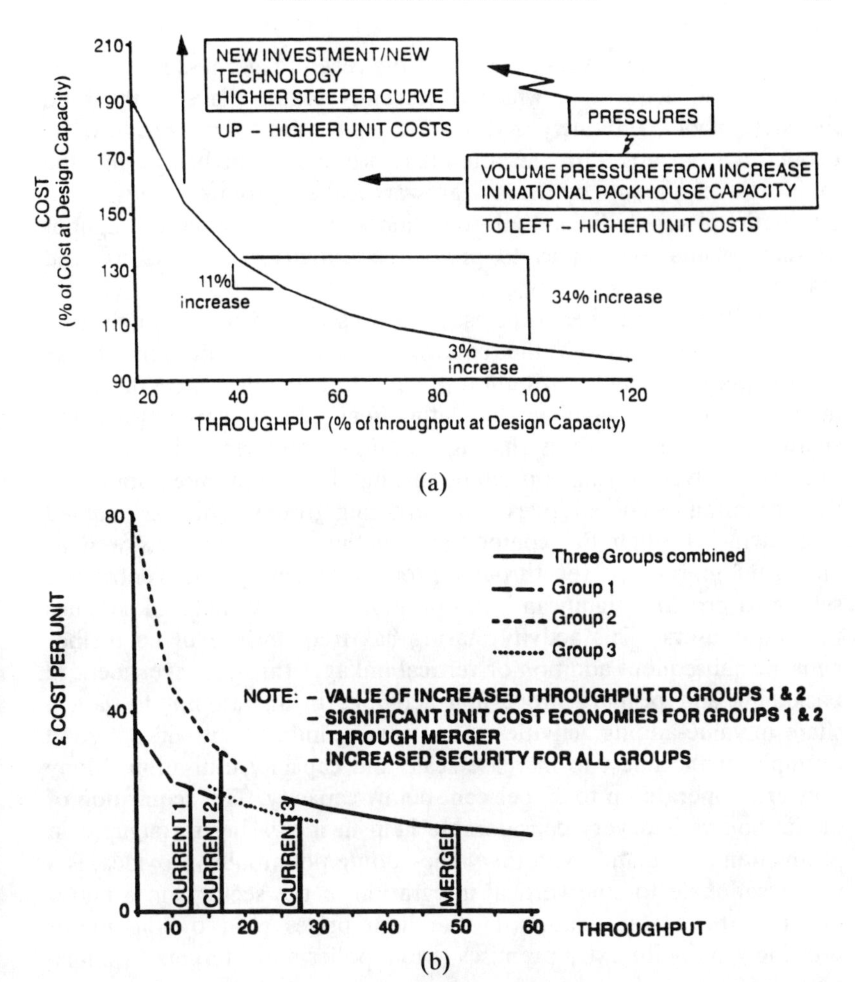

Fig. 12. Packhouse unit costs. (a) Unit cost pressures. (b) The merger/acquisition advantage.

results in increased volume moves the firm to a more secure and less price-sensitive segment of the curve; they know that if volume does fall and prices increase this comes right off the bottom line unless it can be passed back along the chain. The benefits of scale are seen clearly in Fig. 12b which illustrates the benefits to three parties of merger, reduction in unit costs for Firms 1 and 2 and improved volume/price security, and buying and selling power for all three.

The above analysis says a lot about the recent eagerness for merger, takeover and acquisition which is currently taking place in the food industry, about the increased handling of third party produce by cooperative groups with the attendant problems that brings and the major development by many grower/packer groups to establish international supply links, both to obtain a more aseasonal throughput of their plants as well as to provide continuity to outlets to hold markets.

In addition, a number of measures are used by firms to obtain cost competitiveness through the control of capacity utilisation. These include peak load or contribution pricing, line extension, ceding share at peak periods or leaving fluctuating segments to competitors and sharing activities. Activity-sharing amongst producers has most frequently involved buying and selling through horizontal integration and the formation of cooperatives and marketing groups. This has enabled cost saving through the competitive purchase of inputs, as well as enhanced market power through product assembly and centralised selling to provide 'significant' and programmed raw material supplies for major users. This activity-sharing has frequently evolved further with the subsequent addition of vertical linkages through investment in packaging and primary processing activities in an attempt to gain a share in value-adding activities. The brewing industry provides a good example of the quest to increase scale and capacity utilisation. Many breweries operate up to 30 per cent below capacity. The acquisition of public houses is a very considerable help in increasing throughput in production plants and, with the value-adding potential it provides, is a major rationale for the vertical integration in the sector and a major factor in the determination of the high prices paid by the major brewing groups for extra premises (Monopolies and Mergers Commission, 1989).

(ii) *Procurement.* Procurement costs minimisation is increasingly important at all stages in the food supply chain. The increasing sophistication and high cost of the equalisation processes involved in marketing has meant that grading variable product to requirement is no longer viable. The author recently analysed the outturn of six 'well run' packhouse operations where outgrades ranged between 7 and 39 per cent with an average of 25 per cent. Twenty-five to thirty per cent outgrades are not uncommon across the produce sector and are not sustainable in the light of the expense of storage and grading and the increasing demand for constant volumes of tightly specified raw materials by users.

Meeting this demand for raw materials at the highest realisation price is the challenge for producers, and obtaining it at the cheapest price compatible with longer term supply assurance the challenge for the buyer. These conflicting objectives lead to ever-increasing efficiency in service operations. To a degree, resolution of the price conflict depends on a gentleman's agreement and is an area of brinksmanship. Each party pleads poverty, the producer attempting to obtain a higher back-to-farm price but with the fear of jeopardising unwritten longer term supply relationships, the buyer exerting price pressure to improve his margins but always knowing that he cannot go too far as he wants an industry which maintains a range of significant suppliers to reduce his supply risks and maintain competition. Thus, at all stages in the supply chain efforts are being made to tune raw material specifications more precisely to needs and to select only those suppliers and channels that can meet these requirements.

Reconfiguring the Value Chain

This is a further means of increasing competitiveness and involves consideration of the efficiency of each activity: can it be performed differently or eliminated?; can a group of activities be reordered or regrouped?; or, can coalitions with other firms lower or eliminate costs?

Fundamental to this reconfiguration exercise is 'differentiation analysis', i.e. does the value of the unique benefits provided through the differentiation of a product or service more than offset the extra costs? An integral part of this exercise is examination of the technology strategy. This involves examination of existing and new technologies at each stage in the supply and marketing chain, identifying relevant technologies including new ones from a technical and commercial viewpoint, and finally selecting a strategy which reinforces competitive advantage. Increasingly in the food industry this differentiation strategy involves investment in research and development to create new products and processes (Blanchfield, 1983).

Implications

Commercial strength therefore depends upon cost competitiveness and the understanding of the need for continuous differentiation and repositioning of the business, its products and service package. Effective repositioning has also to take account of the influence (beneficial or adverse) of any affects on, and reaction from, the ecosystem and the social system. Increasingly, assessment of political and consumer acceptability is a prerequisite to product and process development as well as to enhancement of market power and can be used as a promotional and differentiation attribute.

The market dynamics and business development pressures described above give rise to structural and relational changes throughout the food chain system. Thus, the aggregate marketing system is not structured in frozen immobility but is dynamic, with a constant interplay and mutual redefinition between individual entities and social institutions (Burns and Stalker, 1961). To understand these changes in a systems sense is the key to effective strategic decision-making in business which, in turn, is the key to the survival of the business.

PRACTICAL MANIFESTATIONS

The Produce Sector

Western Europe

The produce (fruit and vegetable) sector (Malcolm, 1983; PSL, 1987; Street, 1988) provides a good example of the system adjustments which have taken place under the above influences and pressures.

In volume terms the market is growing only very slowly. An increasing proportion of the supply has come from imports. Imports into the EEC originate from many parts of the world and the EEC is seen as a prime target export market by many. The result is that there is oversupply of most products in most seasons, few unfilled product windows remain and downward price pressure is severe.

Alongside this have been major changes in the composition of demand in terms of the product mix and the value-adding operations. Wants have been developed and tastes are broadening, there is a much greater propensity to experiment with new products and there is an expectation that products will be available in all seasons. The most important factor which applies across Europe is the rising expectation of product quality, virtually no demand existing for produce of low or mediocre quality. Consumers expect products which are fresh, clean, attractive, undamaged and unblemished. There have been associated major technical developments, principally involving a revolution in transport methods and the widespread adoption of cool-chain distribution from farm to retail.

In addition, there has been a trend on a Europe-wide basis to the development of the convenience of one-stop shopping through the development of supermarkets and hypermarkets with a decline in independent retailers. The trend in Europe is to fewer and larger stores owned by multiple retail groups (Jones, 1987) and a consequential concentration of buying power. The share of the fruit and vegetable market controlled by the retail multiples may be estimated at about 50 per cent for Western Europe as a whole and rising rapidly. This figure is approximately correct for the UK and France and is a little higher in the Netherlands at about 60 per cent. In Western Germany the figure is higher still, estimated between 80 per cent and 90 per cent, resulting from the large number of supermarket groups and less tradition of small independent retailers selling fruit and vegetables. In Sweden, the three leading multiples have a share in fresh produce approaching 70 per cent and in Switzerland one group alone, Migros, has a share approaching 40 per cent followed by the Cooperative with 20 per cent. Thus, throughout Europe the trend towards increased size of outlets, decline of independents and concentration of purchasing power is a general one.

The supply requirements of these multiples are different to those of the independent retailer with the result that the traditional wholesaler is declining in importance. Supermarket supply is based on a highly organised system based on central distribution depots providing stock-refurbishment on an individual day-by-day and store basis. Requirements are increasingly for packed, branded and bar-coded products. To meet this need, delivery to the distribution depot is highly organised, involving exact specification of quantity, time of delivery and quality of product. The multiple wants no part of selection from a wholesale market and instead sets out a specification for a complete service package for suppliers. This applies to domestic suppliers and also operates through importers to overseas suppliers. Thus, very large companies have emerged to handle the distribution operation. Many take no ownership of the product and carry out the distribution operations for a management fee; others are importers and distributors and have either to be involved in or integrate supplies through other packers. So companies such as Scipio in West Germany, Pomona in France, and Geest and Salvesen in the UK have taken on this role and become major players in their own national markets.

The United Kingdom

The UK, which mirrors the European situation, has led developments in some areas. Of the total household expenditure on food, fresh fruit accounts for about 6 per cent and fresh vegetables about 8 per cent. By any standard this is a large market estimated at £3·2 billion at retail selling prices and £2·3 billion at distributor prices (PSL, 1987). Today the multiple retailers' share of the fresh fruit and vegetable trade is growing fast and approaching 50 per cent having expanded from about 20 per cent in half a decade. Shares range between 35 and 55 per cent depending on the product; the multiples are estimated to handle 55 per cent of the exotics but only about 35 per cent of the heavy vegetables. Almost all of this growth has been at the expense of the independent greengrocer with a concomitant reduction in wholesale market activity, now down to about 40 per cent of the total volume. UK produce imports have grown spectacularly and now account for about 20 per cent of market volume and a value share approaching 40 per cent. There has been a marked move away from heavy vegetables to light vegetables and salads, and the season for the latter is continually extended. There is increased demand for novelty and a growth in the market for 'exotics' (Hallam and Molina, 1988).

Today the UK supply industry, despite import and price pressure, is an efficient productive enterprise selling about 45 per cent of its production to the multiple groups, graded to a tight standard and to a large extent pre-packed, pre-prepared, priced and branded. To achieve this there has been considerable rationalisation amongst growers and a large proportion of this produce going to multiples is produced by a dozen or so large producers and producer groups passing through about the same number of marketing cooperatives. Whilst the multiples and wholesale markets remain the major outlets for UK produce an increasing proportion is sold for processing.

Continuous adoption of new technology, changing specifications and market organisation have been a regular feature of this sector. Despite continued accusation of weak farm gate prices and abuse of market power, the major producer groups who can meet the new exacting requirements of the users are still in business but have rationalised, concentrated and merged.

Evolution

In the early 1960s most produce came from the traditional market gardener who supplied local shops or the wholesale market (Fig. 13).

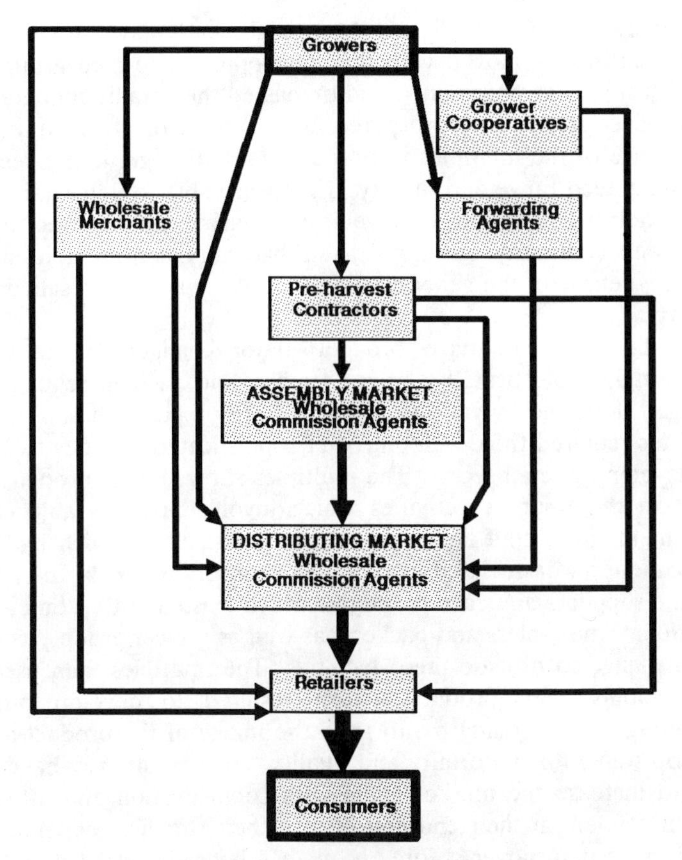

Fig. 13. Marketing channels—traditional system.

The emerging supermarkets found the wholesale market procedure contrary to all their procurement criteria. They started to establish mechanisms for buying direct from growers under contract. Their requirements were simple; they had to ensure that produce of a known quality and volume would be available on a day-by-day basis from a known and guaranteed supplier. They needed a product which was consistent and always available. In this respect produce was no different to a can of beans.

They set out to satisfy their procurement criteria by contracting supplies from three categories of source: producer marketing cooperatives; growers grouped around a marketing agent; and identification of large individual growers. There were, in the early days, premiums for quality and many moved with this development, formed groups and contract supply arrangements, and by-passed their traditional markets. They have grown and prospered as a result of it. In turn, the importance of the multiples in produce marketing grew as a result of the guaranteed range and quality. They were able to differentiate and brand their products and, to develop the market, they offered quality, range and continuity of supply. This had very considerable success with the demise of the general market gardener and wholesale market activity.

The last 20 years have brought major changes to the sector. Increasingly sophisticated storage, grading and packing requirements emerged, followed by various degrees of pre-preparation and processing. This required the development of sophisticated facilities to supply the developing retail sector. The multiples showed little propensity to invest in these supply activities which involved considerable capital investment, fixed staff and management costs, all of which had scale implications and associated increases in exposure and risk. In order to become suppliers and earn premiums which existed at that time, some traditional wholesalers and packers, as well as grower groups accepted the challenge to provide these facilities. The multiples then used the more sophisticated products which resulted to develop produce marketing to new standards, to raise the image of the product and to develop wants for uniformity and quality. As a result they have been able to increase the market again with sophistication and attendant high investment at their end and in turn their suppliers benefited and also increased turnover. More recent has been the increase in third party distribution whereby a managed distribution system is used by the major retailers to assemble and break bulk into in-store loads. Increasingly, these systems are part of automated inventory control operations linked to retail outlets through EPOS (electronic point of sales information).

Alongside the general rationalisation and concentration amongst the multiples has come a concentration in produce retailing. Now six major multiples handle about 70 per cent of produce sales through this type of outlet. The multiples have been largely responsible for an associated rationalisation in the production sector. There are now only about 23000 horticultural holdings. The number is still contracting and has reduced by a further 12 per cent in the last 8 years. During the same time the number of vegetable producers fell by 18 per cent.

There has been both a contraction in the number of growers as well as the area planted, the latter down by about 8 per cent during the same period. More important, production has become highly concentrated. Some 43 per cent of vegetable production is in the hands of 6 per cent of the producers and in fruit production 37 per cent originates from 7·5 per cent of producers. In this process of rationalisation and restructuring, producer marketing groups have become important. There are currently about 50 major producer cooperatives in the vegetable sector and 35 in top fruit, handling 17 per cent and 33 per cent respectively of agricultural products in these categories (Plunkett Foundation, 1987). The turnover of these cooperatives continues to increase. Recent estimates give the disposition of cooperative suppliers as 56 per cent to the multiples, 34 per cent to wholesale and 10 per cent to processing and catering. Already these cooperative groups are evolving and in the last 2 years a number are considering company status or have entered formal integration arrangements with independent packers and distributors. The privately owned packers and distributors have also rationalised; there has been a major change in concentration and ownership through mergers, takeovers and flotations. However, despite this concentration and rationalisation, these groups still remain weak sellers with a very limited market share compared with the retailers.

So today a major proportion of produce is derived through programmed supply arrangements between the multiples, producers and producer groups. Price and delivery term agreements are negotiated between them but product flows between the two via packhouses, frequently grower-owned and managed, and distributors who assemble bulk and then break it into in-store loads. Figures 14 and 15 show the new system and how it works.

With most produce, if the required quality is not achieved the outgrades are almost impossible to sell. If they can be sold, the price rarely covers the packing and marketing cost let alone production costs, and when sold they are often responsible for general price degradation in the sector. Novelty in the market and associated technical developments in relation to the product, processing, packing and distribution methods are the accepted order of things, more often than not being user-led initiatives to which the production sector has to be responsive to hold its market. The sector is lean, fit and evolving rapidly in response to the new challenges.

The momentum of change shows little sign of abatement. The

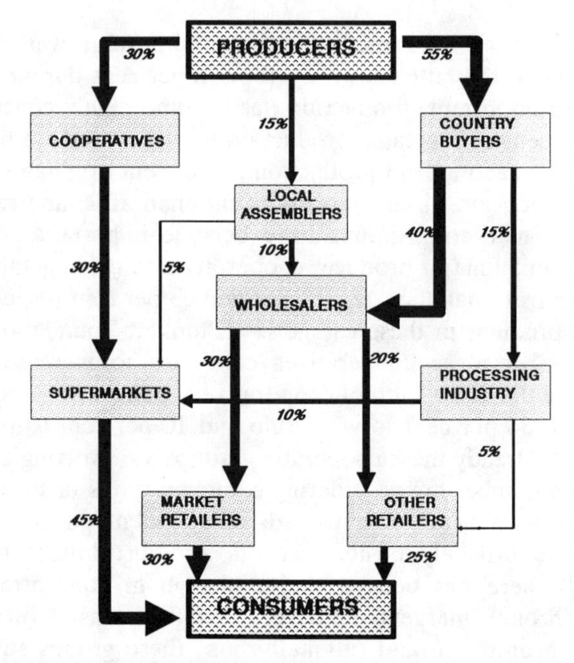

Fig. 14. Marketing channels—the current situation.

intense pressures being fed back from the market, principally from the multiples, and which have to be met by suppliers, are for: greater specialisation; greater standardisation; improved quality; continuity of supply; new packs and labelling; pre-preparation; improved shelf life; improved varieties; daily deliveries; and guarantees of safety. To satisfy these requirements, growers, groups and processors are on an upward spiral of extra investment to retain markets but with increased costs and risks. This is against a background of eroding prices.

Inferences for Other Sectors

The above review illustrates how market, institutional and technical pressures can change the balance of power and the commercial supply and market links. In a systems sense, a system which previously showed considerable homeostasis in terms of products and organisation becomes a highly disturbed one with adaptation and feedbacks all along the chain. Reduced institutional support and technical developments are the key to this adaptation; they permit innovation, product

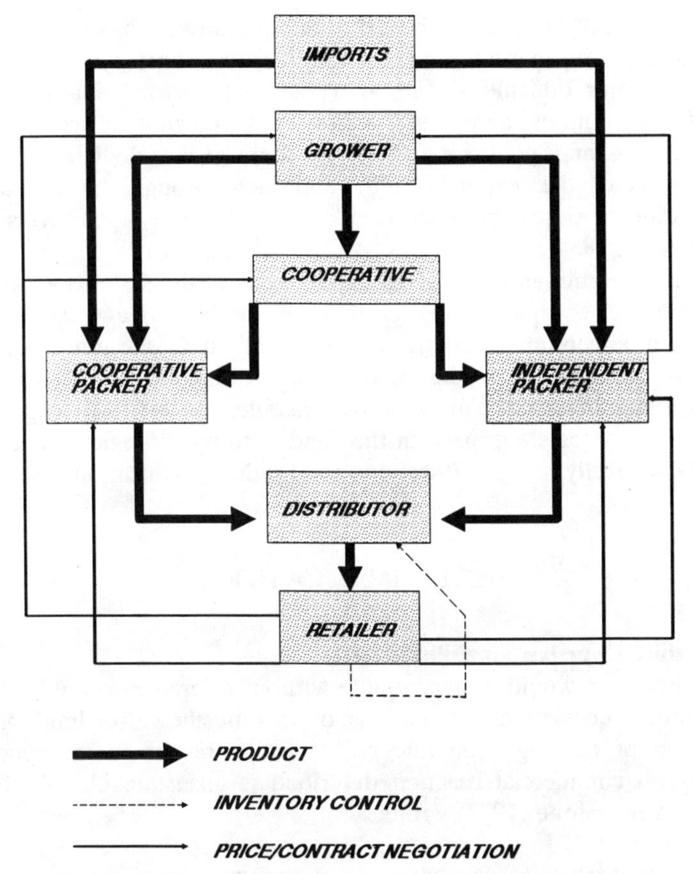

Fig. 15. Contracts, prices and product flow.

development and differentiation. Buyers and users increase control and influence, and formal and informal vertical links follow. The market becomes more dynamic and obsolescence and short life become product development criteria. Those who identify the potential and react appropriately survive largely by the acquisition of the resources and markets of others.

Survival and development require a high level of understanding about what drives the food chain system and in particular how the various sub-systems which comprise it interact. A systems view helps and can undoubtedly improve decision-making. However, formalisation of the relationships involved through quantitative modelling of the development of market linkages and organisational systems is not without major difficulties. Not least, it would involve many behavioural assumptions about the action and reaction of consumers, businessmen and politicians. Perhaps there is a role here for the application of the 'expert system' approach through the structured application of past experience (Dalkey and Helmer, 1963; Weis and Kulikowski, 1984).

What is certain and true of all applications of the systems approach (quantitative or qualitative) is that the holistic view assures an improved perception of the system being studied and awareness on the part of the analyst of the relative importance of the issues and relationships involved. This alone is a justification for further application of the systems approach in this field both by strategic planners as well as generally in agricultural and food industry education.

WIDER IMPLICATIONS

A Conflict Over Sustainability

This discussion would be incomplete without reference to the broader agricultural ecosystem implications of this business treadmill. Continuation of existing economic policy which results in the type of short-term commercial reaction described is unsustainable. As England and Bluestone (1973) wrote:

> *the dilemma facing modern capitalism is that, though a rapid rate of economic growth is necessary to service large corporations and contain class conflict, ultimately this leads to an increasing degree of environmental destruction. Depletion and pollution of the environment are the inevitable byproducts of production and consumption.*

Sooner or later society will have to move towards a steady-state system where stocks are held constant and flows minimised, i.e. where fabrication is for long life, not obsolescence, and wants are increasingly psychic, not material (Daley, 1973). The existing economic philosophy ignores the fact that the earth is as described by Lotka (1957); a closed-loop system of material and energy cycles powered by the sun. Historically, with low population and adequate resources the earth seemed to mankind to be an open system. Now as man presses nearer to terrestrial limits it seems more a closed system with all of the associated constraints (Boulding, 1966). Economics does not distinguish between these two states or between renewable and non-renewable resources and, by reduction of every activity to a monetary value, ignores the finite nature of resources and also the entropy law and the second law of thermodynamics. These physical laws tell us quite uncompromisingly that matter cannot be created or destroyed, that eventually all energy is converted into waste heat and that endless pursuit of the growth path through the exploitation of fossil fuels and other finite resources will result in depletion of resources and thermal pollution. In entropy terms the cost in any economic enterprise is always greater than the product—all activity results in a deficit and consumption of finite resources today with the associated entropy effect reduces the future carrying capacity of the planet.

The economic concepts of continuous relationships and marginality also break down in ecosystem terms. They recognise that each extra unit of resource costs more to extract but conveniently ignore the fact that the extra cost of extraction increases GDP and hence further increases the speed of the treadmill. They also fail to accommodate eutrophic effects and recognise that the incremental unit of energy or pollutant can completely close down the system. Georgescu-Roegen (1971) encapsulates the unsustainability of this economic philosophy by referring to the standard economic textbook which represents the economic process by a circular diagram; a pendulum of movement between production and consumption in a completely closed system possessed with value flows but completely ignoring material flows.

Clearly, there is a system conflict between short-term commercial considerations and longer-term environmental effects. These longer-term effects result from the free-for-all encouraged by the current economic philosophy. This is no longer viable for a society living in a world ecosystem of finite resources and limited ability to handle pollutant affects. Hardin (1968) in his 'The tragedy of the commons' refers to mankind as custodian of the planet and to the problems created by pollution and population growth; he accepts that prohibition is an option not easy to enforce and that:

> *The great challenge facing us now is to invent the corrective feedbacks that are needed to keep custodians honest. We must find ways to legitimate the authority of both the custodians and the corrective feedbacks.*

There is a clear challenge for society to develop a revised economic philosophy which considers energy and resources as stocks and flows in physical as well as value terms and develops weighting factors which take account of inter-generational interests. Since politically an abrupt change from the current situation is likely to be unacceptable, a transition policy must be developed which considers seriously the implications of the second law of thermodynamics and the entropy laws and guides economic policy towards sustainable development. The concept of sustainable development is currently fashionable but we must not forget that this need has long been recognised (Lotka, 1957; Hardin, 1968; Georgescu-Roegen, 1971) and that the problems of sustainability were the subject of many serious systems modelling exercises almost two decades ago at the Massachusetts Institute of Technology (Forrester, 1971; Meadows *et al.*, 1972). This work has continued and, more recently, political reaction to global warming and pollution have popularised this area of research further, e.g. in 1987 through the World Commission on the Environment and Development and more recently by Pearce *et al.* (1989).

Movement towards the steady state will raise some of the most difficult issues mankind has ever had to face. Daley (1973) lists these to include determination of the level of stocks of people and wealth, the optimum level of maintenance throughput, and the time available for transition. A major role exists for economists and systems thinkers to: challenge existing economic and growth philosophies; model material and energy stock and flow systems, taking account of the physical laws discussed above; value material and energy resource use to take account of distribution effects, the demands of future generations and of potential non-continuous and eutrophic effects; and derive improved policy instruments for consideration by politicians in the containment of the adverse effects of the existing system of economic development on the ecosystem.

Finally, one must caution against optimism in this resolve. There are major difficulties in constructing and validating models of the ecosystem and socio-economic system interactions involved, given the paucity of information on many of the inputs, outputs and transformation functions. Hence, the development and testing of effective new policy instruments will pose significant challenges. Perhaps more importantly, there would be major problems associated with the implementation of radical policy. The ecosystem is a 'commons'. Whilst there are exceptions, mankind has manifestly failed in the management of common resources even at the most local level. The management and control systems required are on a global scale and would require the sort of political cooperation between nations that is hard to envisage without some Leviathan being or body controlling the globe.

BIBLIOGRAPHY

Blanchfield, J. R. (1983). Technological change in food manufacturing and distribution, in: *The Food Industry: Economics and Policies,* eds J. A. Burns, J. P. McInerney and A. Swinbank. London: Heinemann.

Boulding, K. E. (1966). The economics of the coming spaceship earth, in: *Environmental Quality in a Growing Economy. Resources in the Future,* Baltimore: John Hopkins Press.

Boussard, J. M. (1988). A French perspective on supply control and management, *Journal of Agricultural Economics,* **39,** 326–39.

Burns, T. and Stalker, G. M. (1961). *The Management of Innovation.* London: Tavistock Publications.

Cottrell, R. C. (1989). Has the science of nutrition benefited from the food industry? *Proceedings of the Nutrition Society,* **48,** 155–8.

Curnock Cook, M. (1989). Ignorance or bliss? What farmers really think about co-operation, in: *Enterprise Farming—A New Approach for the 1990s.* Kenilworth, UK: National Agricultural Centre.

Daley, H. E. (1973). *Towards a Steady State Economy.* San Francisco: W. H. Freeman.

Dalkey, N. and Helmer, O. (1963). An experimental application of the Delphi method to the use of experts, *Management Science,* **10,** 458–67.

Domar, E. (1947). Expansion and employment, *American Economic Review,* March.

Durkheim, E. (1893). *De la Division du Travail Social,* [On the Division of Labour in Society], translated by Simpson G. Glencoe. IL: Free Press.

England, R. and Bluestone, B. (1973). Ecology and social conflict, pp. 190–214 in: *Towards a Steady State Economy,* ed. H. E. Daley. San Francisco: W. H. Freeman.

Eurostats (1977). Luxembourg: Statistical Office at the European Community.

Eurostats (1982). Luxembourg: Statistical Office at the European Community.

Eurostats (1987). Luxembourg: Statistical Office at the European Community.

Forrester, J. W. (1971). *World Dynamics.* Cambridge, MA: Wright Allen Press.

Georgescu-Roegen, N. (1971). The entropy law and the economic problem. Distinguished Lecture Series No. 1, AL: University of Alabama.

Gurr, M. I. (1989). Colloquium on 'Nutrition, marketing and the media'. Has the science of nutrition benefited from the food industry? *Proceedings of the Nutrition Society,* **48,** 159–63.

Hallam, D. and Molina, J. P. (1988). The United Kingdom Market for Exotic

Fruit. Bulletin Number 13, London: Overseas Development Natural Resources Institute.
Hardin, G. (1968). The tragedy of the commons, *Science,* December, 1243–8.
Harrison, A. and Tranter, R. B. (1989). The Changing Financial Structure of Farming. CAS Report 13, Reading: University of Reading.
Howe, W. S. (1983). Competition and performance in food manufacturing, pp. 101–26 in: *The Food Industry: Economics and Policies,* eds J. A. Burns, J. P. McInerney and A. Swinbank. London: Heinemann.
Jones, S. F. (1985). Market Research for Agriculture and Agri-Business in Developing Countries: Courses, Training and Literature. Report G188, London: TDRI.
Jones, S. F. (1987). The European fresh produce market: Marketing challenges and responses for suppliers in the Mediterranean Basin. Paper presented to World Bank and Economic Development Institute Seminar, Oman (June 1987), Newbury: Produce Studies Ltd.
Johnson, W. A. (1973). The guaranteed income as an environmental measure, pp. 175–89 in: *Towards a Steady State Economy,* ed. H. E. Daley. San Francisco: W. H. Freeman.
Lotka, A. J. (1957). *Elements of Mathematical Biology.* New York: Dover Publications.
Malcolm, J. (1983). Food and farming, pp. 66-80 in: *The Food Industry: Economics and Policies,* eds J. A. Burns, J. P. McInerney and A. Swinbank. London: Heinemann.
Manilay, A. A. (1987). Market research for grain postharvest systems, pp. 31–7, in: *Market Research for Food Products and Processes in Developing Countries,* eds R. H. Young and C. W. MacCormac. Ottawa: IDRC.
Marx, K. (1846). Letter to P. V. Ennekov, pp. 401–2, in: *Karl Marx and Frederick Engels, Selected Works,* Vol. 2, London: Lawrence and Wishart.
McDonald, J. R., Rayner, A. J. and Bates, J. M. (1987). Food Consumption Processing and Distribution (Part 2). Department of Economics Paper No. 60, Nottingham: University of Nottingham.
McDonald, J. R., Rayner, A. J. and Bates, J. M. (1989). Market power in the food industry, *Journal of Agricultural Economics,* **40**(1), 101–8.
Meadows, D. H., Randers, D. L. and Behrens, W. W. (1972). *The Limits to Growth.* New York: Universe Books.
Monopolies and Mergers Commission (1989). The Supply of Beer: A Report on the Supply of Beer for Retail Sale in the United Kingdom. Cmd 651, HMSO.
Pearce, D., Markyanda, A. and Barbier, E. (1989). *Blueprint for a Green Economy.* Earthscan Publications.
Plunkett Foundation for Cooperative Studies (1982). *Financial Prospects for Agricultural Cooperatives.* Oxford: St Giles.
Plunkett Foundation for Cooperative Studies (1987). *Statistics of Agricultural Cooperatives in the UK 1986–1987.* Oxford: St Giles.
PSL (1987). *The Co-operative Implications of New Trends in the Distribution of Perishable Food Products.* Newbury: Produce Studies Ltd.
Quilkey, J. (1987). An overview of case studies in market research. In: *Market*

Research for Food Products and Processes in Developing Countries, eds R. H. Young and C. W. MacCormac. Ottawa: IDRC.
Smith, A. (1937). *The Wealth of Nations.* New York: Modern Library.
Spedding, C. R. W. (1975). *The Biology of Agricultural Systems.* London: Academic Press.
Street, P. R. (1988*a*). The balance of power: Who influences the farmer and his choice of products. Paper presented at the FARMSTAT International Conference 1988, Newbury: Produce Studies Ltd.
Street, P. R. (1988*b*). Professionalism in managing change. Paper presented at the PPMA Annual Conference.
Weisskopf, W. A. (1973). Economic growth versus existential balance, pp. 240–51 in: *Towards a Steady State Economy,* ed. H. E. Daley. San Francisco: W. H. Freeman.
Weis, S. M. and Kulikowski, C. A. (1984). *A Practical Guide to Designing Expert Systems.* NJ: Rowman and Allen.

9
Agroecosystems

G. R. CONWAY
The Ford Foundation, New Delhi, India

It is true that any word said by one person to another (or action taken) may have unforeseen consequences, is bound to have some effect and should therefore be pondered most carefully. It also has to be recognised that if this thought renders one speechless or paralysed for fear of doing the wrong thing, one is just as likely to be doing exactly that by inaction. In short, since decisions have *to be taken on a basis of inadequate information, the message must be accepted seriously, but in spirit.*

(Colin Spedding (1975*a*) in *The Biology of Agricultural Systems*, p. 185)

INTRODUCTION

Agriculture is one of the most complex of human activities, requiring numerous day-to-day decisions relating to crops and livestock, family or hired labour, purchased inputs, sales or consumption and so on. Each intervention usually has ramifying consequences and each decision thus involves an implicit or explicit assessment of trade-offs, in both the short and long terms. This is equally true of large industrialised farms in Europe or the USA, and of small subsistence farms in India or Africa. Analysing such complexity in order to improve agricultural performance is a daunting task and we owe a great deal to Colin Spedding for providing us with concepts and tools that make this task much easier (Spedding, 1975*b*, 1979).

The main thrust of Spedding's work has been toward devising means of improving the performance of agricultural systems *per se*—increasing efficiencies and profitability of crop and livestock enterprises. However, in recent years we have come to realise the importance of extending this analysis to the wider environmental and social systems within which agriculture is embedded. As Spedding recognised, agricultural performance is crucially dependent on a great range of biological entities and processes; indeed this topic was the subject of two of his most important books and several papers (Spedding, 1971*a,b*; Spedding, 1975*a*; Spedding *et al.*, 1981). In *The Biology of Agricultural Systems* he devotes a chapter to agricultural ecosystems in order to 'emphasise the ecological approach to whole agricultural systems and not merely to their internal structure and functions'. The chapter's focus is on the interaction between agriculture and the environment and, as Spedding recognised, this interaction is two-way: just as the health of agriculture depends on the proper functioning of environmental processes, so does the health of the environment depend upon a respectful agriculture.

Agriculture and the Environment

Farms in the industrialised countries have become larger and fewer in number, highly mechanised and heavily reliant on synthetic fertilisers and pesticides, many of which are hazardous. Farms have also become more specialised, crop and livestock enterprises being separated geographically so that crop residues and livestock excreta, once recycled, are now wastes whose disposal presents a continuing problem for the farmer. Coincident with these changes, growing urbanisation and population densities, coupled with increasing affluence, have intensified conflicts over land use. Urban populations have become reliant on agricultural catchments for their drinking water, are demanding uncontaminated food and are increasingly valuing the countryside for attributes other than food and fibre production. Amenity, recreation and nature conservation are now important products of the countryside in their own right. Thus, not only is the potential for environmental contamination increasing, so too is the potential value of the damage (Conway and Pretty, in press).

Similar changes are occurring in the developing countries. The advent of new high-yielding cereal varieties as part of the Green Revolution, together with the intensification of export crop agriculture are resulting in a dramatic growth in pesticide and fertiliser use. Pollution problems are already apparent and are likely to grow in importance in the next few years. Although the use of the countryside for leisure is at present confined to very few urban dwellers, many

Third World countries are developing strong conservation movements among whose concerns are the effects of agriculture on wildlife.

Agriculture and Society

Equally we are becoming increasingly aware of a similar two-way dependency between agriculture and human social systems. Agriculture is universally acknowledged, by and large, as the successful provider of food for the world's growing population. If this were not so illness, starvation, civil strife and the collapse of whole societies would become a common occurrence. Agriculture, however, also has more subtle relationships with human institutions. The large, highly mechanised farms of the industrialised countries are a consequence of government policies and, in particular, programmes of subsidies intended to ensure high standards of living for farmers. It is also, though, a process that has led to a considerable reduction in the size of the farming community and a loss of the traditional family farm. In the developing countries similar trends are occurring on those well-favoured lands which have been subject to the Green Revolution, with dramatic changes in the nature and structure of traditional societies. Elsewhere, on the more marginal lands where agriculture is relatively stagnant, there is a growing appreciation of the role of village institutions and family relationships in agricultural development. Here, as in agriculture generally, the role of women in agriculture is finally being acknowledged and acted upon by researchers and extension workers.

The various topics laid out in the preceding paragraphs could be expanded into a series of extensive review articles. In this paper I have chosen to confine myself to two subjects: the first is the role of agriculture as a polluter; the second is the interface between agriculture, the environment and social institutions in the process of development. In the tradition of Colin Spedding's work the focus will be on the methodological challenges that these topics pose for systems analysis.

AGRICULTURE AND POLLUTION

Spedding (1975*a*) has characterised the main elements of an ecological view of agricultural systems as (1) the provision of greater internal detail; (2) a longer term consideration; and (3) a more complete account of external relations. Under the last of these elements is the topic of pollution (Fig. 1).

The traditional polluter of society has been industry. Indeed, agriculture for most of its history has been environmentally benign and this has changed only with the advent of synthetic agrochemicals and the loss of the traditional mixed farm. Farmers and agricultural scientists have long known of the hazards of the older chemicals, such as the arsenates used in pest and weed control, and many were aware of the possible dangers inherent in the new generations of organochlorine and organophosphate insecticides, but it was Rachel Carson's book *Silent Spring* that brought home the dangers to a much wider public (Carson, 1963). Some of her fears were exaggerated but unfortunately many have subsequently proven justified. Today agriculture is recognised as a major polluter not only on a local or regional scale but also globally (Conway and Pretty, in press).

The list of pollution problems caused by agriculture (Table 1) has grown significantly since Rachel Carson's time and, indeed, even in the 15 years that have elapsed since the publication of *The Biology of Agricultural Systems*. There is no room in this paper to review what we now know; instead, I want to concentrate on the problems of analysis of the often complex chains of cause and effect that characterise agricultural pollution problems.

The Assessment of Pollution

Pollution assessment is a complicated process (Conway, 1986*a*; Conway *et al.*, 1988). Ideally it requires a complete understanding of the

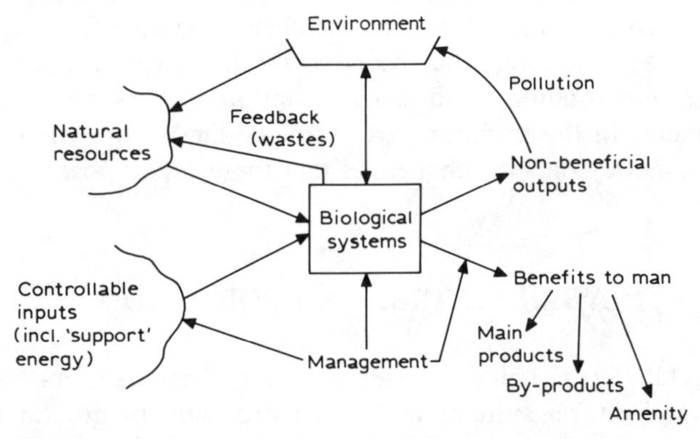

Fig. 1. The role of a biological system in agriculture (adapted from Spedding, 1975*a*.)

TABLE 1
The Principal Pollution Problems Caused by Agriculture

Causes	*Known or possible effects*
Pesticides	Nuisance Damage to human health Damage to wildlife Pest resistance
Nitrates from fertilisers	Damage to human health Eutrophication
Nitrous oxide from fertilisers	Climatic warming Destruction of ozone layer
Livestock excreta	Nuisance Damage to human and livestock health Eutrophication
Silage effluents	Nuisance Eutrophication
Ammonia from livestock and paddy fields	Acid rain
Methane from livestock and paddy fields	Climatic warming
Combustion products from burning forests and from straw burning	Nuisance Climatic warming Acid rain

(Source: Conway and Pretty, in press.)

nature of the components and linkages of a chain that stretches from underlying causes, through circumstances and effects, to perceptions and costs (Fig. 2). Once this is understood then preventative or control actions can be taken against all or some of the components in the chain. Each component and relationship, however, generates its own set of assessment problems and action usually has to be taken before a full assessment can be completed. Inevitably pollution assessment and control is an iterative process, new knowledge—whether acquired from long-term research or the outcome of serious pollution incidents—resulting in piecemeal improvements.

Perhaps the greatest challenge lies in the assessment of the effects of pollution. The problem is that effects are frequently expressed differently depending on the level of system being investigated (Conway, 1986*b*). Moreover, different levels require different kinds of investigation, involving people of differing disciplinary backgrounds and using different tools and standards for assessment. Thus at the cellular or molecular level it is often possible to get very precise results

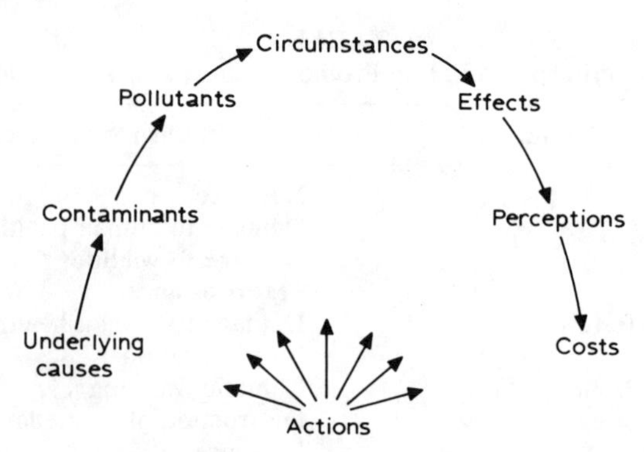

Fig. 2. The chain of pollution assessment (adapted from Conway, 1986*a*).

in terms of the potential toxicity or mutagenicity of a pollutant but this does not reveal whether, in practice, such effects will occur in whole organisms in their environment nor how widespread will the effects be in the range of possible organisms. For example, high doses of aspirin can cause foetal malformations in rodents yet it is widely used by humans with no such ill effects (Wilson *et al.*, 1977).

At the level of whole organisms, laboratory experiments can be used to determine toxicity—usually measured as the lethal dose 50 (LD_{50}), i.e. the amount of active ingredient of the compound required to kill 50 per cent of the exposed population. There are, however, a variety of serious drawbacks inherent in this test (see, for example, Sharratt, 1977; Duffus, 1980). In particular they are very difficult to interpret. For example, animal species differ greatly in their response—the LD_{50} of dioxin for a hamster is 5000 times that of a guinea-pig—so that it is impossible to extrapolate to untested species. The LD_{50} is also a poor basis for extrapolating to a dose at which there is a much smaller, but nevertheless, important effect (Rawls, 1983).

At the population level epidemiological studies seek to establish correlations between observable effects in human or wildlife populations at large and ambient levels of pollutants in the environment. Unfortunately, very high correlations are often readily obtained but these may be misleading or, at the least, incomplete explanations unless other confounding factors are recognised and evaluated.

Organochlorines and Bird Populations

Ideally, the results from the investigations at all three levels should be consistent and reinforce one another as explanations. A good example is the relationship between organochlorine pesticides and the decline in populations of birds of prey.

Birds of prey are usually large and conspicuous creatures so that when their numbers began to decline in the 1950s this soon became apparent. The most dramatic declines were in the peregrine falcon. In the eastern United States the falcon was virtually extinct by the early 1960s and a once flourishing population in southern England had virtually disappeared. In the UK, sparrowhawks, kestrels and barn owls also began to decline. It was at first thought that these changes could be part of the normal cyclic fluctuations that are typical of many birds, particularly raptors. However, a number of scientists began to suspect a link with the use of organochlorine pesticides, and investigation showed lethal levels of dieldrin in the bodies of wild peregrines. It seemed likely that this contamination had been acquired through feeding on grain-eating prey such as pigeons and wood doves (Jefferies and Prestt, 1966). The accumulation of organochlorine pesticides in the food chain is now well established (Fig. 3).

It soon became apparent, however, that this was only part of the story. In the early 1950s Derek Ratcliffe began to notice large numbers of broken peregrine eggs. He began to suspect the role of pesticides and carried out a survey of egg collections in Britain going back to the 1930s. This revealed that from 1947, a year after DDT began to be used, eggshells were significantly thinner by 20 per cent over the

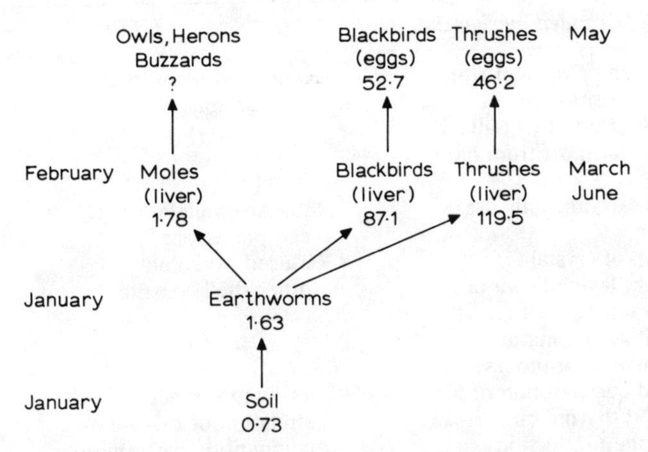

Fig. 3. Food chain accumulation of *pp′*-DDE in an orchard sprayed each April. Total residues in parts per million. (Data from Bailey *et al.*, 1974.)

country as a whole (Ratcliffe, 1958, 1967). Subsequently the relationship between eggshell thinning and DDT content in peregrines was clearly established. The same effects have also been demonstrated in laboratory tests and biochemical studies have uncovered the details of the mechanism of eggshell thinning (Table 2). The thinning results in the cracking of eggs during laying and, in the case of the peregrine falcon, the eggs are often eaten by the parents, leaving the nest empty (Ratcliffe, 1970). The complete story is convincing, although it took over a decade to unravel.

Nitrate Fertilisers and Cancer

By contrast, the evidence for a relationship between nitrate fertilisers and human cancer is contradictory and confusing (Conway and Pretty, in press). In the UK concern was first raised following a study of the population of the town of Worksop where the drinking water had contained over 90 mg nitrate litre^{-1} since at least 1953, the highest level of any borough in the UK (Hill *et al.*, 1973). Over the period 1963–71 the mortality from gastric cancer was shown to be significantly higher than that in nine neighbouring towns, the water supplies of which contained much lower concentrations of nitrate (approximately

TABLE 2

Possible Mechanisms by which Organochlorine Pesticides Cause Eggshell Thinning

Disrupted mechanism	*Effect*
Absorption of calcium from gut; deposition or mobilisation of medullary bone; transport from blood to shell	Reduced availability of calcium ions
Role of carbonic anhydrase system	Reduced availability of carbonate ions
Initiation of crystal growth; laying down of shell matrix; shell growth inhibitors; premature termination of process	Reduced availability of other shell constituents
Reduced consumption of food	Poor health of adult
Disrupted thyroid and hormone production	Stimulation of thyroid or competition with thyroxine

(Source: Cooke, 1973.)

15 mg litre^{-1}). For a variety of reasons this epidemiological relationship has not proven convincing. In particular the presence of a large population of coalminers—an occupation known to have a high incidence of gastric cancer—was not taken into account (Davies, 1980). When the original data were adjusted to take account of the mining population the relationship with nitrate levels disappeared (Table 3).

More recent epidemiological investigations have failed to show positive correlations between nitrate exposure and gastric cancer. Indeed a study of 229 urban areas in the UK produced significant negative correlations between current and historical nitrate levels in drinking water and gastric cancer rates in both men and women (Beresford, 1985). A complementary study which measured salivary nitrate and nitrite levels in people from low and high gastric cancer risk areas of Britain similarly produced a negative correlation (Table 4). Investigations have also been carried out of workers in the fertiliser industry, but have revealed no significant difference in the incidence of gastric and other cancers compared with national averages (Fraser *et al.*, 1982; Al-Dabbagh *et al.*, 1986).

Most developed countries have experienced at least a 30 per cent reduction in mortality from gastric cancer since the early 1950s (because of the high fatality rate, mortality is a good index of incidence). Mortality and incidence rates are much lower in developing countries and are also falling, although in some the rates are unusually high (Conway and Pretty, 1988*a,b*). Chile, for instance, has age-adjusted mortality rates for gastric cancer second only to Japan. It

TABLE 3

Different Estimates of Gastric Cancer Mortality Rates for the Town of Worksop for 1963–71 (national average is set at 100)

	Standardised mortality rate (ratio of observed/expected × 100)	
	Men	*Women*
Estimate of Hill *et al.* (1973)	108	160**
Estimate of Davies (1980) adjusted for social class, mining and more accurate population data	95	131

** $p < 0.01$.

TABLE 4
Mean Nitrate and Nitrite Concentrations (nmol ml^{-1}) in Saliva of Study Populations in High and Low Gastric Cancer Risk Areas

	Low risk (Oxford and southeast)	*High risk (Wales and northeast)*
Nitrate	162·1	106·3
Nitrite	100·2	67·0

(Source: Forman *et al.*, 1985*a*.)

is also the only country with natural deposits of nitrates and has a long tradition of heavy use of fertilisers. There are strong correlations between gastric cancer and cumulative *per capita* exposure to nitrogen and to agriculture as an occupation (Armijo *et al.*, 1981), but there was no correlation with nitrate levels in drinking water, and indeed they are generally low (1·2–15·1 mg $litre^{-1}$) (Zaldivar and Wetterstrand, 1975).

In summary, the epidemiological evidence for a link between gastric cancer and nitrate fertilisers is very weak. (I have not discussed the link between fertiliser applications and nitrate levels in drinking water: the evidence here is much stronger, although still subject to controversy.) The principal evidence for nitrates as a factor in producing gastric cancer comes from biochemical and laboratory experiments with animal populations. The postulated chain of events involves ingestion of nitrate which is then converted to nitrite by bacteria in the gut. The nitrite then combines with amines and amides, derived from food and various environmental contaminants, to produce N-nitroso compounds (nitrosamines and nitrosamides) that are known to be powerful carcinogens in a wide variety of animal species (Magee *et al.*, 1976; WHO, 1978).

On several occasions animals fed with amines and nitrite together or nitrites alone have been shown to develop tumours (Mirvish, 1977; Newberne, 1979; Lijinsky, 1980). Nitrosation has also been demonstrated in several human volunteers, but there is still no direct evidence that N-nitroso compounds cause cancer in humans (Oshima and Bartsch, 1981; Elder *et al.*, 1984).

Expert opinion now is that the chain of events is particularly complex, involving a large number of factors of which nitrates are only one (Fig. 4). Forman *et al.* (1985*a*) and his colleagues summarising the epidemiological evidence conclude that the 'results in general weigh against the idea that environmental nitrates play a major role in

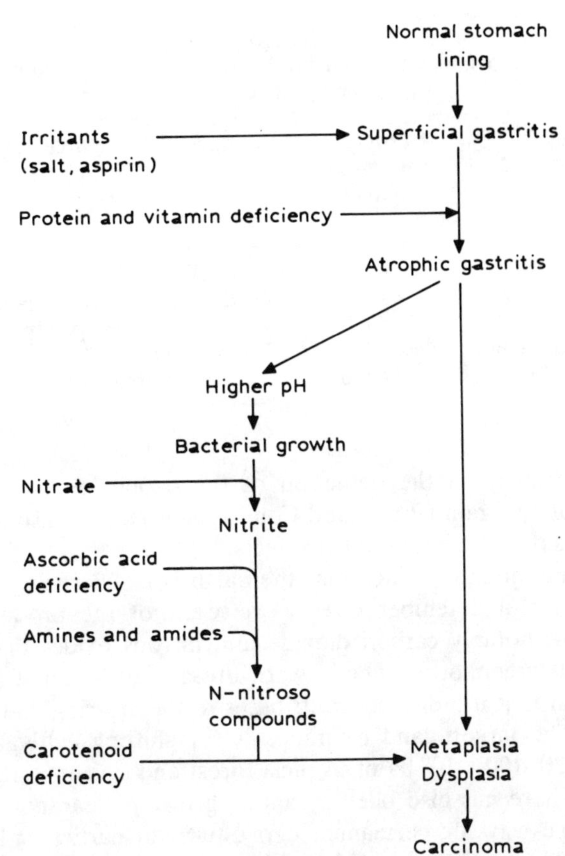

Fig. 4. Simplified version of Correa's model for the development of gastric cancer (adapted from Correa, 1983). The full version includes several other factors.

determining the risk of gastric cancer in Britain' and that expenditure might be better directed to other causative agents. The argument, however, is somewhat academic since strict limits on nitrate levels are being enforced throughout the industrialised countries and there are moves to restrict fertiliser applications.

Agriculture and Global Pollution

While establishing the relationship between nitrate fertilisers and human cancer is complex enough, understanding the role of agriculture in global pollution promises to be an even bigger analytical challenge. In recent years it has become apparent that agriculture is a major source of atmospheric pollution (Table 5) and is thus implicated

TABLE 5
Contribution of Agriculture to Total Production of Globally Important Gases and Smoke

Product	*Proportion produced by agriculture (%)*	*Agricultural source activities*
Methane	40–60	Paddy, livestock, biomass burning
Nitrous oxide	10–25	Fertilisers, biomass burning
Ammonia	80–90	Livestock wastes, paddy
Other combustion gases	60–65	Biomass burning
Particulates and smoke	60–65	Biomass burning

in global warming, in the depletion of the ozone layer and in the production of acid rain (Pretty and Conway, 1989). It is also an actual or potential sufferer from all these effects.

There is mounting evidence that the earth is experiencing a steady rise in mean global temperatures as a result of the production of various gases, notably carbon dioxide and nitrous oxide, that create the so-called 'greenhouse effect'. Agricultural development is partly responsible in that it indirectly contributes to the clearing and burning of forests. The largest land clearance is for shifting cultivation, for which some 20–100 × 10^6 ha of tropical forest and savanna are cleared each year. There has also been a rapidly growing clearance of these lands to make way for permanent agriculture, in particular livestock raising (Seiler and Crutzen, 1980). More directly, nitrogen fertiliser use results in emissions of nitrous oxide to the atmosphere and although at present fertilisers are estimated to account for only 1–4 per cent of emissions, the proportion is likely to increase in the future.

There is also some theoretical evidence that increasing nitrous oxide may be reducing the ozone layer in the stratosphere. Ozone there is broken down by a variety of molecules including nitric oxide, derived from nitrous oxide. Another destructive molecule is the hydroxyl ion, derived from methane and water vapour. Methane is produced by biomass burning and in the guts of cattle and other ruminants. In the troposphere, however, nitrous oxide and methane interact with other gases to increase ozone concentrations. Agriculture is thus simultaneously a causative factor in both the tropospheric increase and the stratospheric decrease of ozone. The former has been shown to harm agricultural crops and reduce yields; the latter by removing some of the protection against ultraviolet B radiation may be increasing the risk of skin cancer.

Finally, agriculture is the major cause of emissions of ammonia to the atmosphere as a result of the volatilisation of nitrogen in fertilisers or in animal excreta. In the industrialised countries livestock and fertilisers are responsible for 80–90 per cent of emissions. On a global scale animal wastes are still the largest source, but biomass burning and losses from fertilized paddies are also important. Losses of nitrogen as ammonia from fertilised paddy fields is usually 5–15 per cent, but sometimes reaches 40–50 per cent. Ammonia in the atmosphere results in the production of ammonium sulphate which leads to increased soil and water acidity (Ap Simon *et al.*, 1987).

The systems involved, however, are highly complex and there is much dispute over the contribution of various factors. Equally controversial are the likely effects, particularly those of global warming, on agriculture. It is commonly and naively believed that increased temperatures will lead to greater crop growth and hence higher yields, at least in temperate regions, but it is likely that the effects will be considerably more complex (Barbier and Oram, 1985; Barbier, 1989; Parry *et al.*, 1988). Especially in the subtropics we are likely to see greater incidence of extreme events—floods and droughts occurring with greater frequency and with devastating effects on agricultural production.

THE NATURE OF AGROECOSYSTEMS

It is now very clear that the process of agricultural pollution presents a challenge for systems analysis as great as, if not greater than that of agricultural production and efficiency. Equally an argument can be made for a comparable challenge posed by the interactions between agriculture and human social systems and institutions. As Spedding (1975*a*) pointed out, agricultural systems lie at the intersection of economics, the social sciences and biology. Spedding's approach to this complexity was to define systems strictly in terms of their 'purpose' and then to define boundaries, components and processes accordingly (Spedding, 1979). This has turned out to be a highly practical and powerful approach. However, a case can equally be made, particularly in the light of our increased knowledge of environmental and social systems, for defining systems not only in terms of human purpose but also by their unique structure and dynamics.

In part this rests on a recognition that natural ecosystems are the basis of all agricultural systems, even if the link in some systems is very tenuous. A swamp or flood plain, for example, is transformed into a ricefield (Fig. 5). A bund is built which creates a clear biophysical boundary and the great diversity of the original ecosystem is reduced to a restricted assemblage of crops, pests and weeds—although still retaining some of the natural elements, fish and predatory birds, for instance. The basic renewable ecological processes remain: competition between the rice and the weeds, herbivory of the rice by the pests and predation of pests by their natural enemies (and of the fish by the predatory birds). These are, however, greatly modified by the agricultural processes of cultivation, subsidy (with fertilisers), control (of water, pests and diseases), harvesting and marketing. The resulting system is usually termed an agricultural ecosystem or agro-ecosystem (Lowrance *et al.*, 1984; Spedding, 1975*a*; see also the majority of papers in the journal *Agro-Ecosystems*).

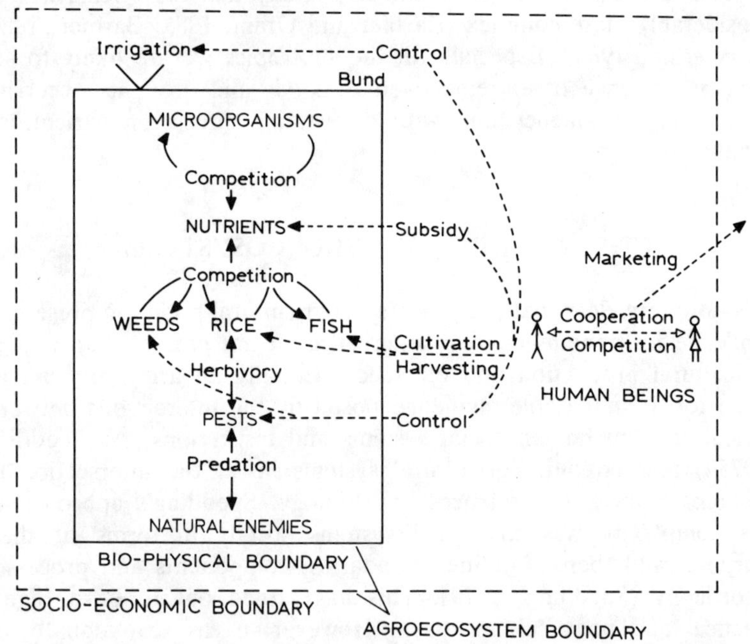

Fig. 5. The ricefield as an agroecosystem (adapted from Conway, 1987).

However, this is only a partial picture of what the transformation entails. The agricultural processes are the results of human decisions which derive from human goals. They are determined by the dynamics of human social and economic cooperation and competition, as embodied in a variety of human institutions. The resulting system is thus as much a socio-economic system as it is an ecological system, and has both biophysical and socio-economic boundaries. This new complex agro-socio-economic-ecological system, bounded in several dimensions, I have defined as an agroecosystem (Conway, 1987).

Properties of Agroecosystems

Four important consequences flow from defining agroecosystems in this way. First, the conceptualisation of systems as being distinctive in their structure and dynamics makes it possible to characterise systems in terms of a set of properties (Conway, 1987). I have suggested that for agroecosystems there are four such properties—productivity, stability, sustainability and equitability. They may be briefly defined as follows;

(1) *Productivity* is the output of valued product per unit of resource input. Common measures of productivity are yield or income per unit area, or total production of goods and services per household or nation, but a large number of different measures are possible, depending on the nature of the product and of the resources being considered. Yield may be in terms of kilograms of grain, tubers, leaves or of meat or fish or any other consumable or marketable product. Alternatively, it may be converted to value in calories, proteins or vitamins or to its monetary value at the market. In the last case it is measured as income as a function of expenditure or profit. But frequently, the valued product may not be yield in conventional agricultural terms. It may be employment generation or an item of amenity or aesthetic value or one of a wide range of products that contribute, in ways that are difficult to measure, to social, psychological and spiritual well-being.

(2) *Stability* is the constancy of productivity in the face of small disturbing forces arising from the normal fluctuations and cycles in the surrounding environment. Included in the environment are those physical, biological, social and economic variables that lie outside the agroecosystem under consideration. The fluctuations, for example, may be in the climate or in the market demand for agricultural

products. Productivity may be defined in any of the ways described above and its stability measured by, say, the coefficient of variation in productivity determined from a time series of productivity measurements. Since productivity may be level, rising or falling, stability will refer to the variability about a trend.

(3) *Sustainability* is the ability to maintain productivity, whether of a field or farm or nation, in the face of stress or shock. The stress may be growing salinity or erosion or debt, i.e. a frequent, sometimes continuous, relatively small predictable force having a large cumulative effect. A major event such as a new pest or a rare drought or a sudden massive increase in input prices would constitute a shock, i.e. a force that was relatively large and unpredictable. Following a stress or shock the productivity of an agricultural system may be unaffected or may fall and then return to the previous level or trend, or settle to a new lower level, or the system may collapse altogether.

(4) *Equitability* is the evenness of distribution of the productivity of the agricultural system among the human beneficiaries, i.e. the level of equity that is generated. Once again, the productivity may be measured in many ways but commonly equitability will refer to the distribution of the total production of goods and services or the net income of the agroecosystem under consideration, i.e. the field, farm village or nation. The human beneficiaries may be the farm household or the members of a village or a national population. Equitability may be measured by a Lorenz curve, Gini coefficient or some other related index.

Socio-biological Interaction

Second, the definition of agroecosystems in terms of both their biophysical and their socio-economic components helps to foster a genuine interdisciplinary approach to agricultural system analysis. It should be noted that the system properties are defined in both biophysical and socio-economic terms. Productivity, for instance, may be in terms of biomass or yield or equally measured as net economic return or some social indicator of human value. Similarly sustainability may refer to either ecological or institutional sustainability.

Too often, while lip service is paid to multidisciplinary analysis of agricultural systems, social and biological scientists tend to work separately, only coming together to write some final synthesis of their work. Yet experience suggests that many, if not most, of the crucial questions for agricultural development lie not in one province or the other, but in their intersection. In Fig. 5 the critical dynamics arise precisely where the socio-economic processes interact with the ecological. Two examples illustrate this point.

The dynamics of international economic markets, together with the high research and development costs entailed in the production of new pesticides, drive multinational chemical companies to seek new compounds which will have wide applicability in terms of the range of countries, crops and pests that can be killed. By producing high-volume sales the investment costs can be recouped and profits maximised. Development agencies and government policy-makers also perceive pesticides to be an easily manipulated instrument of agricultural development and accordingly provide direct or indirect subsidies to encourage their use. Farmers in turn perceive pesticides as a powerful way of controlling pests and tend to use them to excess. However, the reality of the ecological dynamics of pest control is that many pests are well regulated by natural enemies—the brown planthopper on rice, for instance, is regulated by wolf spiders and other predators. Heavy use of broad-spectrum pesticides differentially kills off these predators making the pest problem worse, as has happened in the case of brown planthoppers in Southeast Asia (Kenmore *et al.*, 1984; Joyce, 1988).

Traditional rice varieties are tall and uneven in height. In Java they are harvested by cutting each individual panicle with a small knife called an *ani-ani*. The customary arrangement is for the landless people of the village, and particularly the women, to go from field to field harvesting the grain which is then handed over to the farmer, less some 10 per cent which is retained by the harvester. With the advent of short-strawed varieties this practice has ceased. It is easier to harvest the new varieties with sickles and gangs of men are now contracted as harvesters (Collier *et al.*, 1973). The custom was also for women to mill the rice at home, again retaining a percentage as a payment for the work. This no longer occurs because of the introduction of mechanical rice mills in the villages (Timmer, 1973). One consequence of these changes is that the poor can now only obtain food for cash and hence have to try and find formal employment.

Hierarchies

The third consequence of defining agroecosystems in this way is that it naturally leads to the further concept of a hierarchy of agroecosystems. The importance of such a hierarchic view in agricultural and environmental analysis has already been emphasised above in the discussion of pesticide pollution. There I referred to a hierarchy that went from cells through tissues and organs to whole animals and populations. In agroecosystem terms the hierarchy is one which begins with an agroecosystem that consists of an individual plant or animal, its immediate micro-environment and the people who tend and harvest it (Fig. 6). Examples where this exists as a recognisably distinct system are the lone fruit tree in a farmer's garden or the milk cow in a stall. The next level is the field or paddock. The hierarchy then continues upwards in this way, each agroecosystem forming a component of the agroecosystem at the next level. The higher up the

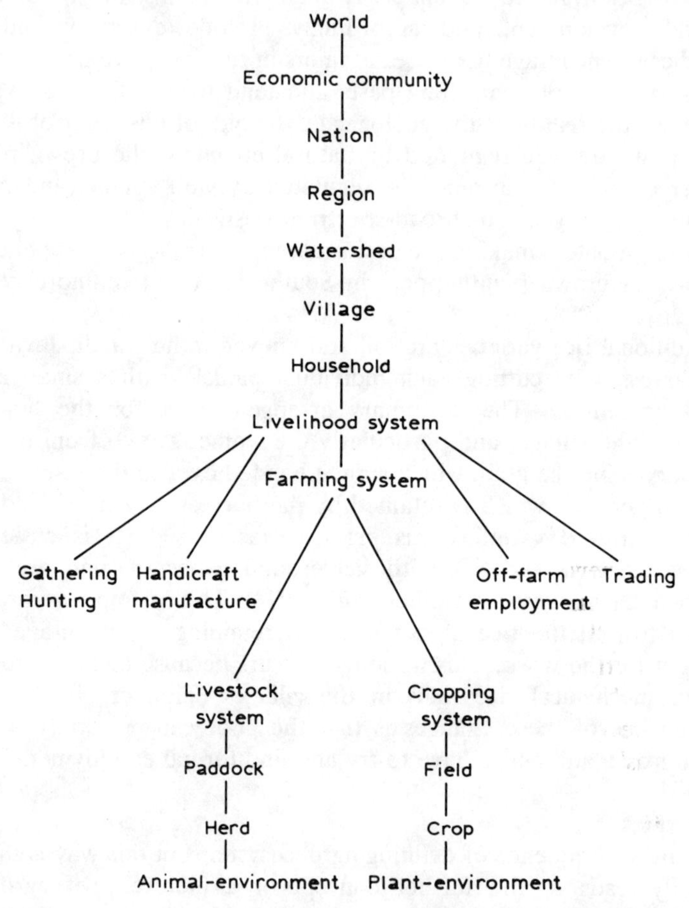

Fig. 6. The hierarchy of agroecosystems (adapted from Conway and Barbier, 1990).

hierarchy the greater is the apparent dominance of socio-economic processes, but ecological processes remain important and, at least in sustainability terms, crucial to achieving human goals (Conway, 1987).

Trade-Offs

Perhaps the most important of the consequences, however, is the recognition of trade-offs in agricultural development between the agroecosystem properties. There are trade-offs, for instance, between productivity on the one hand, and sustainability and equitability on the other. The trade-offs occur within agroecosystems and also between agroecosystems in the hierarchy. Moreover they are particularly associated with the intersection of biophysical and socio-economic processes. The use of pesticides to control the brown planthopper, for instance, represents a case of higher productivity at the expense of sustainability while the changes in harvesting and milling of rice in Indonesia are examples of higher productivity at the expense of lower equitability.

SYSTEMS ANALYSIS

Agricultural Ecosystem Analysis

For the most part the analysis of agricultural ecosystems or agro-ecosystems has focused on discrete component processes. A good example is the work on the dynamics of pest populations and their natural enemies (Croft *et al.*, 1976; Hassell, 1978; Flint and Van den Bosch, 1981; Conway, 1984). This has been highly productive, combining field and laboratory work with analytical and computer simulation models to uncover the natural dynamics, the effects of using pesticides and the potential for biological control. In a limited number of cases the analyses have been extended to encompass socio-economic processes using such tools as expert systems (Mumford and Norton, 1984; Norton, 1986).

There has also been some work on the analysis of whole agro-ecosystems. This has tended to focus on the flows of energy and cycling of materials through agricultural ecosystems (Hart, 1980, 1981). Topics have included the ecological basis for agricultural productivity and the dynamics of nutrients and organic matter, as well as the role of information flows (Lowrance *et al.*, 1984).

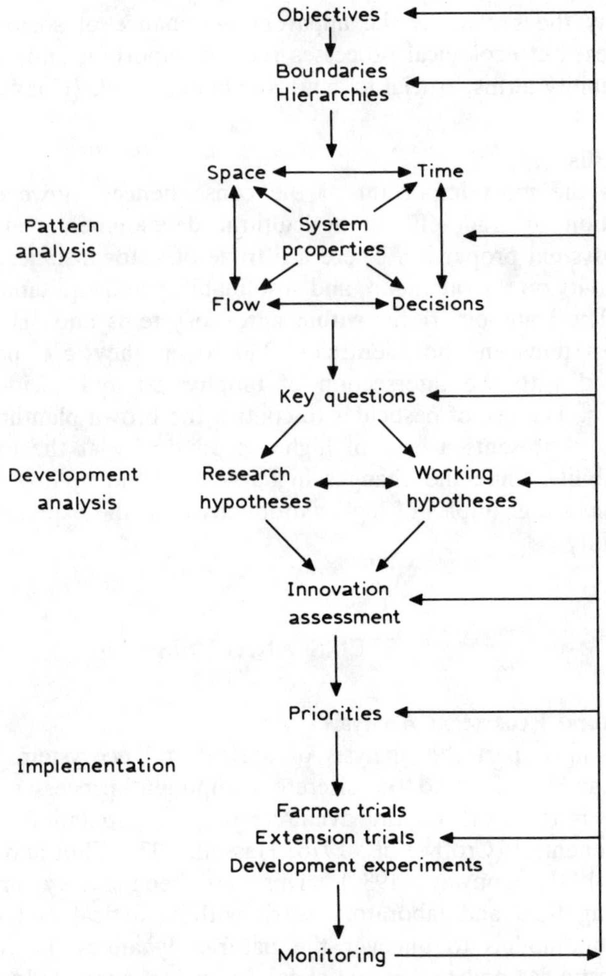

Fig. 7. The procedure of agroecosystem analysis (adapted from Conway, 1986*c*).

Agroecosystem Analysis

The development of techniques to analyse agro-ecosystems, as defined above, began in 1978 at the University of Chiang Mai in Thailand but drew on earlier efforts to analyse natural ecosystems (Walker *et al.*, 1978). In essence, agroecosystem analysis (AEA) draws on the concepts of agroecosystems, agroecosystem hierarchies, agroecosystem properties and their trade-offs as a basis for analysis, both in the field and in a multidisciplinary workshop setting. It relies heavily on a great variety of simple descriptive diagrams, prepared in the field from direct observation and through interviews with farmers. These are then used in the workshop to facilitate communication between the different investigators and to identify accurately the critical development problems and opportunities facing the farmers (Fig. 7). The outcome is a series of key questions and hypotheses which lead to either further research or to development action (Gypmantasiri *et al.*, 1980; Conway, 1985, 1986*c*).

Following its development at Chiang Mai the method was taken to Khon Kaen University in the northeast of Thailand, where it was adapted to the problems of analysing the semi-arid agroecosystems of northeast Thailand (KKU-Ford Cropping Systems Project, 1982*a,b*) and thence to Indonesia where it was applied to the analysis of the research needs of, respectively, the uplands of East Java, the tidal swamplands of Kalimantan and the semi-arid drylands of Timor (KEPAS, 1985*a,b,* 1986). More recently AEA has been used as a method for determining development priorities for the Aga Khan Rural Support Programme in the northern areas of Pakistan (Conway *et al.*, 1985) and for the Ethiopian Red Cross Society in Wollo province in Ethiopia (Ethiopian Red Cross Society, 1988).

Lake Buhi

An example of the application of AEA is its use in an assessment of the environmental and social problems created by construction of a small dam for irrigation purposes in the Philippines (Conway and Sajise, 1986; Conway *et al.*, 1989). The dam had been partially finished in 1986 but was already showing adverse effects on the fishing and agricultural livelihoods of the lakeshore inhabitants, the transportation and domestic water supplies of a lakeside town, and the long-term productivity and stability of the lake itself.

Multidisciplinary teams from the University of the Philippines at Los Banos spent some two weeks in the area, interviewing the inhabitants and preparing a series of maps, transects, seasonal calendars and other diagrams. This was then followed by a 5-day workshop involving government officials, development agency personnel and representatives of the farmers and fishermen. The workshops followed the structure developed for AEA, proceeding from an intensive discussion of the diagrams to produce a set of key questions for research and development. These were then prioritised by the whole workshop in

TABLE 6
Innovation Assessment for Lake Buhi

Improvement/ innovation	*Product- ivity*	*Stability*	*Sustain- ability*	*Equit- ability*	*Cost*	*Time to benefit*	*Priority*
		(Major effects)					
1. Lake level control	+	+	+	+	XXX	XXX	1
2. Fishery optimization	+		+		XX	XX	1
3. Pollution control	+		+		X	XX	2
4. Fish-cage location	+		+		XX	XXX	2
5. Long-term planning			+		X	X	4
6. Fishing technology	+				XX	XX	3
7. Multi-purpose management		+	+	+	XX	XX	1
8. Tourism, etc.	+			+	XX	XX	3
9. Market/credit				+	XX	XX	2
10. Health	+			+	XX	XX	2
11. Water supply		+	+		XX	XX	1
12. Compensation				+	XXX	XXX	1

XXX Low cost or short time to benefit.
XX Medium cost or time to benefit.
X High cost or long time to benefit.
(Source: Conway *et al.*, 1989.)

terms of the agroecosystem properties to produce a plan of action (Table 6). Subsequently, despite considerable political difficulties, a number of the key questions have been answered and recommended improvements implemented.

Rapid Rural Appraisal

Agroecosystem analysis is but one example of the approach known as rapid rural appraisal (RRA) which has been developed over the last decade (McCracken *et al.*, 1988; Chambers, in press; Khon Kaen University, 1987); see also papers in *Agricultural Administration* (1981), vol. 10). RRA may be defined as a systematic but semi-structured activity carried out in the field by a multidisciplinary team, and designed to acquire quickly new information on, and new hypotheses about rural life.

Studies of local rural situations in developing countries have often concentrated on only one set of conditions, relying on extensive data collections involving many researchers over a long period of time and costing large sums of money. The obvious logistical problems of such an approach are frequently accompanied by other more serious shortcomings. Local inhabitants are seldom consulted or at best through fixed and formal channels, for instance, by means of a written questionnaire with the questions determined beforehand. The context of the target data is frequently ignored; 'averages' are sought, while significant variations are often missed. This gives little opportunity for new features of the system to be revealed or for insights to be gained other than those which could have been learnt at the start from the local people. RRA is designed to rectify these shortcomings.

There is no single standardised methodology for RRA. In each situation this depends on the objectives, local conditions, skills and resources. However, there is a suite of techniques in existence which can be used in various combinations to produce appropriate RRA methods. The suite includes:

Secondary data review
Direct observation
Diagrams
Semi-structured interviews
Analytical games
Portraits and stories
Workshops

Secondary data consists of reports, maps, aerial photographs and so forth that already exist and are relevant to the project. Direct observation includes measurement and recording of objects, events and processes in the field, either because they are important in their own right or because they are surrogates for other variables that are important. Diagrams have already been described.

One of the most important of RRA techniques is semi-structured interviewing which is a form of guided interviewing where only some of the questions are predetermined and new questions or lines of questioning arise during the conduct of the interview in response to answers from those interviewed. The information is thus derived from the interaction between the knowledge and experience of the interviewer and the interviewee(s). The latter may be groups, for example, of village leaders or key informants such as school teachers or local government officials, or the farmers themselves selected on one or more criteria.

Analytical games consist of dialogues with farmers which take the form of a game, i.e. they follow certain simple, but mutually agreed rules. One example is 'preference ranking' where farmers are asked to choose between pairs of crop varieties. A set of choices is prepared and farmers are presented with them in every pairwise comparison. Each time they are asked to indicate which they prefer if they could only grow one of the pair and to give the reasons.

Portraits and stories are simple written essays on families and their livelihoods that illuminate their present conditions and the manner of their decision making.

The various techniques described above will be used in various combinations depending on the objective of the RRA. Very broadly there are four principal classes of RRA, which ideally follow one another in the sequence of development activity:

(1) *Exploratory RRA*—to obtain initial information about a new topic or agroecosystem. The output is usually a set of preliminary key questions and hypotheses. (Agroecosystem analysis is an example of an exploratory RRA.)

(2) *Topical RRA*—to investigate a specific topic, often in the form of a key question and hypothesis generated by the exploratory RRA. The output is usually a detailed and extended hypothesis that can be used as a strong basis for research or development.

(3) *Participatory RRA*—to involve villagers and local officials in decisions about further action based on the hypotheses produced by the exploratory or topical RRAs. The output is farmer-managed trials or a development activity in which the villagers are closely involved.

(4) *Monitoring RRA*—to monitor progress in the trials and experiments and in the implementation of the development activity. The output is usually a revised hypothesis together with consequent changes in the trials or development intervention which will hopefully bring about improved benefits.

Institutionalisation

In Thailand agroecosystems research and the use of RRA techniques are now well established at several universities and in government institutes, having evolved into a variety of distinctive styles (Prince of Songkla University, 1988; Conway, 1988). A masters programme in agricultural systems has been running at Chiang Mai University since 1986, being offered initially in Thai and now in English. Undergraduate teaching at both Chiang Mai and Khon Kaen universities also contains course work stressing a systems approach and draws heavily on Spedding's textbooks together with various materials produced by the faculty based on their field research. An association of Southeast

Asian workers in agro-ecosystem research and development, known as SUAN (the Southeast Asian Universities Agroecosystem Network) has been in existence since 1981 and has sponsored a series of workshops, seminars, training programmes and faculty exchanges. It has recently expanded its activities to Laos, Vietnam and China.

A similar process of institutionalisation is also proceeding at other universities in Southeast Asia, and is now beginning at various universities in eastern India. RRA techniques, particularly those emphasising participatory approaches, are presently being rapidly adopted by non-government development organisations in countries as diverse as Kenya, Ethiopia, Zimbabwe, Pakistan, India and the Philippines (Chambers *et al.*, 1989; also see *RRA Notes* series produced by the International Institute for Environment and Development, London).

The rapid and widespread adoption of AEA and RRA in recent years is a reflection of the absence of a bureaucratically defined approach and the positive encouragement of workers to experiment and develop their own techniques suited to their particular circumstances. In my view these are signs of the beginning of a revolution in the way agricultural research and extension workers go about development activities. The hope is that this will result in a process of agricultural development that not only produces sufficient food and fibre for our rapidly expanding world population but accomplishes this in a manner that is environmentally and socially sound.

REFERENCES

Al-Dabbagh, S., Forman, D., Bryson, D., Stratton, I. and Doll, R. (1986). Mortality of nitrate fertiliser workers, *British Journal of Industrial Medicine*, **43,** 507–15.

Ap Simon, H. M., Kruse, M. and Bell, J. N. B. (1987). Ammonia emissions and their role in acid deposition, *Atmospheric Environment*, **21,** 1939–46.

Armijo, R., Ortellana, M., Medina, E., Coulson, A. H., Sayre, J. W. and Detels, R. (1981). Epidemiology of gastric cancer in Chile: I—Case control study, *International Journal of Epidemiology*, **1,** 53–6.

Bailey, S., Bunyan, P. S., Jennings, D. M. and Norris, J. D. (1974). Hazards to wildlife from the use of DDT in orchards: A further study, *Agro-Ecosystems*, **1,** 323–38.

Barbier, E. B. (1989). The global greenhouse effect: economic impacts and policy considerations, *Natural Resources Forum*, February 1989, 20–32.

Barbier, E. B. and Oram, P. A. (1985). Sensitivity of agricultural production to climatic change, *Climate Change,* **7,** 129–52.

Beresford, S. A. A. (1985). Is nitrate in the drinking water associated with the risk of cancer in the urban U.K.? *Journal of Epidemiology,* **14,** 57–63.

Carson, R. (1963). *Silent Spring*. London: Hamish Hamilton.

Chambers, R. (in press). Rapid and participatory rural appraisal, *Appropriate Technology*.

Chambers, R., Pacey, A. and Thrupp, L. A. (Editors) (1989). *Farmer First: Farmer Innovation and Agricultural Research*. London: Intermediate Technology Publications.

Collier, W. L., Wiradi, G. and Soentoro (1973). Recent changes in rice harvesting methods, *Bulletin of Indonesian Economic Studies,* **9,** 36–42.

Conway, G. R. (Editor) (1984). *Pest and Pathogen Control: Strategic, Tactical and Policy Models*. Chichester: John Wiley & Sons.

Conway, G. R. (1985). Agroecosystem analysis, *Agricultural Administration,* **20,** 31–55.

Conway, G. R. (Editor) (1986*a*). *The Assessment of Agricultural Pollution*. London: Centre for Environmental Technology, Imperial College.

Conway, G. R. (1986*b*). Agricultural pollution, pp. 24–40 in: *The Assessment of Pollution,* ed. G. R. Conway. London: Centre for Environmental Technology, Imperial College.

Conway, G. R. (1986*c*). *Agroecosystem Analysis for Research and Development*. Bangkok: Winrock International.

Conway, G. R. (1987). The properties of agroecosystems, *Agricultural Systems,* **24,** 95–117.

Conway, G. R. (1988). *Agroecosystem Research and Development in Thailand: A Review and Bibliography*. London: International Institute for Environment and Development.

Conway, G. R. and Barbier, E. B. (1990). *After the Green Revolution: Sustainable Agriculture for Development*. London: Earthscan.

Conway, G. R. and Pretty, J. N. (1988*a*). Fertiliser risks in the developing countries, *Nature,* **334,** 207–8.

Conway, G. R. and Pretty, J. N. (1988*b*). *The Risks of Fertiliser Use in the Developing Countries: A Review*. London: International Institute for Environment and Development.

Conway, G. R. and Pretty, J. N. (in press). *Unwelcome Harvest: Agriculture and Pollution*. London: Earthscan.

Conway, G. R. and Sajise, P. E. (1986). *The Agroecosystems of Buhi: Problems and Opportunities*. Los Banos: Program on Environmental Science and Management, University of the Philippines.

Conway, G. R., Alam, Z., Husain, T. and Mian, M. A. (1985). *An Agroecosystem Analysis for the Northern Areas of Pakistan*. Gilgit, Pakistan: Aga Khan Rural Support Programme.

Conway, G. R., Gilbert, D. G. R. and Pretty, J. N. (1988). Pesticides in the UK: Science, policy and the public, pp. 5–32 in: *Britain Since 'Silent Spring'; An Update on the Ecological Effects of Agricultural Pesticides in the UK,* ed. D. J. L. Harding. London: Institute of Biology.

Conway, G. R., Sajise, P. E. and Knowland, W. (1989). Lake Buhi: resolving conflicts in a Philippine development project, *Ambio,* **18,** 128–35.

Cooke, A. S. (1973). Shell thinning in avian eggs by environmental pollutants, *Environmental Pollution,* **4,** 85–152.

Correa, P. (1983). The gastric precancerous process. *Cancer Surveys,* **2,** 437–50.

Croft, B. A., Howes, J. L. and Welch, S. M. (1976). A computer-based extension pest management delivery system, *Environmental Entomology,* **5,** 20–34.

Davies, J. M. (1980). Stomach cancer mortality in Worksop and other Nottinghamshire mining towns, *British Journal of Cancer* **41,** 438–45.

Duffus, J. (1980). *Environmental Toxicology.* London: Edward Arnold.

Elder, J. B., Burdett, K., Smith, P. L. R., Walters, C. L. and Reed, P. I. (1984). Effects of H2 blockers on intragastric nitrosation as measured by 24 hour urinary secretion of N-nitrosoproline, in: *N-nitroso compounds: Occurrence, Biological Effects and Relevance to Human Cancer,* eds I. K. O'Neill, R. C. von Borstel, C. T. Miller, J. Long and M. Bartsch. IARC Science Publication no. 57. Lyon: IARC.

Ethiopian Red Cross Society (1988). *Rapid Rural Appraisal: A Closer Look at Rural Life in Wollo.* London: Ethiopian Red Cross Society, Addis Ababa and International Institute for Environment and Development.

Flint, M. L. and Van den Bosch, R. (1981). *Introduction to Integrated Pest Management.* New York: Plenum Press.

Forman, D., Al-Dabbagh, S. and Doll, R. (1985). Nitrates, nitrites and gastric cancer in Great Britain, *Nature,* **313,** 620–5.

Fraser, P., Chilver, C. and Goldblatt, P. (1982). Census-based mortality study of fertiliser manufacturers, *British Journal of Industrial Medicine,* **39,** 323–9.

Gypmantasiri, P., Wiboonpongse, A., Rerkasem, B., Craig, I., Rerkasem, K., Ganjapan, L., Titayawan, M., Seetisarn, M., Thani, P., Jaissard, R., Ongprasert, S., Radnachaless, T. and Conway, G. R. (1980). *An Interdisciplinary Perspective of Cropping Systems in the Chiang Mai Valley: Key Questions for Research.* Chiang Mai, Thailand: Faculty of Agriculture, University of Chiang Mai.

Hart, R. D. (1980). *Agroecosistemas: Conceptos Basicos.* Turrialba, Costa Rica: Centro Agronomico Tropical de Investigacion y Ensenanza.

Hart, R. D. (1981). An ecological systems conceptual framework for agricultural research and development, pp. 45–58 in: *Readings in Farming Systems Research and Development,* eds W. W. Shaner, P. F. Philipp and W. R. Schmehl. Boulder, CO: Westview Press.

Hassell, M. P. (1978). *The Dynamics of Arthropod Predator–Prey Systems.* Princeton: Princeton University Press.

Hill, M. J., Hawksworth, G. and Tattershall, G. (1973). Bacteria, nitrosamines and cancer of the stomach, *British Journal of Cancer,* **28,** 562–7.

Jefferies, D. J. and Prestt, I. (1966). Post-mortems of peregrines and lanners with particular reference to organochlorine residues, *British Birds,* **59,** 49–64.

Joyce, C. (1988). Nature helps Indonesia cut its pesticide bill, *New Scientist,* 16 June.

Kenmore, P. E., Carino, F. O., Penez, C. A., Dyck, V. A. and Gutierrez, A. P. (1984). Population regulation of the brown planthopper (*Nilaparvata*

Stal.) within ricefields in the Philippines, *Journal of Plant Protection in the Tropics,* **1,** 19–37.

KEPAS (1985*a*). *The Critical Uplands of Eastern Java*: *An Agroecosystem Analysis.* Jakarta, Indonesia: Agency for Agricultural Research and Development.

KEPAS (1985*b*). *Swampland Agroecosystems of Southern Kalimantan.* Jakarta, Indonesia: Agency for Agricultural Research and Development.

KEPAS (1986). *Agro-ekosistem Daerah Kering di Nusa Tenggara Timur.* Jakarta, Indonesia: Agency for Agricultural Research and Development.

Khon Kaen University (1987). *Proceedings of the International Conference on Rapid Rural Appraisal.* Khon Kaen, Thailand: Khon Kaen University.

KKU-Ford Cropping Systems Project (1982*a*). *An Agroecosystem Analysis of Northeast Thailand.* Khon Kaen, Thailand: Faculty of Agriculture, Khon Kaen University.

KKU-Ford Cropping Systems Project (1982*b*). *Tambon and Village Agricultural Systems in Northeast Thailand.* Khon Kaen, Thailand: Faculty of Agriculture, Khon Kaen University.

Lijinsky, W. (1980). Significance of *in vitro* formation of N-nitroso compounds, *Oncology,* **37,** 223–6.

Lowrance, R., Stinner, B. R. and House, G. J. (eds) (1984). *Agricultural Eco-systems*: *Unifying Concepts.* New York: John Wiley.

Magee, B. N., Montesano, R. and Preusmann, R. (1976). N-nitroso compounds and related carcinogens, pp. 491–625 in: *Chemical Carcinogens,* ed. C. F. Searle. Washington, DC: American Chemical Society Monograph no. 173.

McCracken, J. A., Pretty, J. N. and Conway, G. R. (1988). *An Introduction to Rapid Rural Appraisal for Agricultural Development.* London: International Institute for Environment and Development.

Mirvish, S. S. (1977). N-nitroso compounds, nitrate and nitrite: possible implications for the causation of human cancer, *Progress in Water Technology,* **8,** 195–207.

Mumford, J. D. and Norton, G. A. (1984). Economics of decision making in pest management, *Annual Review of Entomology,* **29,** 157–74.

Newberne, P. M. (1979). Nitrite promotes lymphoma incidence in rats, *Science,* **204,** 1079–81.

Norton, G. A. (1986). *Pest Management and World Agriculture—Policy, Research and Extension.* Papers in Science, Technology and Public Policy, no. 13. London: Imperial College of Science and Technology.

Oshima, H. and Bartsch, H. (1981). Quantitative estimation of endogenous nitrosation in humans by monitoring N-nitrosoproline excreted in the urine, *Cancer Research,* **41,** 3658–62.

Parry, M. L., Cater, T. R. and Konijn, N. T. (Editors) (1988). *The Impact of Climatic Variations on Agriculture, Vols 1 and 2.* Dordrecht: Kluwer Academic Publishers.

Pretty, J. N. and Conway, G. R. (1989). *Agriculture as a Global Polluter.* London: International Institute for Environment and Development.

Prince of Songkla University (1988). *Farming Systems Research and Development in Thailand*: *Illustrated Methodological Considerations and*

Recent Advances. Haad Yai, Thailand: Prince of Songkla University.
Ratcliffe, D. A. (1958). Broken eggs in peregrine eyries, *British Birds,* **51,** 23–6.
Ratcliffe, D. A. (1967). Decrease in eggshell weight in certain birds of prey, *Nature,* **215,** 208–10.
Ratcliffe, D. A. (1970). Changes attributable to pesticides in egg breakage frequency and eggshell thickness in some British birds, *Journal of Applied Ecology,* **7,** 67–107.
Rawls, R. L. (1983). Dioxins: Human toxicity is most difficult problem, *Chemical and Engineering News,* **61,** 37–48.
Seiler, W. and Crutzen, P. J. (1980). Estimates of gross and net fluxes of carbon between the biosphere and the atmosphere from biomass burning, *Climatic Change,* **2,** 207–47.
Sharratt, M. (1977). Uncertainties associated with the evaluation of health hazards of environmental chemicals from toxicological data, pp. 105–23 in: *The Evaluation of Toxicity Data for the Protection of Public Health,* eds W. S. Hunter and J. G. P. M. Smeets. Oxford: Pergamon Press.
Spedding, C. R. W. (1971*a*). An ecological approach to agriculture needs synthesis rather than analysis, *Agricultural Institute Review,* **26,** 3–7.
Spedding, C. R. W. (1971*b*). Agricultural ecosystems, *Outlook on Agriculture,* **6,** 242–7.
Spedding, C. R. W. (1975*a*). *The Biology of Agricultural Systems*. London: Academic Press.
Spedding, C. R. W. (1975*b*). The study of agricultural systems, pp. 1–19 in: *Study of Agricultural Systems,* ed. G. E. Dalton. London: Applied Science Publishers.
Spedding, C. R. W. (1979). *An Introduction to Agricultural Systems*. London: Applied Science Publishers.
Spedding, C. R. W., Walsingham, J. M. and Hoxey, A. M. (1981). *Biological Efficiency in Agriculture*. London: Academic Press.
Timmer, C. P. (1973). Choice of technique in rice milling in Java, *Bulletin of Indonesian Economic Studies,* **9,** 57–76.
Walker, B. H., Norton, G. A., Conway, G. R., Comins, H. N. and Birley, M. (1978). A procedure for multidisciplinary ecosystem research: with reference to the South African Savanna Ecosystem Project, *Journal of Applied Ecology,* **15,** 481–502.
WHO (1978). *Nitrates, Nitrites and N-nitroso Compounds*. Geneva: World Health Organisation.
Wilson, J. G., Ritter, E. J., Scott, W. J. and Fradkin, R. (1977). Comparative distribution and embryotoxicity of acetylsalicylic acid in pregnant rats and rhesus monkeys. *Toxicology and Applied Pharmacology,* **41,** 67–78.
Zaldivar, R. and Wetterstrand, W. H. (1975). Further evidence of a positive correlation between exposure to nitrate fertilisers ($NaNO_3$ and KNO_3) and gastric cancer death rates; nitrates and nitrosamines, *Experientia,* **31,** 1354–5.

10
Understanding and Managing Changes in Agriculture

C. T. de Wit

Department of Theoretical Production Ecology, Agricultural University, Wageningen, The Netherlands

PRODUCTIVITY AND SUSTAINABILITY

Agriculture has to be productive and sustainable (TAC/CGIAR, 1989). From the ecological viewpoint, this implies that renewable resources are maintained, non-renewable resources are used with foresight and the intrinsic value of the natural environment is recognized, and from the socio-economic viewpoint that farm families make a decent living and that increasing and changing demands for agricultural products are satisfied at affordable prices. These are each reasonable goals and boundary conditions, but together they constitute a world in which conflicting productivity and sustainability demands have to be satisfied at the same time and thus form a continuing source of political debate. The outcome of this debate depends on ideological and political views, socio-economic conditions, technical possibilities and past performance.

For most of the history of mankind, the possibilities of increasing agricultural production by increasing the production per hectare have been very limited, so that increase in demand was in general met by expanding the area under cultivation. In some situations, sustainable farming systems were developed. Examples are shifting cultivation in tropical forests and rainfed bunded rice in Asia. However, in most situations productivity demands overruled sustainability demands, so that soil resources were over-exploited. This is witnessed by the vast areas of once good agricultural land that have been lost or damaged by wind and water erosion, exhaustion or salination. Well-known examples are the bare hills in the Mediterranean region, the saline soils in the Middle East, the mining for nutrients of the commons in Western Europe and the destructive dust storms in the 1930s in the USA.

THE FIRST GREEN REVOLUTION

This disastrous development came to an end in the industrialized Western world during and shortly after World War II. At that time the linear rate of yield increase per unit area soared in terms of wheat equivalents from an average of about 2 kg ha^{-1} $year^{-1}$ to more than 50 kg ha^{-1} $year^{-1}$ (de Wit *et al.*, 1987; de Wit, in press), while the rate of growth of the population declined. This 'first green revolution' was characterized by innovative combinations of old and some new techniques, such as the use of more sturdy varieties with shorter straw, better water management and timeliness made possible by soil amelioration and mechanization, improved soil fertility by the use of inorganic fertilizers, and better control of weeds, pests and diseases through the use of biocides.

At the same time, the growth of the population slowed down, but large discrepancies between supply and demand were kept under control by an increased use of luxury products, like meat, vegetables and flowers, and by taking land out of production in marginal regions. When everything else failed, overproduction was accepted against prices with a social function and subsequently dumped on the world market. Such price subsidies were made possible because of the rapid decline of the number of people working in agriculture due to mechanization. This is in sharp contrast to the past when the agricultural sector was a dominant sector of the economy and was not only used as a source of agricultural products but also of labour and capital to develop other sectors of the economy.

This increased production per unit area and per man required an increased use of external inputs from the industrialized sector of the economy, and large-scale reconstruction and reallocation of land in regions with agricultural perspectives. This has led to irreversible changes in the landscape that affected diversity of the environment and quality of natural resources. At the same time the social and environmental fabric was jeopardized by marginalization in agriculturally poorly endowed regions. These developments also reflected conflicts between productivity and sustainability, but now conflicts arise out of affluence and the lopsided use of technical means, no longer out of poverty and lack of technical possibilities.

EFFICIENCY OF RESOURCE USE

Referring to the so-called law of diminishing returns, it is often taken for granted that the large increase in yield that has been witnessed required a much more than proportional increase of means of production. Accordingly, when yields per hectare are related to, for instance, nitrogen use per hectare, in the course of time one would expect a decreasing return. However, it appears (de Wit, in press) that at present at the high end of the yield range the efficiency of nitrogen use is the same as at the low end, so that any sign of diminishing returns is absent.

The reason for this apparent contradiction is that the law of diminishing returns only holds when increasing amounts of a certain production factor are used under otherwise the same conditions. However, a feature of intensification of production is that it is not the improvement of one growing factor that is decisive, but the improvement of a number of them. This leads to positive interactions that imply that the total effect of the measures that are taken is larger than the sum of the effects of each of them separately. Liebscher (1895) acknowledged this in the last century by formulating his so-called law of the optimum that states that a production factor contributes the more to production, the closer the other production factors are to their optima. The law reflects the fact that the agricultural production process in low-yielding situations, where many limiting and partly unknown factors interact, is not very well understood and therefore difficult to manage, whereas in high-yielding situations it implies better control so that inputs may be better timed and adjusted to demand. Of course, this does not exclude the possibility that some inputs are so cheap that there is little incentive for the farmer to economize on their use, so that more of them are used than is necessary to reach the production goal. A recent analysis (de Wit, in press) confirmed this law of the optimum: no production resource has to be utilized less efficiently upon the approach of optimal growing conditions and many are used more efficiently. Therefore, where it is profitable to use an input mix, it is more profitable to use each of the inputs of this mix at optimum intensity. Yield increases are thus determined in an autonomous process that is fuelled by public and private research, extension and marketing. System analysis and simulation studies (Buringh, *et al.*, 1979) have shown that in many regions of the industrialized Western world there exists still a considerable yield gap between what farmers actually produce and could potentially produce. This makes any optimum a moving target for the farmer.

DIVERGING DEVELOPMENTS

Any market-oriented policy that seeks to adjust agricultural demand and supply by downward adjustments of prices has to come to grips with this phenomenon of autonomous yield increases. It is true that the supply will be reduced by marginalizing those regions that are less suitable for agriculture. However, at the same time yields continue to increase in regions where agriculture is considered profitable.

This autonomous increase of yields is on average about 1·5 per cent per annum in the European Community (EC) (Meester and Strijker, 1985) and, since land that is abandoned has a below average yield, it is necessary to take at least 2 per cent of the land out of regular production each year to compensate for the yield increase on the land that stays in production. Until the year 2000 this would amount to at least 20 per cent of the agricultural land or about 15×10^6 ha in the EC alone. This is equal to the surface of all the agricultural land in Great Britain. Such marginalization of agriculturally less-endowed regions is not a new phenomenon but has been going on for quite some time in the industrialized world, as witnessed by the vast tracts of abandoned land in, for instance, France and the eastern USA.

It is sometimes suggested that the internal demand for agricultural products within the EC should be increased by replacing imported concentrates by local products. This would compensate for a production increase of about 4 years (de Wit *et al.*, 1987). However, any additional import restrictions at the borders of the EC are internationally unacceptable. There are also attempts to increase the demand by agro-industrialization, for instance the production of ethanol for cars, pulp for paper and vegetable oils for lubrication. However, the production of ethanol requires huge subsidies whereas the energy balance of the process is either not or barely positive. The cultivation of fibre crops for pulp production is technically feasible but is unlikely to be able to compete with imports for a long time to come. Most attractive may be the cultivation of oil crops for the production of biodegradable lubricants to replace environmentally unacceptable mineral oils (Slettenhaar, 1989). If one could succeed in capturing every year an additional 2 per cent of this market of 5×10^6 t year^{-1}, this would account for 10^5 ha year^{-1}, but this is less than 10 per cent of the land that was calculated to go out of regular production every year. Such figures illustrate the magnitude of the problem.

Well and less-endowed agricultural regions are geographically unevenly distributed within the EC (de Wit, 1988). Therefore any policy of price liberalization will come to a deadlock unless the more prosperous countries and regions, out of proper self-interest, are willing to support the less-endowed regions in creating development possibilities that aim at a social and economic structure that can replace the agricultural structure to a large extent. Where this is not feasible, social programmes should be supported that enable the younger part of the population to move to more promising regions and the older part to fade away in some grace. This requires more authority, leadership and money in the EC, but without this the common market in agricultural products is put at risk by any market-oriented policy. Such a partition into regions where agricultural production continues to increase and those where agriculture disappears as an important source of income has divergent consequences for the sustainability of agriculture.

Land that is taken out of regular production in the less-endowed regions will in some cases be used for ranching, forestry and the establishment of nature reserves. The profitability of such land uses is so small that much of the land will be left to run wild. Such extensification could be considered a positive development but for the fact that the environmental and social fabric formed by traditional farming systems is destroyed in the process. Hence, there may be valid reasons to put the brake on such destruction by supporting agriculture in regions that are marginalized. However, conserving a way of life in which 30 per cent of the population is supposed to make a living in agriculture is in the long run not only very expensive but also unacceptable to the population as a whole.

Support of agricultural development in potentially suitable regions of new member states is something else and may be necessary for political reasons. However, this requires that in other suitable regions land is taken out of production for other purposes such as, for instance, the strengthening of the ecological infrastructure. Such developments would occur at the expense of efficient resource use, but this may be very well worth it.

Pollution prevention is best served by concentrating farming in the most favourable regions. The need for energy, fertilizers and biocides per unit product is then the lowest. This relieves the overall burden on the environment; even so, environmental standards continue to be threatened locally because of the increased use of resources per unit area in regions where agriculture continues to be practised. Also, smaller farmers are caught between falling prices for their products and an increasing general level of prosperity. Hence, the number of farmers in well-endowed regions continues to decrease and the size of the farms continues to increase and with this their level of mechanization and automation.

PROBLEMS OF SUSTAINABILITY

Merciless exploitation of comparative advantages results in regional specialization and narrowing of crop rotation. Especially in combination with heavy mechanization, this threatens sustainability. Fields are then barren for too large a part of the year and subject to loss of soil structure and to unacceptable rates of nitrogen leaching and of wind and water erosion. Another problem of specialization is the development of pests, diseases and weeds that are difficult to control. Therefore much of the crop husbandry research is directed towards widening of the crop rotation and the development of soil cultivation practices and machinery that conserve the structure of the soil and the reintroduction of cover crops.

As summarized elsewhere (de Wit, in press), high-yielding crops may be more susceptible than low-yielding crops to obligatory parasitic pests and diseases, like aphids, mildew and rusts, mainly as a result of higher nitrogen concentrations in the attacked tissues. This promotes insurance spraying, i.e. spraying without first establishing the risk of damage to the crop. Integrated pest management methods have been developed to reduce such high use of biocides. These rely on varieties with a broad resistance, observation-driven chemical and biological control systems, avoidance of over-fertilization with nitrogen and other cultural practices. With such methods the average number of spray applications to wheat crops is kept at 2·5 in the Netherlands, compared to 8·5 in the UK and 7 in the northwest of the FRG (Rabbinge, 1987). On the other hand, it has been found that higher-yielding crops may be less susceptible to non-obligatory parasitic diseases like septoria, fusaria and verticillium and to nematodes. Fast-growing crops are also better able to suppress weeds than low-yielding crops, so that herbicides are less needed under high-yielding conditions at later stages of growth. This is contrary to the widely held belief that high-yielding crops require in general more biocides for their protection.

Corrected for inflation the price of nitrogen is at an all-time low in the EC in spite of the large increase of the price of energy in the 1970s. This has promoted considerable wastage, especially in animal husbandry. An environmental tax at least to double the price of inorganic nitrogen from the factory has been suggested (e.g. German Council of Experts for Environmental Problems, Rat Umweltfragen, 1985) to reduce such wastage. To obtain some idea about the long-term effects of such a tax, a comparison may be made with the tax on gasoline; in countries with high taxes on gasoline, small, energy-efficient cars have been developed that do the job of transporting the driver and his occasional passenger just as well as the big, gas-guzzling cars in countries with low gasoline taxes. The proceeds of such a tax could be guided into an EC fund for structural improvements in the less-endowed regions of the Community (de Wit, 1988). If need be, the tax burden on the farmers in well-endowed regions, who use most of the nitrogen, could be compensated for by price support within the framework of the Common Agricultural Policy. The costs are then passed on to the consumer, like costs associated with structural differences that supposedly make European agriculture more costly than agriculture in some other economic blocks. That is the price that has to be paid if one wants to maintain a viable agriculture in the European Community and to meet long overdue environmental standards.

A practically unavoidable consequence of greater uniformity is increased genetic erosion. By means of gene banks and conservation *in situ* attempts are being made to slow down this process, hopefully to such an extent that the rate of genetic impoverishment is counterbalanced by the increased ability for gene manipulation. There may be some comfort in the thought that any random sample of plants may already contain most of the individual genes in the plant kingdom albeit not in the combination that would induce desirable characters.

THE SECOND GREEN REVOLUTION

Of course there are far too many poor people in developing countries, but in contrast to disheartening expectations at the end of World War II, the medical doctor did not steam ahead of the agriculturalist. The Law of Malthus has not finally been proved to be true but agricultural production has kept abreast of the strongly increasing demand for food. Despite an almost three-fold increase in Third World population since 1945, global food shortages have been averted and famines have been reduced in both frequency and size. This required an increase in production of over 3 per cent per annum. The demand for food will continue to increase at least at these rates in the coming decades because of a continuing increase in population and income.

Roughly two-thirds of the increase of production was achieved by increasing yields per unit area and the other third by extending the surface under cultivation during the last 40 years, but regional differences cover the whole range from zero increase in yields to zero area expansion. There are still large areas that could be reclaimed in some parts of Africa and of South America, but in many regions of Asia most of the land that is suitable for some form of agriculture is already in use.

The Rockefeller and Ford Foundations were aware of this at an early stage (Baum, 1986) and founded at the beginning of the 1960s the International Rice Research Institute (IRRI) with the objective of transferring the yield-increasing techniques that were developed in the Western world to the cultivation of rice in Asia. This transfer appeared not too difficult and the results were readily adopted in regions where farmers were guided by some extension and attractive terms of trade.

Famous examples are the unprecedented yield increases in the Punjab and on Java. On the latter island, the yields of sawah rice increased only at a rate of about $2{\cdot}5\ \mathrm{kg\ ha^{-1}\ year^{-1}}$ before 1968, so that it was necessary to import increasing quantities of rice for the increasing population. Since then, yields have increased at a rate of $125\ \mathrm{kg\ ha^{-1}\ year^{-1}}$ (de Wit *et al.*, 1987), so that the growth of the population has been overtaken. The average yield increases over the whole of Asia were so large that the godowns filled up with rice and the prices went down. This solved the problem of hunger due to physical scarcity, but problems of hunger due to poverty and problems of equity remained.

The possibilities for increasing production are still so large that well-endowed regions are likely to be able to meet demands at relatively low prices in the coming decades in most countries of Asia (Buringh *et al.*, 1979). As in the developed world, this will marginalize less-endowed regions because their terms of trades will become less favourable. Scarcity of funds and the lack of political power of the population of marginal regions makes it very unlikely that the transfer of money that will be needed to reverse such marginalization processes will occur in the foreseeable future. Most farmers in these regions are thus condemned to subsistence and to exploitation of natural resources to satisfy immediate needs or to moving elsewhere. Agricultural research which is particularly directed towards improvement of least endowed regions may open up new possibilities in some situations. However, in many cases its results are more readily applied in regions that are better off. The comparative advantage of these least endowed regions is then not increased and their marginalization continues.

Seed stock, industrial fertilizers and biocides can be traded in small amounts, so that, in areas where application of new techniques is attractive, the smaller enterprises will eventually follow the larger ones. However, this is of little or no consolation to the smallest farmers. In spite of increasing yields, they cannot maintain their aspired level of income because they are caught between falling prices for their products and an increasing general level of prosperity with the consequence that they become impoverished. Experience in the Western world and in Japan shows that this is an unavoidable process unless one accepts permanent and increasing income transfers from the other sectors of the economy to the agricultural sector. However, this would drain scarce resources that are so badly needed to create more diversified regional economies with job opportunities outside the agricultural sector for small farmers and landless labourers.

MORE SUSTAINABILITY PROBLEMS

This second green revolution manifested itself only 20 years after the first green revolution and is therefore a striking example of a rapid transfer of techniques. This did not come about without problems.

For instance, pests and diseases can develop easily to epidemic proportions in the tropics because first principles of crop rotation are violated by growing the same species side by side during the whole year in various stages of development and by cultivating crops in mixtures instead of in rotation. Change of crops and of seasons then ceases to form a natural control on epidemic developments. Blanket spraying, however, may give rise to more problems than are solved because resistances develop and natural enemies are decimated. An example is the epidemic increase of brown planthoppers on Java in the middle of the 1970s that for some years caused a set back in the rate of increase of rice yields.

Even more so than in temperate climates it is therefore necessary to develop integrated methods to control pests, diseases and weeds, but this draws heavily on the infrastructure for research, teaching and extension. Therefore, epidemics and adverse consequences of their control may very well remain a serious problem for some time to come, and with it a risk of calamities during manufacture, transport and use of biocides.

In contrast to biocides, minerals and nitrogen occur in the natural environment but this is no excuse to waste them. However, inspired by many bad examples in the developed world, injudicious methods of fertilizer use have been promulgated and are now in wide use. For instance, the uptake of nitrogen fertilizer by irrigated or rainfed rice is often considerably less than 20 per cent of the amount that is applied (van Keulen, 1977). Better methods of application would enable an increase in this recovery to 50 per cent or even more. Such improvements of efficiency would make the use of fertilizer attractive in many more regions than at present and reduce the adverse environmental impact of over-fertilization in other regions.

There are many irrigated areas in Asia but these are often subjected to aging and inappropriate management. Sometimes the upper parts of the watersheds are reclaimed or deforested in such ways that soil erosion rapidly silts up reservoirs and channels. In other cases, there is lack of drainage, so that lower lying parts of the irrigation systems become saline. It is also disheartening that the importance of a proper social, economic and administrative infrastructure is often underestimated, so that many projects that were technically promising turned out to be failures. These problems apply far less to rainfed cultivation of rice as practised in many regions of South-East Asia. This may very well be one of the most sustainable farming systems that has been developed.

URBAN BIASES

The situation in South America is in sharp contrast to that in Asia. This continent is sparingly populated and has potential for increased production both by more intensive use of the land that is already in production and by reclaiming more land with due concern for the maintenance of indispensable tropical rainforest reserves. The coexistence of these vast underutilized resources for production and large numbers of poor small farmers is evidence of the geographic and socio-economic inaccessibility of this land. The commercial farm sector controls most of the better endowed but underutilized land resources, while the large peasant sector often occupies and overuses marginal lands (FAO, 1988). Problems of soil degradation and deforestation originate from inappropriate agricultural policies that perpetuate this limited access of small farmers to land and provide tax shelters for labour-extensive development of new areas where extensive systems of arable cropping and cattle ranching are not sustainable.

The key to improvement should have two levers: land reform should make it possible for the small farmer to expand onto agricultural land that is at present controlled by the large absentee landowners and agricultural policy reforms should make it attractive to intensify such utilization so that further degradation by overexploitation comes to an end. These are revolutionary changes that go against the immediate interests of the powerful urban middle class in many South American countries.

Such urban biases (Lipton, 1977) have led to a policy of neglect of the rural population in Africa. Dump prices on the world market, low exchange rates, lopsided trade policies and taxes, price control and subsidies promoted food imports, kept food prices low and thus discouraged food production. This enabled the maintenance of low wages in urban areas, so that investments in urban regions were favoured at the expense of those in rural areas.

This policy of disregard of rural development is running into a dead end for several reasons: there are less surpluses available against dump prices, industrial products of developing countries are often not attractive for the spoiled Western consumer, foreign debts are surging, local currencies are devaluing and imports have to be curbed. These changes make it more attractive to bring local agricultural products to the urban markets, to invest in the development of rural regions and to use fertilizers and other external means of production. This would enable farmers again to grow more of the food where the mouths are. Such renewed urban–rural linkages are essential for a more sustainable and equitable development. It requires a policy that is geared towards strengthening of the infrastructure of roads and markets, of agro-industrial activities in both urban and rural regions, and of agricultural research, teaching and extension.

A THIRD GREEN REVOLUTION?

In spite of the present dismal situation in most of Africa, there are regions with considerable potential, for instance the extended savanna regions, in which growing seasons of about half a year are characterized by a good level of rainfall and radiation and by relatively warm days and cold nights. Soils may be chemically poor but they respond very well to the use of fertilizer and are easy to cultivate. A number of these regions are suitable for the development of productive family farms that are also viable in the long run. To compete successfully with life and work in urban regions it would also be necessary to improve the productivity of labour and to mitigate the drudgery of farming. However, the soils are vulnerable to erosion, so that amelioration measures to control surface run-off are needed and farming systems have to be developed that keep the soil as much as possible covered with crops and mulches. Agricultural systems as developed in the Western world pose far too many risks of erosion. They require also large imports in the form of machines, spare parts and fuel and generate too little employment to be socially attractive.

However, a number of African countries are mainly in the humid lowlands that are covered with tropical rainforests. In some circumstances, part of their demands could be met by expansion of agricultural land use at the expense of natural forests, but in others it would be preferable to save these forests and to meet increasing demands for food on existing agricultural land. Since demand is at least going to double in the coming 25 years, this would require that the yields on the land that is already in agricultural use have to increase at a rate of about 50 kg ha^{-1} $year^{-1}$. These are as high as green revolution yield increases in the Western world and Asia, but they have to be achieved under far less favourable agro-ecological and socio-economic circumstances. Because of the limited potential for irrigation and the necessity of erosion control, attempts are made to meet such increasing needs by the development of techniques that build upon traditional systems of shifting cultivation. Systems are then visualized in which fallow periods are shortened by growing annuals and woody perennials in combination and green mulches are used to keep the soil covered. Even biotechnology is not able to create green miracles. Hence, none of these systems are likely to meet increasing needs without an increase in the use of external inputs comparable with those of the first and second green revolutions. This would require a third green revolution, but this is not sufficiently acknowledged because too often the efficiency of traditional subsistence systems is underestimated and the carrying capacity of natural resources is overestimated.

The Sahel may serve as an example. The pastoralists exploit the excellent pastures in the north of the Sahel at the border of the Sahara during the short rainy season by a yearly transhumance and survive the rest of the year in the south of the Sahel, where little feed but sufficient drinking water is available during the dry season. This herding system is labour intensive, but measured by off-take per unit surface area is the most productive system that can be visualized under such agro-ecological circumstances (Breman, 1982). Any attempts to improve production by the introduction of other techniques would fail without the introduction of external production resources. This holds also for a better integration of transhumance in the north with arable farming in the south if this were socially acceptable to begin with.

SOIL VERSUS OIL

About half of the soils on earth may be considered suitable for some form of agriculture. Half of this half is already in use and the other half requires either large efforts to reclaim it, or is very fragile or occupied with ecosystems such as tropical forests that are worth conserving. The only way to avoid a rape of the earth and at the same time to meet the needs of an increasing world population is by the use of intensified agricultural systems that produce increasing amounts of food on land that is already reclaimed and by making use of fossil fuels for the manufacture and use of the external inputs that are needed. By doing so the use of one unrenewable resource is substituted by the use of another.

The wisdom of such a substitution depends on the relative scarcity of soil and oil and the possibilities for substitution (TAC/CGIAR, 1989). Although it is possible to grow valuable crops in water culture on bare rocks, soils are an irreplaceable constituent of natural ecosystems and without them it is practically impossible to maintain agricultural systems for the production of the bulk of food and feed. The proven and surmised reserves of fossil oil are about 100 times the present annual use, whereas huge amounts of fossil energy are available in the form of natural gas, coal, tarsands and others. Unfortunately, their indiscriminate use leads to an increase in the carbon dioxide (CO_2) concentration of the air the adverse effects of which are considered to outweigh the beneficial effects. There are, however, substitutes in the form of solar and wind energy, and nuclear energy out of fission and, possibly, fusion that contribute far less to CO_2 production. Bio-energy recycles the CO_2 but its use is nevertheless counterproductive, because it requires agricultural exploitation of large areas of land.

Obviously, the way we continue to live is not sustainable in the long run, but in the meantime society is likely to be better served by forms of agriculture that save soil at the expense of oil than the other way round.

REFERENCES

Baum, W. C. (1986). *Partners Against Hunger*. Washington, DC: The World Bank.

Breman, H. (1982). L'aménagement des pâturages, in *La Productivité des Pâturages Sahéliens*: *Une Étude des Sols des Végétations et de l'Exploitation de cette Ressource Naturelle* (With extended English summary), eds F. W. T. Penning de Vries *et al.* Pudoc, Wageningen: Agricultural Research Report no. 918.

Buringh, P., Heemst, H. D. J. van and Staringh, G. J. (1979). Computation of the absolute maximum food production for the world, in *MOIRA. Model of International Relations in Agriculture,* eds H. Linneman *et al.* The Netherlands: North-Holland Publishing.

FAO (1988). Potentials for agricultural and rural development in Latin America and the Caribbean. *Main Report and Annexes*. Rome: Food and Agriculture Organization.

Keulen, H. van (1977). Nitrogen requirements of rice with special reference to Java, *Contr. Centr. Res. Agric. Bogor,* no. 30.

Liebscher (1895). Untersuchungen uber die Bestimmung des Düngerbedürfnisses der Ackerböden und Kulturpflanzen, *Journal für Landwirtschaft,* **43**, 49.

Lipton, M. (1977). *Why People Stay Poor: A Study of Urban Bias in World Development.* London: Temple Smith.
Meester, G. and Strijker, D. (1985). Het Europese landbouwbeleid voorbij de scheidslijn van zelfvoorziening. *Voorstudie WRR V46.* The Netherlands: Staatsuitgeverij.
Rabbinge, R. (1987). Implementation of integrated crop protection systems, in *Pest and Disease Models in Forecasting, Crop Loss Appraisal and Decision Supporting Crop Protection Systems,* eds D. J. Royle *et al.* Intern. Org. Biol. Control/Org. Intern. Lutte Biol. Bulletin.
Rat Umweltfragen (1985). Umweltprobleme der Landwirtschaft. *Der Rat von Sachverständigen für Umweltfragen.* Stuttgart en Mainz, FRG: Kolhammer GMBH.
Slettenhaar, G. (1989). Gebruik plantaardige smeermiddelen spaart milieu, *Nieuws uit Wageningen,* no. 55, November.
TAC/CGIAR (1989). *Sustainable Agricultural Production: Implications for International Agricultural Research.* FAO Research and Technology paper no. 4. Rome: Food and Agriculture Organization.
Wit, C. T. de (1988). Environmental impact of the CAP, *European Review of Agricultural Economics,* **15,** 283–95.
Wit, C. T. de, Huisman, H. and Rabbinge, R. (1987). Agriculture and its environment: are their other ways? *Agricultural Systems,* **23,** 211–36.
Wit, C. T. de (in press). On the efficiency of resource use in agriculture, *Agricultural Systems.*

11
Agricultural Sector Modelling for Policy Development

D. R. HARVEY

Department of Agricultural Economics and Food Marketing, and Centre for Land Use and Water Resource Research, University of Newcastle-upon Tyne, UK

Vicar: With God's help, you have made this garden a thing of beauty.
Gardener: You should 'ave seen it when 'e 'ad it to 'isself!

INTRODUCTION

That systems analysis has become a traditional phrase, if not technique for looking at problems of agricultural development and policy in the UK is due in large part to the activities of Colin Spedding. He has spoken, it seems with increasing authority, from his platform of the Chair in Agricultural Systems and, perhaps more importantly, as Director of the Centre for Agricultural Strategy at Reading University to agricultural and rural policy-makers on behalf of the agricultural and rural sectors and the research professions which seek to serve those sectors. His authority, however, derives not from the platforms on which he stands, but from his ability to 'see things in the round' and to relate apparently disparate facts, theories and pieces of applied research into an integrated holistic picture.

> *The understanding of whole agricultural systems thus requires a synthesis of several biological disciplines, management and economics. The risks of oversimplification and superficial treatment are obvious. The most common solution, of studying the constituent parts separately, leaves the essential synthesis to be undertaken by those engaged at the level of*

> *enterprise studies . . . (but) all relevant disciplines are required to integrate the results of both research and practice into useful models.*
>
> (Spedding, 1975)

Without resorting to the details or vocabulary of applied systems analysis, the basic principle behind the approach is very simple: that things are as they are not by accident or random chance occurrence but because some systematic patterns of behaviour and mutually responsive interactions occur within and between people and their natural surroundings. Understand these patterns and interactions, especially the feedback linkages, and the world will be a more understandable and therefore predictable place. Furthermore, with an accurate, if simplistic and incomplete understanding of why things are the way they are, it becomes possible to project the consequences of doing things differently and thus to make suggestions about policy changes to better achieve increasingly complex objectives. The principle applies from the management of individual enterprises at the farm or firm level, through the operation and planning of the whole firm to the development of consistent and effective policies at the regional, national and international levels.

Cynics or pessimists might object that the world and things do not operate by clockwork mechanisms but rather are cloudy and uncertain. A systems approach does not deny this, though it has to be admitted that the incorporation of uncertainty and noise makes many systematic models of the world both complex and ambiguous, to say nothing of the increased problems of verification and validation. Numerical approximation and simulation of models then become the only practical methods of exploring system properties; analytical abstraction and analysis is no longer sufficient. The rapidly expanding power of modern computers makes such exploration increasingly available, as does the emergence of major research funding exercises, at least in some areas, especially on questions associated with the rural and natural environment at the present time.

Nevertheless, there are dangers associated with such a rapidly developing subject; depending on the way in which such models are constructed, tested and used, we may be in danger of inventing policies and practices which suit the models more or less perfectly but are, in fact, totally divorced from the real world—we may be exactly wrong rather than roughly right. Can we be sure that our systems have the earth going round the sun and not *vice versa*? While the differences may not be dramatic for practical options or predictions close to historical experience (that the sun will 'come up' tomorrow), they become vital when considering futures which may be radically different from our past (like landing people on the moon or sending space probes throughout the planetary system). Enthusiastic practitioners of a systems approach to socio-economic behaviour are urged to read Isaac Asimov's *Foundation* saga (1951–68), which explores the implications of an ultimate systems theory termed 'psychohistory' in which all human development can be predicted. Fortunately, we are a very long way from such science fiction, though the present rate of change may well give us pause for thought.

Enough has already been said to indicate that many of the fundamental problems associated with trying to model complex human/natural systems are philosophical in nature rather than technical or single-discipline theoretical, intractable though the latter may seem to practitioners. However, before returning to these questions, it is necessary to lay out the ground of agricultural sector modelling for policy purposes.

The rest of the paper deals with, first, the basic elements of the economic theory of the behaviour of markets and the integrated general equilibrium in which this theory culminates. Readers familiar with economics may profitably skip this section, which is included to give non-economists a flavour of the theoretical economic system which to a greater or lesser extent informs applied economic modelling efforts. The third section turns to an illustrative survey of some of the types of models to which this theoretical framework gives rise. The fourth section deals with the implications of the previous sections for the development of a comprehensive modelling system of the agriculture/environment complex in the light of current activities in this area at both the Centre for Agricultural Strategy, Reading and the Centre for Land Use and Water Resource Research, Newcastle. The fifth section emphasises some major conflicts and issues associated with these developments. Thus, this paper is not a comprehensive survey of systems analysis and agricultural sector modelling. This has already been adequately covered in the literature, for instance, as a classic example, Johnson and Rausser (1977) which in spite of its age is still a major and relevant review of the problems and issues, and more recently Bauer (1989*a*). Rather this paper is a distillation of personal experience with trying to model aspects of the agricultural and food system for policy analysis purposes and trying to understand what other models can tell us about policy effectiveness, efficiency and development.

ECONOMIC THEORY—AN ABSTRACT ECONOMIC SYSTEM

Economic models, unlike some scientific counterparts, are heavily dependent on the underlying theory. Although there has been a long debate among economists as to the relative importance of theory versus empirical evidence (see Lipsey (1983), Chapter 1, for an elementary discussion and Blaug (1980) for the definitive version), the role of theory is still extremely important in most cases, especially in the case of policy analysis. There is a simple reason for this. Economic systems, as a sub-set of social, political and personal systems, are generally open and extremely complex and also largely incapable of controlled experimentation (notwithstanding recent work on this, see e.g. Roth (1988) for a review). Without a general methodology which allows for the separate identification of component parts of this system, social scientists are forced to proceed on the basis of 'thought experiments'—suppose people behave according to the simplest motives of profit maximisation (in the case of productive enterprises) and utility maximisation (in the case of consumers, where utility is at least an ordinal and generally monotonic function of consumption levels), what would the resulting economic system look like?

The answers, coupled with the assumption that people behave rationally (they actually strive to achieve these objectives in their behaviour), allow the construction of an elegant and sophisticated theory about economic systems. Of course, this model is an abstraction from and simplification of the real world, as is inevitable in all models of the real world. Nevertheless, so long as it captures the essential elements of human economic behaviour it may be relied upon. Empirical evidence is then assembled which serves two important but not always complimentary purposes: first, the evidence is intended to *test* the theoretical hypotheses, though often many of them are incapable of being tested since the whole framework is entailed in the hypotheses (see Blaug, 1980); second, the empirical evidence is used to *estimate* the theoretical relationships, in order to use the resulting empirically calibrated model to predict and project the future, where predictions (as unconditional forecasts of the future) are less common as well as less reliable than projections (as forecasts conditional on certain fixed values of exogenous variables and relationships). In logic, testing of the hypotheses should precede the estimation of relationships. However, until the recent promulgation of statistical causality tests (stemming from Grainger, 1969), estimation of relationships has been treated as equivalent to testing hypotheses, subject to the usual immunising stratagems (Blaug, 1980, pp. 17–20) used to protect the current paradigm of thought and protect the researcher against the journal and peer review process.

The essential difference, therefore, between socio-economic systems and their natural or physical counterparts is the lack of experimental evidence on the nature and characteristics of fundamental relationships between variables and states (as collections of variables at particular levels). Characterising a system at its simplest possible level, as a sequence of inputs (I), transformation functions (Tf) and outputs (O), Table 1 provides one classification of systems models.

It stands to reason that complete lack of knowledge about any of the essential elements of the system prevents modelling of the system and forces the investigator to restrict attention to exploring the system to improve knowledge. Simpler systems often approximate type 1 characteristics while increasing complexity often leads towards type 7, though, of course, the very application of such a classification implies some subjective knowledge or presumption about the way in which the system works. The identification (knowledge) of inputs implies that some ideas about the transformation function and its outputs is also presumed. In the policy arena, at least, policy-makers are forced by their occupations to believe that the instruments under their partial control actually influence their objectives (target outputs) through some systematic transformation function, even if objective knowledge of the transformation functions and outputs is lacking.

Economic models are generally of type 2 or 5, i.e. knowledge of the transformation functions is lacking. It is here that the formulation of the theory is important, to provide surrogate knowledge about the transformation functions, and hence about the inputs and outputs which are of most concern.

Economic theory pictures individuals as pursuing their own interests, within constraints established by legal and institutional provisions as well as those imposed by their own circumstances and fortunes, in response to signals about the relative social values of goods and services determined in the market place. Two further influences are

TABLE 1
Knowns and Unknowns and their Consequences for Systems Analysis

	Knowns	*Unknowns*	*Solution existence?*	*Nature of solution*
1	I, O, Tf	—	Yes	System description
2	I, O	Tf	Yes	System identification and estimation: input/output relationships examined, models hypothesized, parameters estimated and model tested.
3	I, Tf	O	Yes	System convolution and forecasting: output predictions on basis of simulation of input and transformation function over probability ranges of inputs and noise.
4	O, Tf	I	Yes	System deconvolution: inverse of (3), inputs reconstructed within probability bounds and retrospective forecasts compared with actual events.
5	O	Tf, I	Yes (?)	ARMA system identification and estimation: cannot differentiate between system Tf and inputs.
6	Tf	I, O	Yes (?)	Pure simulation
7	I	Tf, O	No	Frequent policy control situation, where the system mechanisms and its outputs are indeterminate

(Source: Bennett and Chorley, 1978, p. 71.)

important: the technological conditions determining the relationships between inputs and outputs in the production of goods and services, which at least constrain the nature of the economic transformation function; the influence of government (at all levels) which alter prices and terms of trade within the various markets as well as partly determining the institutional practices and regulations which govern these trades, thus altering the inputs and endogenous variables within the system. Within each market (as a collection of people willing to demand and supply a particular good or service) it is expected that the process of trading will result in an equilibrium in which the total quantity supplied will equal the total quantity demanded at the same (equilibrium) price. Such 'partial equilibria', so called because they deal independently with separate markets, are expected to respond to changes in incomes, tastes and technology, and to shifts in related markets, adjusting to new equilibria to reflect these changes. Thus, each market viewed as a system is expected to exhibit homeostasis.

The pure theory of market behaviour can be extended to produce a general equilibrium in which all markets are simultaneously in balance. It can also be shown that, given the distribution of ownership of factors of production (land, labour, capital, management and entrepreneurial ability) amongst the population, the final mix of production and consumption in this equilibrium is such that no one person can be made better off (according to his/her own judgement) without making someone else worse off—a condition known as Pareto optimality after the economist who formalised this criterion against which improvements in economic welfare are generally judged. This is the basis of Adam Smith's dictum that the invisible hand of market competition produces an optimum wealth of nations. However, this optimum is also dependent on a series of conditions which define perfect competition within all markets.

Among the most important of these conditions are: that individuals are the best judges of their own welfare; that they have perfect (or competitively available) information about the present from which to make informed judgements about the future; that one person's consumption or one firm's production does not affect other people's consumption or production other than through the market mechanism; that no individual or group of people on either side of the market can separately or individually influence the market outcome; that all goods and services potentially or actually demanded by the population are rival in consumption (that is, my consumption denies others' consumption of the same unit) and that people are excluded from consuming or benefiting from goods unless they have paid for them.

Although it is not expected that all of these conditions will be met all of the time, a variant of the market ideology argues that existence and encouragement of competitive market mechanisms will encourage the emergence of new techniques, goods and services to satisfy new demands. From this point of view, over time a system of workable competition, in which these conditions are approximately met, is expected to result in optimum economic welfare.

How can the obvious problems associated with the development of agriculture and the countryside be reconciled with this rather optimistic and perhaps complacent view of the operation of market mechanisms (transformation functions)? In the simplest terms, this story is one of the *efficiency* of markets, in which the allocation of resources to the production of goods and services is optimal given the prevailing distribution of resources amongst the population. However, the story suffers from three serious difficulties as a practical prescription for social organisation. First, the mechanism has nothing to say about the distribution of income and wealth, and hence economic opportunity, throughout society. In fact, the operation of the market mechanism can be expected to lead to some concentration of income and wealth among those whose original endowment of skills, ability and inherited wealth is greater than others, or whose particular comparative advantages more closely match the demands for goods and services as expressed through markets. Yet a number of religions and philosophies (see, as the most prominent modern example, Rawls (1971)) emphasise the notion of equity and argue strongly that sustainable societies require an equitable, though not necessarily equal distribution of economic welfare and opportunity. Hence governments are needed to ensure redistribution to the worse-off at the expense of the better-off. A major difficulty associated with such intervention is that it almost certainly breaks the conditions under which competitive markets result in an optimum allocation of resources to the production and consumption of goods and services.

Second, markets often take some time to adjust to new circumstances and in so doing give rise to difficulties for those people faced with adjustment problems. There is often a case, therefore, for public intervention to assist with the adjustment processes. Third, markets do not work competitively in practice. In particular, a number of important 'market failures' can be identified which prevent the achievement of welfare optimisation by the market mechanism and which therefore require social (political) intervention.

The first and second of these deficiencies of the market mechanism have been used in the past to justify substantial intervention in rural areas, especially in agriculture. In the process of development from a predominantly agrarian to an industrial economy, income growth is accompanied by a less rapid growth in demand for food. As a consequence, agriculture grows less rapidly than the rest of the economy and, according to the market mechanism, should release resources, especially labour and capital, to the rest of the economy. Such a process is often accompanied by technological change which reinforces and even dominates the market tendency. The market signals which encourage resource release mainly consist of reduced incomes and earnings for resources in agriculture relative to the rest of the economy. Those with good opportunities elsewhere will be encouraged to leave while the remainder (with limited or no alternative opportunities) remain in the sector and remain relatively poor as a consequence. Prior to the development of other economic activity in rural areas, the relative decline in agriculture is equivalent to rural decline while support of agriculture also supports rural areas.

The economic prescription for government intervention in this case is to assist with the adjustment process. There are examples of such policy intervention, such as the Rural Development Areas scheme in the UK, which not only encourage adjustment of rural activity other than agriculture but also are intended to redistribute economic activity from urban to the rural areas in the interests of more balanced geographical development. Agricultural policy, on the other hand, has an explicit objective of supporting the incomes of farmers together with other objectives which are not necessarily consistent with it. That is, agricultural policy has often been associated with protecting the sector from adjustment which would otherwise occur through the operation of market forces. The distributional benefits of farm policy within the agricultural sector have rarely been studied (Josling and Hamway (1972) is an exception) although a much greater body of analysis has shown that the main distributional effect of the Common Agricultural Policy is to effect a major redistribution from consumers to farmers and landowners (see, for example, Buckwell *et al.*, 1982). The government responsible for taking such decisions is also subject to political pressure from its subjects. This may result in interest groups being able to 'capture' government support in their own rather than society's interests. There is a growing, though predominantly American literature which develops this theme through the application of theories of public choice to agricultural policy. Mueller (1979) and Buchannan and Tollison (1984) provide the general overview to this body of theory while a recent agricultural application is forthcoming (de Gorter and Tsur, 1989).

Aside from distributional and adjustment problems, it should be clear from the above outline that markets are said to fail if conditions necessary to achieve economic welfare optimisation are not met. Of particular importance under this heading are: (i) circumstances where private costs and benefits reflected in prices and costs do not properly include the external effects (externalities) of production or consumption decisions (pollution is a classic example) or where market structure (concentration of firms, or less often consumers, in a market) allows individual decision-makers to affect prices or costs so as to exploit the other side of the market and thus divorce private prices and costs from their social counterparts; (ii) the existence of an important class of goods known as public goods. The latter are goods which, in pure form, are neither rival in consumption nor excludable. Thus, my consumption of a beautiful landscape does not reduce another's consumption of that same landscape. Similarly, it is difficult to exclude people from the benefits of a pretty countryside. These characteristics make it impossible for private markets to provide public goods. They have to be provided through or by the public sector if they are to be provided at all. Harvey and Whitby (1988) explore these issues in more detail, including the important related concept of property rights.

Market failure relies heavily on the value-judgement necessary to determine optimal economic welfare, which is conventionally the Pareto criterion. Applied over successive generations, this criterion, restrictive though it is, is sufficient to guarantee sustainable development of both economic and environmental resources as defined by Pearce *et al.* (1989) though, as they point out, this does require some accounting framework within which economic and environmental goods and bads can be compared and, as necessary, traded. This does not deny that even perfectly functioning markets may fail to meet other equally legitimate and perhaps more important social objectives, particularly equitable distribution. Other value-judgements than the Pareto criterion are necessary to decide on appropriate distribution and on the strength to be accorded to other motives and objectives than those included in the conventional definitions of economic welfare. Public choice theory seeks to explore this phenomenon through application of the pervasive economic motivation of self-interest to the formation and persistence of policies, characterising the political process as a market place in which votes are traded for policies, with politicians and policy-makers supplying policies in response to the electorate as a collection of more or less interested pressure groups. There are some reasons to question this approach, but this paper is not the place to explore them. Suffice it to say that, in the opinion of this author, economic welfare optimisation can be regarded as a necessary but not sufficient condition for personal and social bliss in a wider consideration of all relevant social and political conditions, which is at once a sobering thought for economists and a comforting one for everyone else.

The conclusions from this brief review of economic theory are that: (a) inputs and outputs from such systems are all that can be directly observed—the transformation functions (at least as far as the socio-economic elements are concerned) can only be inferred from these observations; therefore, (b) economic theory is a necessary but incomplete and abstract formulation of the transformation functions of the economic parts of an agricultural system; (c) even given the accuracy and reliability of the economic theory, the resulting characterisation of agricultural systems from an economic perspective can ultimately be only a partial and incomplete picture, however general the economic formulation might attempt to be. Consequently, the proper use of such system models for policy purposes can only be advisory, never dictatorial. Considerable judgement and democratic discussion of the results and their policy implications is absolutely necessary if they are to achieve any advance in the human condition.

THE DEVELOPMENT OF AGRICULTURAL SECTOR MODELS

A brief paper such as this cannot hope to provide a comprehensive survey of the wealth of literature dealing with agricultural sector models. All that can be offered is an outline of the recent development illustrated with examples. As such, it is inevitable that the following survey is somewhat subjective and idiosyncratic, coloured by the experience and interests of the author. The caveat should be borne in mind for the rest of this paper.

The formal modelling of economic systems in which policy intervenes is a complex task, requiring a substantial reflection of the complete production–marketing system in order to derive useful and relevant policy information. The underlying lags in response and linkages between commodities, based on the technical aspects of agricultural production, have to be combined with economic behaviour of farmers to determine supply and associated input and resource use, in response to both market signals and the often extremely complex policy interventions. The latter affect behaviour both through altered market signals and also through quantitative regulations and restrictions, which include the activities and mechanisms of producer marketing boards and state trading and marketing institutions. In addition, especially in the United States, there are complex rules about participation in those cases where the programme is voluntary, compliance (where participation in one programme is dependent on meeting the conditions under another, for example the set-aside and deficiency payment/loan rate system—Gardner, 1981) and eligibility limits which restrict participation to certain sizes of farm or business. Overlaying these interactions are the transformations of products and inputs in time, form and space. The resulting structures, including feedback linkages and expectation formation mechanisms, can be formidable. A useful, though orientated towards United States, and now rather dated survey of methods used to tackle these problems is provided by Judge *et al.* (1977) while Rausser (1982) also provides discussion of possible approaches which is again heavily directed towards North America.

Table 2 shows a somewhat artificial classification system of the sorts of models which have been developed. In this table the definitions of the various methods and approaches which have been used to model the agricultural sector are necessarily arbitrary. 'Econometric' implies that relationships are estimated on the basis of historic (usually time series) data using econometric techniques of varying levels of sophistication and that, after testing their validity, the resulting model is

TABLE 2
Agricultural Sector Model Classification

Method/approach:	*Econo-metric*	*Spatial/resource equilibrium*	*Simulation or mixed*	*General equilibrium*	*Public choice*
Scope					
1 'Single' commodity/ One side of market:	1	na	na	na	na
2 Single commodity/ Single country	2	3	4	na	5
3 Multicommodity/ Single country	6	7	8	9	10
4 Single commodity/ Multicountry	11	12	13	na	14
5 Multicommodity/ Multicountry	15	16	17	18	19
6 Agricultural/ Rural environment					
a Regional	na	20	21	22	23
b National	na	24	25	26	27
c Multinational	na	28	29	30	31

perturbed to reflect changes in policy or market conditions and the results used to project the consequences of these changes. The 'Spatial/resource equilibrium' class includes two major sub-types of model: 'Spatial' models where the programming technique is either used to specifically identify the market spatial relationships (including transportation and sometimes processing costs), or 'Resource' models which utilize the programming approach to impose more structure on the underlying production relationships and their interactions with the underlying resource base and structure. 'Simulation' models do not generally involve the separate estimation of relationships, rather they borrow other estimates of the basic relationships necessary to drive the model. 'Mixed' models include those which employ a mixture of techniques and approaches to capture the structure of the system. 'General equilibrium' models are designed to capture the theoretical interaction of all markets in the economy and typically involve either the separate modelling of the rest of the economy, including the labour, capital and money markets, or the borrowing of these formulations from existing macroeconomic models of the economy. 'Public choice' models (the most recent approach) are specified to capture the costs and benefits to specified interest groups and to relate these to the ways in which the political process reconciles competing and conflicting interests.

The scope classification should be self-explanatory, except to note that type 1 models include those which examine only one side of the market but for a varying number of commodities, including a large number of sophisticated demand studies which deal with the whole of food demand, and a number of supply side studies which capture the whole of agricultural supply. The resulting classification (abbreviated through the combination of different approaches under a single column heading in two cases) provides for 31 different types of model, after eliminating those nine cases where the approach and scope are incompatible.

The typical development of modelling expertise and research capacity both within countries and globally has involved progression from the north-west corner to the south-east corner of this table. Thus, as yet there are no substantive examples of cells 23 or 26–31 to this author's knowledge, while there could well be some dispute about whether cell 23 (regional agricultural/rural environment systems modelled as general equilibrium systems) is a legitimate approach, given that regional markets cannot be isolated from the rest of the economy. The entry here reflects the supposition that the rest of the country (other regions) can be captured in minimal detail to 'close' the model of a specific region.

However, this picture of the development of agricultural models is unduly simplistic. There have been cycles and branches. Thus a major expansion in econometric sector models (type 6) succeeded the development of partial models of commodity sectors (type 2), especially in North America (e.g. Fox, 1965; Chen, 1977; Scherr, 1978, Agriculture Canada, 1980), which were used primarily for forecasting, though inevitably also for policy simulation. However, these have since fallen into disuse and disrepair, though Agriculture Canada's FARM and Food and Agriculture Policy Research Institute (FAPRI) are still extant (Meyers *et al.,* 1987; CARD, 1988). Some of the reasons will be returned to below. In Western Europe, less emphasis has been placed on forecasting models and earlier sector models were more eclectic (e.g. Thomson and Buckwell (1979), a farm-based programming model of the UK; McFarquhar *et al.* (1971), a mixed econometric/input–output model of the UK agricultural sector), especially designed to capture the effects of the major policy shift consequent on the UK's accession to the European Community.

Since then, modelling agriculture has pursued rather different paths. On the one hand, computable general equilibrium models have been seen as necessary to place agriculture within an overall picture of economic development (seen as especially important in developing countries, where the World Bank has applied the technique to several countries, following Adelman and Robinson (1978)). The same approach has been employed for multinational modelling (Parikh and Rabar, 1981) in the most comprehensive and ambitious agricultural modelling exercise yet attempted, involving the development of consistent and detailed national models for 12 participating countries, 33 collaborating institutions from around the world and a large number of researchers and associated resources, and also to developed countries (e.g. Stoeckel, 1985).

On the other hand, national concerns about regional distribution of economic activity and appropriate resource allocation have encouraged the development of programming models to identify spatial distributions and incorporate explicit structural and technical relationships, especially in Australia (Walker and Dillon, 1976; Wicks *et al.,* 1978), the US (Hertel *et al.,* 1989) and also in Canada. A third branch of modelling activity has been encouraged first by the commodity boom of the early 1970s and latterly by the GATT negotiations on agricultural trade. The commodity boom persuaded policy analysts that international market conditions were important in conditioning if not determining domestic policy, while the distortions of the world market consequent on domestic agricultural support policies encouraged both the United States and the European Community to place agricultural trade at the top of the agenda in the Uruguay Round of GATT negotiations begun in 1986. The development of multinational models of world trade (type 17) has both preceded and followed such policy interest and will be dealt with in more detail below.

Within the European Community, policy interest has centred on aggregate trends and directions of change, and the national distribution of agricultural activity and on the consequences of the Common Agricultural Policy. A number of policy models have been developed in this context (see, for example, Baur and Henrichsmeyer (1989) and European Commission (1988)). Of these, the model most well-connected to the European policy-making machinery is the Sectoral Production and Income Model (SPEL) built around a comprehensive social accounting framework and recently given wider publicity in Henrichsmeyer (1989) and the related papers in the same source.

Meanwhile, the emergence of environmental concerns has encouraged the development of model systems to examine the relationships between agriculture as the major land user and the natural environment. Once again, the programming approach (type 24) has proved popular in this development and will also be returned to below. Alongside such model development trends have been individual efforts to examine specific policies in more detail and depth, increasingly adopting more rigorous specifications of theoretical models and more constrained estimation techniques to impose a pre-defined structure on the model system and to test for the causal links in an attempt to reduce the dangers of spurious correlations and to circumvent the underlying problem of a lack of detailed experimental work which can provide information on the transformation functions.

It is clearly impossible to review all examples of these various model types, while restriction of attention to examples which have been constructed for the UK would also be inappropriate. To focus discussion on the major differences the next section will concentrate on four major types of model: large-scale econometric models of national agricultural systems (type 6 in Table 2); programming models of national sectors (type 7); multinational multicommodity models (type 17); general equilibrium models (types 9 and 18). Before that, however, some general issues of policy modelling need to be considered.

Application of economic models to policy analysis and evaluation requires a minimum of (i) specified target variables considered important by policy-makers (the output); (ii) specified instruments available to policy-makers (the inputs); (iii) a system of behavioural, identity and technical relationships which link (i) and (ii), the transformation functions, (iv) controllability between the inputs and outputs. Construction and use of such models involves capturing the essential elements of dynamic, stochastic, uncertain and open socio-economic systems that characterise agriculture; establishing (or more usually presuming) causality linkages; generating conditional forecasts (projections) of consequences of policy options; interpretation of results in the light of omitted variables, relationships and interactions; persuading policy-makers of the credibility and validity of these projections and the implications for appropriate policy choice. In the last resort, the acid test of a policy model is its usefulness to, and thus use by policy-makers and their critics. Many intended policy models have failed this test in the past and have consequently fallen into disrepair.

Rausser and Just (Chapter 22 in Rausser, 1982) explore some of the possible reasons for this phenomenon and identity ten principles which need to be followed to achieve a successful policy model. The principles and implications are also discussed by Thomson and Rayner (1984); de Haen (1979) deals with the same problem. The following outline follows the Rausser and Just framework.

(i) *Identify the target policy variables:* which seems all too obvious. However, the distributional consequences of policy alternatives are often left on one side as policy analysis concentrates on the allocative efficiency of policy alternatives. Josling (1969) pointed out the importance of economic transfers generated by policies compared with the net social welfare gain which is the focus of textbook economics, and followed this up with an analysis of the distributional consequences of the Common Agricultural Policy compared with the UK system of deficiency payments (Josling and Hamway, 1972). Multiple and often conflicting objectives raise the question of identifying trade-offs, the associated concept of policy efficiency as the rate of achievement of policy objectives per unit cost (Thomson and Harvey,

1981) and, possibly, establishing the appropriate weights so as to maximise social welfare. In addition, concentration on distributional issues requires that the effects of policies on asset values should be explored (Harvey, 1990), while the identification of potential conflicts between short- and long-run objectives is also important, implying at least a dynamic aspect to the model if not a fully specified dynamic system. De Haen argues the same principle from a rather different perspective, that the 'congruency between the domain of competence of economic theory and the range of responsibilities of practical policy . . . does not exist' (p. 26.2) and goes on to argue for a wider range of interest in policy models than the traditional preserves of economics, namely economic efficiency and growth, to include 'more equitable distribution of income, more ecologically sound patterns of growth and an increased awareness of the impact of domestic policy on international relations' (p. 26.3).

(ii) *Exploit the experimental role of policy models:* which involves recognition that there is no simple answer to the question of the appropriate policy. First, multiple objectives require multiple instruments for their satisfactory achievement. Second, unknown, uncertain and changing weights on various elements in the multi-objective function lead to variations in the appropriate policy set depending on unobserved social/political valuations. Third, the applicability of policy prescriptions and evaluations depends critically on belief in the essential elements of the system by the policy-makers. On all three counts, close involvement of policy-makers with both the construction and use of policy models is to be encouraged, while policy modelling itself must be seen as a process of continual experimentation and improvement rather than a once-and-for-all activity.

(iii) *Verification and validation is critical:* Rausser and Just argue that neither classical nor Bayesian statistics are sufficient to capture the validity and utility of policy models since the former ignores the influence of extraneous but 'known' information while the latter ignores the costs of gathering and processing information. They argue instead for the explicit inclusion of the costs of complexity and of inaccuracy, which implies experimentally designed sensitivity analysis as a major component of policy models in order to identify the key variables and relationships and to concentrate limited effort on these at the expense of more peripheral or unimportant relationships. The dictum that it is better to be roughly right than exactly wrong is crude but apt; similarly the economic principle that the marginal benefit should exceed the marginal cost for an activity to be worth pursuing is also pertinent to the research effort devoted to the construction of policy models.

(iv) *Model structure is important:* All models are abstractions and simplifications of the real world, otherwise they would not be useful. Major areas in which simplification and abstraction can cause difficulties are: omitted variables, mis-specified variables, inaccurate (often aggregate) data and simplified functional forms. Estimation or construction of transformation functions under these conditions can lead to both equally plausible but different functional relationships and also to parameters which vary over time. The result may well be two or, usually, many more equally credible and apparently reliable model structures which give rise to different policy consequences under the same initial conditions. An example of the consequences of different structural specifications is provided by Meilke and Zwart (1979).

Econometric methods for dealing with time-varying parameters without necessarily tackling the underlying problem (on the basis of principle (ii) above) are outlined in Rausser (1982). In addition, heavier weight on more recent data can help to overcome parameter variability while the frequent updating and re-estimation implied by this principle reinforces the notion that policy modelling is a process rather than simply the creation of answers. At a more general level, the principal difference between forecasting models and their policy counterparts is the absolute necessity for a formal structure in the latter. Good predictors are useless for policy analysis unless there is a direct and explicit linkage between policy control variables and their intended targets. Hence the importance of controllability within the model system (Bennett and Chorley, 1978, p. 73 *et seq.*).

(v) *Data availability and model structure are inter-related:* Where policies represent marginal changes on previous history and where there exist long runs of data econometric approaches are appropriate, but when policy changes are outside the range of historic data or substantially change the institutional or market mechanisms then estimation of reliable models based on historic data is not possible. The development of the Newcastle micromodel (Thomson and Buckwell, 1979) is an example of the latter. The mix of policy evaluations required often encourages the development of hybrid or mixed models, or at least links between different models, for example, between the CAS (1989) land use allocation model (LUAM) and the

Manchester model (MAPS) (Allanson, 1988) where the spatial distribution and detailed resource implications of national changes in policy are identified in the LUAM while aggregate effects of policy changes within historical experience are better modelled through the Manchester econometric model and used as constraints in the LUAM programming model.

(vi) *Partial analysis is of limited policy value:* this principle has led in the past to the development of extremely large and complex models of the agricultural sector (Agriculture Canada's FARM has over 500 equations) in order to capture the inter-relationships and feedbacks between commodities. However, not all of these relationships and variables are of policy significance. Hence, it is often possible to concentrate attention on the 'reduced form' of the model, in which all endogenous variables are expressed purely as functions of the exogenous or lagged endogenous variables, or even on the 'final form' in which even the lagged values of endogenous variables are eliminated, which can be considered as equivalent to the ARMA version of the model as outlined in Thomson and Rayner (1984). The danger of compressing the structure of the general structural model into the reduced and final forms is that information on the key structural parameters may be lost, so that the ability of the modeller to explain to the policy-maker why certain consequences follow from particular policy changes is limited to non-existent. However, this principle also leads to the emphasis on maintaining comprehensive and consistent data sets, as opposed to complex models, where the concept is of a general purpose database from which specific policy models can be developed quickly as the need arises. In many cases the accumulation and cleaning of data represent a major research resource cost which often results in the structure of the model becoming dependent on easily available data rather than the structure determining the data required.

(vii) *The incorporation of prior information and judgement is essential:* Indiscriminate use of data correlations without necessary theoretical structure or expert judgement about the transformation functions must be avoided. Policy models have to be credible to users, which implies either the incorporation of their major preconceptions about the behaviour of the world or perhaps more often their education as to the 'real' behaviour. As Thomson and Rayner (1974) phrase it, the incorporation of all relevant extraneous information is needed to avoid '*ex post* confrontation between beliefs and results'.

The FARM (Agriculture Canada, 1980) makes explicit use of commodity expert judgement in the preparation of forecasts, working in a formally defined manner with the results of the model itself, according to the principles developed by Rausser and others for the incorporation of such prior and expert information (see Johnson *et al.*, 1982). Just (1990), in a paper which in spite of the title deals with general principles, argues strongly that both theory and *a priori* information, including expert judgement and intuition, are necessary to impose sufficient structure on the model to obtain reliable and consistent estimates which are understandable, a prime requirement for a successful policy model. On the other hand, such an approach can amount to 'feeding the model with preconceived answers' in which case the models become little more than accounting devices (de Haen, 1979). Associated with the need to impose structure on the model is the imperative to acquaint the users with the rationale, logic and implications of this structure and obtain their acceptance, which is no easy task.

Rausser and Just conclude their list with a tenth principle (the eighth and ninth have been concatenated with the fourth and sixth respectively in this discussion)—that policies should generally be formulated with learning in mind, that is policy experiments should be carried out with the objective of generating data and as continuous variations on existing policy rather than as discrete jumps. This appears to be an extremely contentious argument leading to policy-makers serving the policy analyst rather than *vice versa* and is not one this author can recommend. However, from the perspective of policy development it can be observed that policies tend to evolve rather than revolve and that therefore the process of policy development itself should exhibit some systematic elements. If these elements can be identified, perhaps through the public choice approach, then these can and should be built into the analysis and evaluation of policy.

Application of these principles is honoured more in the breach than in the acceptance in practice. In view of the stringent requirements for successful policy modelling, it comes as little surprise that there is not more of it about. In fact, the surprise is that there is so much. Almost by definition 'good' policy models, which attract the attention and use of policy-makers, also attract those responsible for their construction into the policy debate and thus into day-to-day policy response activities. As this happens, so the time and resources necessary to continue with policy modelling and evaluation are reduced. On the basis of casual historical evidence, a major, often the only output of policy models is a group of well-trained policy analysts with a learnt experience of policy interactions and alternatives derived from the modelling exercise. This expertise is then either used to continue the education process through the development of different models or is attracted into the policy-making machinery directly. Governmental 'in-house' policy modelling efforts frequently dissolve into front-line policy response groups with rapidly dwindling research capital, while research contracts and grants with non-governmental institutions generate models of less or ephemeral applicability, highly dependent on their designers and constructors maintaining positive links with the policy process. Career movement, governmental change and other pressures and interests often conspire to break these links. In other words, the formal analysis of principles of policy modelling is not the only explanation for the apparent failure of many research efforts in this area; there are behavioural and social explanations too. The reader is left to choose or combine these explanations as (s)he sees fit in the light of personal experience and knowledge of particular policy model efforts.

AGRICULTURAL SECTOR MODELS: MAJOR APPROACHES

The following classification and discussion may lead one to suppose that the various approaches are competitive. However, it has frequently been pointed out that the several approaches are complementary (e.g. King, 1975). Some elementary thoughts on practical methods of integrating different approaches are given in the next section with reference to agricultural/environmental modelling.

Large-Scale Econometric Models of National Agricultural Systems (Type 6, Table 2)

In the 1970s this approach was extremely popular, especially in the US, but since then, with some notable exceptions, has become much less fashionable. Agriculture Canada (1980) provides a comparison in tabular form of four major econometric models of US agriculture being used on a regular basis for both forecasting and policy analysis purposes at the time of the development of the FARM model. Three of these were run by commercial consultancy houses, Wharton EFA, Data Resources Incorporated (DRI) and Chase Econometrics, and the fourth by the USDA. Since that time only FARM is still in positive existence and is being used regularly; the remaining models have been mothballed. In addition to the FARM model, two other current examples deserve mention under this heading: the CARD/FAPRI system (CARD, 1988) as probably the most well-developed and extensive example of this genre and the MAPS (Allanson, 1988) model of UK agriculture as a local though as yet preliminary and less ambitious example.

FARM was developed on the basis of, though separately from a variety of commodity-specific models previously used for both forecasting and policy analysis in Agriculture Canada (Agriculture Canada, 1978). FARM is primarily used as a forecasting device in support of Agriculture Canada's regular Outlook programme and has been designed and constructed specifically with this in mind. Market Commentaries are published on a quarterly basis for animals and crops and annually for horticulture and farm inputs. The proceedings of the annual Canadian Agricultural Outlook Conference are also published annually. The incorporation of detailed and specific structure necessary for serious policy analysis has been accorded secondary status while the development of formal mechanisms and techniques for regular and consistent updating, incorporation of expert opinion and revising model structure to improve forecast performance have all been given high priority. Two of the leading North American experts in model design and econometric techniques as applied to agriculture, S. R. Johnson and G. C. Rausser, were closely involved in the development of this model and their influence is demonstrated in the operational version. Agriculture Canada accords a high priority to the maintenance and regular use of the model and the necessary resources are provided and protected. An early report on the implementation of this model can be found in Johnson *et al.* (1982).

The model is treated as a piece of operational capital by the department and is now in 'production' mode rather than being used as a major research tool. Hence, the outputs are a set of regular forecasts embedded in the Outlook publications of Agriculture Canada and a large and well-maintained database, rather than a stream of journal articles. The ambition to produce a compact version which could be used more frequently and which might be tailored to suit policy analysis has not yet to this author's knowledge been completed. An obvious potential policy use of such a model would be the elaboration and evaluation of various possible outcomes of the GATT negotiations on Canadian agriculture. However, it has proved necessary to construct a smaller model for the purpose of trade policy evaluation for Canada. Thus this highly sophisticated and successful model does not readily serve as a policy tool.

The CARD/FAPRI model, on the other hand, has been developed as both a forecasting and a policy analysis tool. Thus results of baseline and policy analyses are routinely supplied to the US Senate and House agriculture committees, while congressional briefing sessions on policy issues are also provided on the basis of the model system. Biennial Agribusiness and Outlook Conferences are held at which FAPRI baseline projections are presented (CARD, 1988). Meyers *et al.* (1987), give details of the results of a FAPRI-based analysis of several policy alternatives under the US Food Security Act, 1985.

The model is actually a system of related sub-models: domestic crops (annual, 160 endogenous variables, 200 exogenous variables); domestic livestock, currently being extended to include dairy and eggs, (quarterly, 38 endogenous variables, 57 exogenous variables); trade, explicitly modelling major importing and exporting countries (22 for wheat, 14 for feed grains and 12 for the soybean/oilseed complex); and a set of country modelling systems, which provide more detailed multicommodity models for individual countries (Zambia, Jamaica, Indonesia and Argentina) developed for specific projects in these countries.

The CARD/FAPRI system is clearly being developed as a multi-purpose system supporting applied research and policy analysis as well as a forecasting role. There is little doubt that the research capital represented by both the model system and the research teams supporting their development represents one of the most sophisticated and extensive institutions on the planet. This view is reinforced by the parallel development of a large programming model of US agriculture in the same centre (see below). It is impossible for an outsider to assess the value of this modelling system either in the space or the time available in this paper, but, applying the acid test of use, both in terms of publications and provision of results to the policy-making process the system must be judged a success. However, comments by Meyers *et al.* (1987), Abbott (1987) and Clayton (1987) reveal less than complete satisfaction with the policy evaluation capabilities of the FAPRI system. The major criticisms are that the system does not deal with a key policy issue, namely the distribution of income and of policy benefits among farmers, and that the structure of the model represents a particular view of the way in which the world works, which 'may be known to some (but) is simply a black box to others'. The first is valid, but rather unfair; it is not possible for a single model to deal with all issues and it is possible to develop a distributional model driven by the aggregate models results, though this takes both time and resources. The second, however, is more telling. It is a point which will be returned to below.

The MAPS model (260 equations) is rather different from either of the previous two models in that it is much less ambitious and has only just been delivered to the sponsors, the UK Ministry of Agriculture, Fisheries and Food (MAFF) and hence has not yet seen wide application. The terms of reference for the model development included the requirements that it be directly related to the annually published national accounts for the UK agricultural sector (the Departmental Net Income Calculation, DNIC) reported in the Annual Review of Agriculture white papers and that it should be suitable for policy simulations. The first requirement means that the input and resource use components of the agricultural system were to be explicitly represented in the model, unlike the FAPRI system. It also presented some difficult data problems. DNIC data is on a calendar year basis, while agriculture does not work like this and many data relating to specific sectors are not so reported. On the other hand, since the focus is on output and input use, the demand and trade sides of the market system can be treated as exogenous, entering through the prices of inputs and outputs, which simplifies the model structure considerably compared with the previous models. Demand and trade modules could, in principle, be added at some later date, though there are no present plans to extend the system in these directions. Limited structural restrictions are employed in the estimation of the model, with heavy reliance being placed on the accuracy of the estimated equations. The general modelling approach is developed from Colman (1983).

Allanson (1988) reports the results of three dynamic simulations with the present model: an *ex post* forecast for the period February 1972 to February 1987 using actual values of all exogenous variables and predicted values of lagged endogenous variables; an *ex ante* forecast of the period 1988–95 to explore the extrapolative robustness of the model; a policy exploration of UK green rate of exchange policy between February 1979 and February 1987. The *ex post* forecast provides evidence on the interpolative robustness of the model. On the whole the model performs reasonably well, though error statistics are not provided so that a brief general assessment is not possible. The *ex ante* forecast makes dismal reading for the agricultural sector, with gross output, gross inputs and gross product all falling in real terms, while farm incomes (the residual in the DNIC and the model) fall by nearly one-third. Of course, these projections depend not only on the structure of the model but also on the values chosen for the exogenous variables (especially that product prices rise by 3 per cent per annum while input and labour prices rise by 6 per cent per annum and interest rates are set at 14 per cent throughout the period). The policy simulation suggests that agrimonetary policy over the period both improved and stabilized farm incomes compared with the base policy of zero monetary compensatory amounts (MCAs) throughout the period.

Programming Models of National Sectors (Type 7)

Norton and Schiefer (1980), in a major review article, identify four major deficiencies of econometric models which have led to the development of programming (usually linear) models of the agricultural sector: (i) estimation of inter-relationships (cross elasticities) between outputs and inputs is difficult because of lack of degrees of freedom; (ii) relationships estimated on historical data may not be valid if future or possible conditions to be examined fall outside the historical data range; (iii) inequality constraints, especially associated with available resources, primarily land, but also labour and capital, cannot be easily incorporated in econometric estimation; (iv) differences in regional and farm size responses cannot or are not usually identified by econometric methods but are often important for policy purposes.

Thus the advantages of the programming approach include the ability to include detailed description of technical relationships between inputs and outputs, differentiation of the production sector by farm business type, size and location, the direct imposition of capacity constraints and the reflection of their relative importance through the determination of shadow prices associated with their relaxation.

Disadvantages include the rigorous application of optimization behaviour, discontinuous response surfaces and associated corner or extreme solutions, the lack of formalised calibration or validation procedures and the associated lack of classical statistical confidence parameters, and the restrictive form in which input–output relationships can be expressed (Bauer, 1989*b*). Considerable effort has been made to overcome these difficulties and also to overcome the problems of aggregation and the lack of dynamics (Hazell and Norton, 1986). For example, modern models often include the incorporation of additional structure and constraints which reduce the scope of the optimization procedure, including integer activities and bounds; the introduction of risk through either quadratic programming or its linear equivalent; recursive techniques; multi-goal objective functions (Rehman and Romero, 1984; Romero and Rehman, 1985).

However, extension of the basic modelling approach to include these refinements often results in extremely large models and often, especially in the case of recursive systems and heavily constrained models, the introduction of arbitrary brakes on the model behaviour. Thus, the CARD programming model system, Agricultural Resource and Inter-regional Modelling System (ARIMS) (CARD, 1988), which is specified as a cost minimizing model, incorporates 105 production regions, 31 marketing regions, 34 ecological regions and ends up with between 120 000 and 140 000 production and marketing activities and 6000–8000 constraints. This size rather than innate complexity highlights another problem with such models, namely that the feasible solution surface of the model is restricted to include only a sub-set of available activities determined by the number of separate constraints imposed on the system. This problem also afflicts another class of programming models which have been used to model spatial market equilibria and which are also potentially valuable in examining temporal and form (processing distribution and retailing) equilibria–quadratic programming (QP) models. Hence in QP trade models, trade flows are restricted to be one less than the total number of importing and exporting regions (countries) in the model. ‘Any trade flows in excess of (the optimal solution) represent departures from the global welfare maximising spatial allocation of resources (given the representation of competition and constraints within the model), and their existence should lead the analyst to seek alternative explanations for why they should exist’ (Thompson, 1981). Thus, in the CARD system, many of the activities are likely to be effectively redundant as far as any useful solutions are concerned. CARD (1988) notes that it has been necessary to construct individual and more compact state models ‘for shorter term policy analysis than is appropriate for ARIMS’.

However, Hertel *et al.* (1989) and Preckel and Hertel (1988) outline a procedure for obtaining continuous differentiable approximations of the basic model's solution surface through econometric estimation based on a range of simulation results from the programming model. This is a promising approach for very large-scale models, though the dangers of concealing the reasons and logic of certain responses in yet another 'black box' should not be overlooked. In spite of the detail included in the ARIMS model, Hertel *et al.* (1989) note that there appear to be very limited possibilities for input substitution, resulting in implausible responses to input price changes. This is a serious criticism for those interested in the consequences of market or policy induced changes in these prices to reflect either increasing scarcity of fossil fuels or concerns about their environmental impacts. Furthermore, production and input use adjustment tends to take place between regions as a consequence of limited alternative activities within regions. Again, this is a serious drawback for policy purposes since geographical shifts may be taken more seriously in the policy process than is justified, while the potential for farm and local adjustment of production practice and input mix have important policy implications which can be overlooked.

While the ARIMS illustration of this type is a sector model, in which the production activities are divorced from their farm base and the sector is treated as a single firm, albeit behaving as a competitive aggregate, an alternative specification involves modelling representative farm types explicitly as farms and then aggregating these within the body of the model through the farm use of inputs and resources, and production of output. A UK example is reported in Thomson and Buckwell (1979) while Hanf (1989) also discusses this type of model from a methodological perspective. The aggregation problems associated with this type are especially severe though not intractable (Norton and Schiefer, 1980), as are the problems of reflecting farm size adjustment and structural change in the sector. However, it does avoid the difficulties of identifying sufficient mutually exclusive and genuinely competitive alternate production activities to represent the heterogeneity of the farm population, which plague sector programming models as illustrated in the comments above.

De Haen (1979) is doubtful of the usefulness of programming models as policy tools. He argues that, because of lack of empirical content, use is restricted to (a) input–output systems for checking data consistency and disaggregating national data on the basis of existing production relationships; (b) testing sensitivity for short-term fluctuations in yields or prices; (c) comparative analysis of equilibria under alternative exogenous conditions to determine new policy goals and targets; (d) determination of shadow prices under alternative allocation patterns. However, Norton and Scheifer (1980) conclude that 'imperfect and abstract though they may be, models sometimes provide the policy makers' best—and most consistent—picture of reality'. They go on to argue for the development of efficient multilevel algorithms which would permit the simultaneous or recursive solution of descriptive models with policy optimisation counterparts (normative models) allowing for the incorporation of intersectoral linkages and the endogenisation of price determination in more realistic and fruitful ways (see, for example, Schatzer and Heady, 1982). They also conclude that the problems of adequate reflection of land tenure systems, factor markets (land, labour, management and capital) and integrated farm household decisions are the most important unsolved problems with this class of model.

Multinational Multicommodity Models (Types 15 and 17)

The development of this type of model has been fostered largely by the recognition that the 'small country assumption' that domestic trade flows do not affect world prices is an inadequate reflection of reality, especially for the EC and US. Given the massive research resources required to estimate independently reliable and realistic behavioural models for the world and also the abundant, though often disparate estimates of the components of the world supply and demand systems available in the literature, the technique has been to borrow these estimates, typically as matrices of price elasticities, and construct a simulation model which is a quantitative reflection of the theory of market behaviour (McCalla and Josling, 1985). Under the impetus of increasing international political awareness of the importance of multinational price and trade relationships, demonstrated by the OECD trade mandate and the Punta del Este declaration at the beginning of the Uruguay round of GATT negotiations, several models of this type have now been developed. Among the most important are the OECD Trade Mandate Model (TMM) (OECD, 1987; Harley, 1989), the USDA's Static World Policy Simulation Model (SWOPSIM) (Roningen, 1986) and Tyers and Anderson (1988), whose model (TA) was used to provide the analysis of agricultural world trade problems in the World Bank Development Report in 1986. An early version of this approach was adopted by

Buckwell *et al.* (1982) in the construction of a model of the Common Agricultural Policy, discussed in Thomson (1987). A useful review of the economic consequences of agricultural support, with which these models are primarily concerned, is provided by Winters (1987) in which the results of these and other models are compared.

There are three major problems with this approach. First, the results are absolutely dependent on the elasticities used to reflect producers' and consumers' responses to price changes and on the levels of consumption, production and prices used to calibrate the base run of the model. Ideally, experimentally designed systematic sensitivity analyses should be conducted with these models around the degrees of uncertainty, possibly subjective, associated with the estimates and values used. This has not been done, at least in published form, for any of the models cited although, as Thomson notes, a limited amount of such analysis on the elasticities of the Newcastle CAP model appeared to show that changes in these do not substantially affect the contrast between the rather large producer, consumer and taxpayer transfers resulting from the policy, though they do affect the net of these flows as the measure of net economic welfare. Changes in the base levels used to calibrate the model had much more important effects. The TA model employs a stochastic component which reflects the production variability through probability distributions, which may avoid some of these problems. The SWOPSIM model attempts to overcome the difficulties of ensuring that the synthetic elasticities are reasonable by imposing the theoretical symmetry and homogeneity restrictions on the demand matrices and the analogous restrictions developed from multi-output production theory on the supply matrices (Roningen and Dixit, 1989). Similar though more *ad hoc* procedures were employed in the CAP model. This, at least, ensures some theoretical consistency though does not answer questions about their real world accuracy.

The TA model is rather more sophisticated in that the elasticities are replaced with synthetic supply and demand functions representing the reduced forms of a set of structural equations for both sides of the market. These functions include partial adjustment terms which distinguish between the short and long run, which enables the model to represent some elementary dynamic properties. In addition, the relationship between livestock production and feed grain use is modelled more explicitly than simply through cross-price elasticities by including input–output coefficients, while the relationships between domestic and world prices are incorporated through directly estimated price transmission equations, which thus incorporate the protection and stabilisation effects of domestic and trade policies. The price of this added sophistication is increased complexity and a less transparent structure to the model, which makes interpretation more difficult.

Second, the representation of policies and policy changes is generally restricted to price wedges between supply and demand schedules as represented by constant elasticities, while policies in practice are much more complex than this and have rather different effects on trade flows and market prices depending on their mix (de Gorter and Harvey, 1990). Some effort has been made with the CAP model to incorporate more sophisticated representations of policies, especially the dairy quota system of the European Community (Thomson, 1987), but the SWOPSIM model is subject to the criticism that price wedges derived from estimates of Producer Subsidy Equivalents (as aggregate measures of agricultural protection and support) are inexact and misleading reflections of the complexity of support measures, especially in the United States for the grains sector (Roningen and Dixit, 1989). Similarly, the TA model condenses complex policy mixes into variations on the price transmission equations while liberalization of domestic policies is reflected through changes in nominal rates of protection (domestic prices relative to world prices).

Third, both the SWOPSIM and CAP models are truly comparative static, i.e. they compare one equilibrium solution with another and say nothing about the time paths of possible changes, which are often extremely important to policy-makers. The TA model, on the other hand, does incorporate a considerable amount of elementary dynamics, requiring the projection of changes in both exogenous variables (incomes, populations, productivity growth rates, exchange rates) and parameter values (elasticities and policy responses). Given the stochastic element, which allows the stability of alternative scenarios to be examined, and the dynamic properties, the TA model appears to provide a more comprehensive set of results of interest to the policy-maker. However, once again the resulting lack of transparency of cause and effect makes interpretation more difficult.

The power of this type of model derives from its essential simplicity. The influence, however, depends on the credibility of the results for which there are no established formal tests. At one extreme the results are simply a quantitative expression of the theoretical arguments in favour of free trade and are no more or less convincing than the underlying assumptions. At the other extreme they provide realistic assessments, given the current state of knowledge and the large costs of obtaining better information and understanding, of the effects of real policy decisions. Between these two extremes these models do provide insights into the interactions of even relatively simple constructs of the ways in which policies and markets interact. These insights can be more easily though less rigorously obtained through numerical analysis and quantitative experimentation rather than formal mathematical analysis. In the more realistic cases (McCalla and Josling, 1985) the latter may not even be possible. As a final comment, this type of model has apparently been rather more productive in at least generating policy debate if not influencing policy decisions than any of the other types discussed here. In that sense they have already earned their keep.

General Equilibrium Models (Types 9 and 18)

The history of this approach begins with the conception of the whole economy as a system encapsulated in an input–output table (Leontief, 1941) in which each sector or industry in the economy is characterised as producing a certain known final output using a collection of intermediate inputs derived from other sectors and flows of services from the stock of natural and man-made resources (including labour) in the economy. The characterisation of the economy is restrictive in its assumptions about the technological relationships between inputs and outputs as fixed coefficient processes and about fixed (exogenous) final demands, factor or resource supplies and prices. However, it has proved a productive base model of the economy. Agricultural applications generally need to decompose the usually highly aggregated agricultural sector into its component parts and often to regionally disaggregate the national matrix, involving an inclusion in the matrix of regional trade flows.

Keyzer (1989) briefly outlines two rather distinct developments of this basic framework into general equilibrium models, which seek to capture the theoretical construct of general equilibrium indicated above. Computable General Equilibrium (CGE) models replace the fixed technical coefficients with price-dependent cost functions and feedback relationships while the model is driven by final demand generated through utility maximising consumer behaviour in which utility is a function of final demand quantities, which are effectively maximised subject to an income constraint and which involve the endogenous determination of product prices since expenditure depends on both prices and quantities. Incomes in turn are generated through the returns to the factors of production assumed to be owned by the consumers. Both trade and governments are typically included as additional sectors of the economy.

Activity Analysis General Equilibrium (AAGE) models formulate the economy as a non-linear programme with weighted sums of consumers' utilities as the objective function and production functions and commodity balances as constraints. The objective function weights are then adjusted through a '*tatonnement*—feedback' system of relations so that income (budget) constraints are satisfied. Shadow prices on the commodity balances then represent the market-clearing prices associated with the general equilibrium solution. The non-linear system can then be linearised and solved as a linear programming problem.

The most ambitious application of this type of model to date has been undertaken as an international cooperative effort through the International Institute of Applied Systems Analysis (IIASA) under the Food and Agriculture Programme (FAP). Keyzer (1989) explains that the approach adopted in this model system is a hybrid between the CGE and AAGE approaches, replacing the input–output block of a CGE model with a net revenue-maximising LP which allows the incorporation of trade and transportation charges, capacity constraints (as opposed to equalities in the I/O block) and priority orderings on policy instruments such as trade quotas, stock holding and guarantee schemes. This basic framework was applied to detailed models for 12 participating countries, including the European Community as a single block, the US and Canada. An early exposition of this model system is provided by Parikh and Rabar (1981) while a more recent account can be found in Keyser (1986) and Parikh *et al.* (1988). Since the end of the FAP under IIASA, the major researchers involved in this work have dispersed, so that the model framework now exists as a network with CARD as the North American hub and the Centre for World Food Studies, Amsterdam, as the European hub. A working version of the Basic Linked System (BLS) is maintained at each location, where the BLS consists of a set of standardised national models together comprising a model of world agriculture.

Several other general equilibrium models with agriculture as a major component are reviewed by Winters (1987) with particular emphasis on the results of these models on the effects of liberalizing agricultural trade through the reduction or elimination of domestic support policies. In this context, many of the existing models, with the possible exception of the IIASA model system, have a very limited reflection of the complexity of agricultural policies and are faced with the same problem as those under the previous heading, namely reflecting complex policy instruments as price wedges. Furthermore, not many people can do general equilibrium analysis in their heads, so that interpretation of the results, especially those which appear counterintuitive, is difficult. Most commentators have to treat the models very much as black boxes, since almost by definition it is not possible to decompose the solutions into the consequences of separate blocks, components or particular structural forms used in the models (see, for example, Tweeten (1988)).

It is salutary to consider what potential advantages the complexity and scope of general equilibrium models, often associated with a lack of specific detail on agricultural sectors, offers for agricultural policy analysis. No doubt, given careful formulation and solution to the several difficult data and estimation problems, such an approach is theoretically and academically respectable. No doubt, too, the potential influence of macroeconomic variables such as interest rates, wage rates, exchange rates, employment and income levels and inflation all have a significant influence on the agricultural sector. In turn, the level of agricultural activity may have noticeable effects on the rest of the economy, most obviously through agricultural trade balances, government support spending and food prices. However, whether it is sensible to model these interactions within the same overall system is another matter. In particular, what is the rationale for using a modelling approach which is, generally, quite different from the approach used to model the macro economy? Well-maintained, frequently updated and generally available models of the macro economy exist (see Wallis (1987) for an assessment of the current UK models). In general, this author would find an analysis which links these models with a separate agricultural sector model both more convincing and easier to comprehend. While the iterations necessary to achieve a stable outcome might be seen as a drawback, the strength of the interactive linkages, at least at the national level, is not likely to be so strong as to make this a serious problem in practice, particularly given the confidence limits around the outputs of each model.

These comments refer to developed countries. The situation for developing countries is rather different, as de Haen (1979) notes. The greater the share of agriculture in total resources, the less realistic the assumptions of zero or infinite elasticities of intersectoral supply and demand and the more important is the construction of a full national model, especially since intersectoral resource flows and foreign trade can hardly be analysed adequately with an isolated agricultural sector model. Although the results of such models are intended to capture all of the knock-on effects throughout the economy, and to incorporate all the feedbacks to the agricultural sector, the assumptions and preconceptions built into the structure of such models means that they are more subject to criticism of being quantifications of theory calibrated to fit current circumstances than most other sorts of model, with an empirical content which is often out of balance with their scope. While the multiplier effects of changes in agricultural activity can be important measures of the overall cost and benefit of agricultural policies, estimates of these effects can be achieved through the application of agricultural sector model results to multipliers derived from existing well-tested macroeconomic models with at least as much transparency and confidence as can be obtained through the construction of an elegant and sophisticated general equilibrium model.

LESSONS FOR MODELLING THE AGRICULTURE–ENVIRONMENT COMPLEX

The problems associated with trying to model agricultural–environmental interactions in order to evaluate policy alternatives raise all of the problems encountered in other policy analyses, with the added complications that the inputs, transformation functions and outputs are often even less well-defined than for agricultural sector modelling work. In addition to the traditional requirements for agricultural sector policy analysis, environmental considerations also require that the effects of production and land use decisions on the natural environment be included in the model system. Given that the social environment is also of public (and therefore policy) interest, the interactions between agriculture and the rural economy and the possible separate development of the latter also need to be considered.

This section briefly outlines some approaches to this set of problems currently being developed under several different projects with which the author is closely connected. Dubgaard and Nielsen (1989) provide a collection of papers on environmental aspects of agricultural production, which includes some reports of other modelling activities elsewhere in Western Europe. Haxsen (1989) describes a programming-based model for Germany developed to examine the possibilities and consequences of decreased production intensity but without a spatial dimension and simply concentrating on the overall levels of food production versus dietary needs. Dosi and Stellin (1989) outline a programming-based spatial model of a specific region in Italy designed to illuminate the problems of nitrate pollution associated with fertilizer and manuring practices. Drake (1989) reports on the current development of a non-linear sector model of Swedish agriculture, based on regional profit and production functions separately estimated for four different regions and eight commodity groups, which is intended to give special emphasis to land use and environmental effects. Elsewhere, Bauer (1989*b*) reports on a model system developed for forecasting agricultural sector change and analysing policy measures for West Germany involving the integration of programming and econometric approaches, the Dynamic Analysis and Prognosis System (DAPS), though this system has not yet incorporated any substantial environmental component. Tamminga (1989) describes a linear programming approach to modelling Dutch agriculture and the environment. This model lays greater emphasis on the direct representation of interactions between crops and livestock and less on the regional differentiation of production activities than the LUAM described in Harvey and Rehman (1989) and CAS (1989) but otherwise is closer to the approach being adopted by the present author than work elsewhere.

LUAM is to be used to analyse changes in agricultural practices, and the environmental and ecological consequences associated with them, which are likely to result from changes in agricultural policies or market conditions. In addition, the model is also capable of reflecting the agricultural consequences (including the implications for policy and market incentives) of particular specifications of environmental and ecological characteristics which are required for particular areas or types of land. The basic focus of the model structure is the agricultural system in England and Wales, as reflected in agricultural land use. The model is a linear programme, with agriculture being modelled as a set of production activities, differentiated by intensity of input use per hectare and by land class as defined in the Institute of Terrestrial Ecology's Land Classification System (ITE LCS) which is now able to classify every kilometre grid square in the country to one of 32 different land classes, reflecting the geological, geographical and climatological heterogeneity of the land base, related to its productive potential.

Additional activities are included to reflect forestry possibilities and also to reflect leisure and recreational activities which could use resources (land, labour and capital) presently used by agriculture. Agricultural production activities are estimated from the annual MAFF Farm Business Survey (FBS), while total areas of land are determined from the annual June Census of agriculture. National levels of input use and output are defined according to the MAFF Departmental Net Income Calculation (DNIC). The model is driven through an objective function in which net margins (receipts minus input, capital and labour costs) are maximized. This is taken as a reflection of the behaviour of a competitive market system competing for use of land. However, possible extensions could involve the substitution of different formulations to reflect different perceptions of the behaviour of the system.

The major aims of this model are: (i) to reflect the spatial pattern of existing land use and management practices in agriculture according to a national land classification system which reflects the potential of the land base, current farming practice as reflected in the annual Farm Business Survey and in the national statistics contained in the annual agricultural census and the national agricultural accounts of MAFF; (ii) to project consequences of changes in this pattern either through specific policies designed to exclude particular areas or types of land from commercial agriculture through regulation or through changes in market or commodity policy conditions affecting land use practices throughout the country albeit in different ways in different regions (land classes); (iii) utilising the ecological detail embodied in the ITE land classification system to identify the major environmental effects of such changes at a regional and local level.

Several features of this particular approach reflect problems identified in earlier studies in the previous section. The specification of the model at the industry rather than the farm level is intended to avoid the problems associated with aggregation of individual farm businesses and of modelling structural change as farms amalgamate and grow, while the explicit incorporation of farm business survey data as the basis for the production relationships provides the basic model structure with substantial empirical content. The use of national aggregate statistics to calibrate and constrain the model also allows econometric estimates of national responses to be used as constraints on the model since these can be independently simulated through the MAP model (Allanson, 1988), rather than determined within the LUAM. This avoids some of the problems of the discontinuous nature of LP solutions and the associated extreme allocations which can occur. Labour and capital availabilities to the system can also be restricted by type (e.g. owners' vs. hired or debt) either on a land class or national basis, with additional resources only available at higher costs to the agricultural sector, which again can be used to restrict the solution space of the LP and further differentiate the optimal enterprise mix. Reflection of different production systems for the same enterprise allows some reflection of observed heterogeneity within the agricultural system, at least by land class and also by level of intensity. However, at the national level there is clearly a limit to the extent to which the model can adequately reflect local variations.

There are a number of obvious shortcomings to the national programming approach which require separate modelling efforts to overcome. However, this approach is best seen as part of an integrated set of approaches and models which promise to provide a more comprehensive picture of the real system in time. At the aggregate (national) level, a programming approach to modelling rural land use allows for the eventual integration of demand-side models, both for raw materials and foodstuffs as well as environmental goods and services, depending on a related programme to elicit values for non-marketed aspects of the environment (Lowe and Harvey, 1989). Regional input–output models of the national economy can be associated with the input use and output results of the LUAM to explore the general economic interactions between land-using activities and other aspects of the rural economy, which eventually could be linked to national macro models.

Given the limits to regional and local detail which can be incorporated into a national model, there is considerable scope for the development of related models of land use and consequences at the regional and local level, which is currently being undertaken by the Natural Environment Research Council/Economic and Social Research Council (NERC/ESRC) Land Use Programme (NELUP) at Newcastle, outlined in Fig. 1, which is to focus on a single river basin, the Tyne, in the first instance.

Similarly, farm/field level interactions cannot be modelled at the

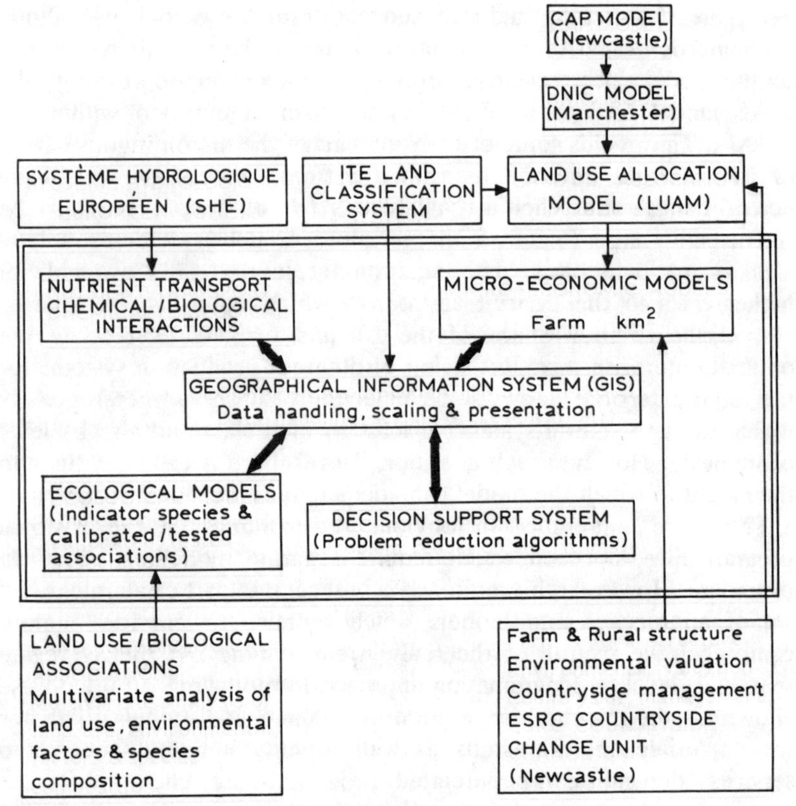

Fig. 1. General representation of NERC/ESRC land use programme (NELUP).

national level but the choice of appropriate farm characteristics cannot be determined without a knowledge of the regional pictures. Thus, following the development of aggregate models it is intended to develop illustrative farm-level models including detailed spatial relationships between productive land use and farm/field boundaries which, in lowland agriculture at least, contain the bulk of ecological interest. This research forms the major focus of a project under the Joint Agriculture and Environment Programme (JAEP), sponsored by the Agricultural and Food Research Council (AFRC), ESRC and NERC, which is about to begin in conjunction with NELUP at Newcastle.

Although programming has a number of advantages at the national level, especially the formal reconciliation of national data with regional counterparts and the direct imposition of strict accounting identities and market-clearing relationships, these advantages are not so relevant at the local level. As a result, the economic components of local and farm level models are likely to use different methodologies. In particular, the determination of land-use patterns and management practices will be modelled through time series analysis of past patterns and trends under NELUP in the first instance, with economic explanations being applied to the resulting time series results. At the farm level the programming approach is a powerful one though it is also restrictive in both the linear structure and the primacy of the objective function, and there are alternatives such as Monte Carlo techniques applied to decision tree models which may prove more useful in mapping the feasible solution surface and exploring the consequences of different goal and constraint sets of owners and operators.

NELUP is conceived as a multidisciplinary programme designed to produce a decision support framework for policy purposes, in which the major hydrological–ecological–economic interactions can be identified and supported with more detailed explanations of each component of the overall system. While it builds on existing research expertise, it is not dependent on the success or otherwise of the related research since the objective is to identify the critical relationships and provide the framework within which information and understanding of these relationships can be integrated.

Returning to the economic modelling aspects of these programmes, the present approach, as illustrated by the LUAM, is subject to five major criticisms.

(i) *Farmer and Landowner Response:* The present approach is heavily dependent on the behavioural responses of farmers and landowners which are characterised as economic maximisation subject to constraints. In practice, the constraints can be formulated to include a variety of non-economic objectives, which restricts the force of this limitation. In addition, the use of the FBS data to define the input–output relationships on a land class basis does allow for different behavioural responses by land class (possibly associated with different agricultural structures, e.g. age of farmers, size of farms, tenure arrangements, family characteristics, extent of off-farm income opportunities). Furthermore, the estimation of these relationships provides evidence of the range of behavioural relationships exemplified in the FBS data. Although there may be some question as to how representative these data are of the real world, they currently form the only repeated and consistent sample of farm behaviour and operational practices. In effect, the current model assumes either that farm structure is unchanging or that any changes in structure will not affect the behavioural response of the farm population. Neither of these assumptions is warranted and the ESRC Countryside Change Programme at Newcastle University and University College, London is examining this aspect (Lowe and Harvey, 1989). In particular, this work will further explore the FBS data, with a view to identifying relationships between the structure of agriculture and the behavioural responses, and will incorporate additional survey work both existing and new.

Sceptics might quarrel with the assumption that landowners and farmers maximise economic returns, arguing that other objectives are also important and may override the economic objective. However, there are two major reasons why this criticism is not so important for the current model. First, the input–output relationships incorporated in the model reflect profit maximisation as it is actually practised at the farm level, at least within the FBS sample, and therefore implicitly include the effects of other objectives on this behaviour modifying the extent to which farmers actually adjust inputs and outputs. Second, the capacity within the model to include non-economic constraints improves the ability of the model to reflect other objectives. It might be added that there is nothing sacrosanct about the current form of the model in using net economic returns as the objective function. It is perfectly possible to alter the objective function to some other criterion or even to extend the model framework to incorporate multi-objectives.

(ii) *Environmental relationships.* In the current version of the model, these are represented on the basis of existing relationships between environmental characteristics and land use and use-practice patterns. There is no guarantee that these relationships will hold as land use and practices are changed. Some of the more obvious discontinuities are already incorporated, such as the impossibility of generating traditional hay meadows overnight simply by reducing intensity. However, the detail of what habitats and species might appear or disappear as intensities and uses are changed from their current levels have not yet been modelled. The art of modelling environmental change in this manner is still in its infancy and it is to be expected that considerable progress can be made in this area under NELUP.

(iii) *Scale of analysis*. The focus of LUAM is the land class. Specific detail on relationships within land classes are only available on a probabilistic basis. Micro information requires more detailed base information and more detailed modelling of relationships. However, since there is already considerable detail available for the sample squares within the ITE LCS, it is possible to develop the micro analogue of the current model to increase the resolution of the model framework.

(iv) *Policy/behaviour interactions*. The state of the rural environment is the result of a complex interaction between people, technology, markets and politics. There are many feedback loops and interactions between these elements such that changes in one will affect the behaviour, response and practice of others. None of this is captured in the present model. Although regular updating of the current model will reflect these changes, they are not incorporated in the model framework so that they are ignored in the projections of the consequences of change. This aspect of modelling socio-economic systems will be returned to in the next section.

(v) *Dynamics*. The current model is 'comparative static'—that is it represents snap shots of the rural environmental system at a particular point, starting with the 1984 situation, and identifies likely changes resulting from a particular set of changing circumstances as a new snapshot of the situation once the effects of these changes have worked through the system. There is a degree of ambiguity about the time scale required for these changes to occur. In part this is resolved in the construction of the model through restricting the amount and direction of change that can occur to reflect short- or long-run scenarios; in part the resolution occurs in the interpretation of the model results. In this sense, the dynamic elements of the system are taken into account in an informal way. The techniques for modelling systems dynamically and the information and understanding of dynamic processes involved which would be necessary for such models are not currently available, although it is possible to make some progress in this direction along the lines of a succession of snapshots replicating the film of dynamic change. However, the critical element of such analysis is the relationships between particular snapshots and their successors, which is of course the dynamic process itself. Without clear and robust specification of these processes, dynamic socio-economic modelling will remain an art rather than a science, however formally or informally modelled.

CONFLICTS, ISSUES AND POSSIBLE DEVELOPMENTS

To conclude this overview of agricultural sector modelling, this section highlights two major conflicts and two outstanding issues arising from current attempts to build models for policy purposes. The conflicts are real in the sense that there are no unambiguous rules or guidelines for their satisfactory resolution; each policy problem requires its own solution which cannot be guaranteed to be unique. Similarly with the issues, though here one might hope that the future development of socio-economic theory, systems theory and the related statistical and mathematical techniques will help. As will be seen, the division of the discussion into these headings is rather arbitrary.

Conflicts

(i) *Forecasting vs. policy analysis:* Policy analysis requires belief about the way in which the system works and hence a policy model requires an explicit structure which incorporates these beliefs. Forecasting, on the other hand, does not require such a formal statement of system structure, merely the establishment of reliable associative relationships, at least if concern is restricted to forecasting events within the range of historical experience. In principle, the separation of forecasting from policy analysis is straightforward; one needs different models for each purpose. However, there are two serious problems with this approach: first, the demonstration that the policy model is credible often requires a satisfactory forecasting performance from the model; second, the parallel use of forecasting and policy models of the same real system often throws up conflicts between the two groups. Hence, in the most well-developed field of model building in economics (macroeconomic models) both functions are required. The conflict therefore in developing a model to be used primarily for policy analysis is the extent to which policy transformation mechanisms have to be compromised in the interests of forecast accuracy. The latter can often be achieved through a black box approach (within which I would include time series analysis since the parables explaining why such a policy instrument has those particular effects are impossible to tell in such models) while the former require an explainable linking and causative structure. Once again, the conflict appears to require a mixture of approaches which can be separated from each other. For example, the Newcastle CAP model was constructed as a policy model but required some forecasting of trends in prices, supplies and demands in order to update the model on a regular basis and to examine the longer run consequences of policy changes. Although the development of an independent forecasting module remains primitive and can certainly be improved, the parallel system has considerable advantages over a more general model which attempts to fulfil both tasks simultaneously.

(ii) *Theoretical rigour and academic respectability versus policy credibility:* A prime example of the ascendancy of the former over the latter is the recent development of sophisticated general equilibrium models (see above). Nevertheless, the difficulties of constructing policy models within an academic or research environment which places heavy weight on the former should not be underestimated. The conflict gives rise to long-run dangers. Theoretical rigour is an absolute requirement for genuine research to improve understanding of the ways in which the world works. However, the understanding exhibited by the policy-makers and advisers is always several years, if not generations, behind the leading research edge. If policy models are driven entirely by the users, then policy research will also seriously lag behind research developments. On the other hand, for current research to be able to play a substantial part in policy analysis, there is a requirement for policy-makers to be regularly updated in their knowledge and understanding of the system. The practical difficulties of this leave open the serious possibility that advances in theoretical and applied understanding of the system will be imperfectly understood by policy-makers and even modellers, and that results of policy analyses will be chosen on the basis of correspondence with preconceived notions rather than on the basis of objective evaluation of the underlying model structure.

Abbott (1987) argues that 'simple models illustrating important relationships reveal more of use to the policy process than (counterintuitive) outcomes of black boxes. The policy debate is enriched when issues are explicitly addressed and models focus on differences in views'. The black boxes to which he refers are those which enclose large-scale and often complex models, in this case the FAPRI econometric model of US agriculture. While to the model builders these model systems are not black boxes, lack of familiarity with the structure and techniques employed can easily lead the potential user and the ill-tutored commentator to view them as such. Those whose careers are helped by, if not dependent on the development of realistic models can be tempted to include more detail and descriptive accuracy than is needed for the articulation of major policy linkages. Furthermore, single integrated models are less obvious and transparent to the potential user than similar systems which incorporate the major transformation mechanisms as separate but related models, which avoids the problem of requiring a similar methodology to all parts of the system. On the other hand, it can be both tedious and inefficient to operate such related systems rather than to incorporate them within a single model.

As de Haen (1979) points out, the interpretation and use of policy models depend critically on their credibility to policy-makers, which in turn depends on their heterogeneous policy/system perceptions. He identifies three principal factors which influence the failure of many policy models to prove useful to the policy process:

Lack of clarity in structure, empirical basis and policy relevance.
Uncertainty about empirical performance.
Inappropriate implementation of model in policy process.

'Quick and dirty' policy analysis often lacks empirical content and excludes relevant facts and relationships in the interests of the pressing need to evaluate policies in terms of their costs and benefits. As a broad and therefore dangerous generalization, the more credible the policy analysis, the more partial and simplified it is. Creating usable and useful policy models requires much more attention to explanation and implementation than is usually catered for in research budgets or operating procedures, activities which also carry few academic or research merit points.

Issues

(i) *Controllability, observability, the nature of the executive and noise:* Socio-economic policies are seldom so well structured as to present clearly defined instruments and objectives, especially in their related forms. Thus agricultural price support has a number of different objectives ranging from stabilising agricultural markets through guaranteeing supplies to consumers and supporting farm incomes. Furthermore, instruments of support vary from deficiency payments through import levies and export subsidies to production or marketing quotas. The controllability of these instruments varies considerably, while the observability of the targets is often open to several different interpretations. The situation is even less structured in the case of environmental objectives and the possible instruments which seek to influence these ill-specified objectives are even more varied and often local in character, such as designation of environmentally sensitive areas (ESA) and sites of special scientific interest (SSSI). The controllability of the instruments is incomplete while the observability of the targets is fraught with difficulty because of the problems of specification.

The nature of the 'executive' or controller of socio-economic systems is clearly bound up with human motivations. Economics simplifies these to the business motive of profit maximisation and the consumer motive of utility maximisation. Given these simplifications it is possible to construct a theoretical transformation function which translates variations in instruments to changes in targets given the behaviour of exogenous elements in the system. There are two important consequences of this state of affairs. First, the relationships between the instruments and targets is preconditioned by the structure of the model (the theory) which is not unique in the sense that there are competing theories which provide similar qualitative relationships between the inputs and outputs but which have different policy implications. The debate about the influence of money and monetary policy on the behaviour of the economy is a case in point. Similarly, the definition of the instruments and targets is also preconditioned by the theory employed. For instance, simple theories of international trade models treat most market interventions as wedges between producer and consumer prices regardless of their actual implementation while more sophisticated or complex theories argue that the details of intervention are important in determining the consequences. Under the conventional theory of market behaviour, farm incomes, as the incomes earned by farmers for their own labour and equity capital, are determined by the opportunities available to the owners for incomes elsewhere in the economy. Thus farm incomes, at the margin at least, are not influenced at all by agricultural market intervention. As a result, defining a policy target which is both consistent with policy-makers' objectives and which is also controllable, influences the nature of the executive which is needed for the model.

Second, the inevitable noise in the data with which these models are calibrated and estimated, as well as in the specification of the transformation functions, cause severe problems in separating the influence of policy instruments from exogenous or state variables. The ways in which this noise is dealt with determine to some extent the model relationships between the instruments and targets. Deterministic models, such as programming models, deal with the problem in a quite different way from stochastic econometrically based models.

The question of the executive or controller raises another issue in socio-economic modelling. The policy process itself is a part of the system being modelled, though it is frequently and traditionally regarded as exogenous. Rational expectations, whereby economic agents are presumed to employ 'optimal' forecasts of the future in making present decisions, subject in more sophisticated versions to the costs of information assembly and processing, imply that present and expected future policy changes will be taken into account in current decisions. Thus, for example, Chaveau and Gordon (1988) demonstrate that rational expectations of world prices would include expectations about policy stances of the major traders. Given this, their responses will be conditioned by both the type (quotas versus price fixing) and also by the sustainability of the policy, where obvious unsustainability either encourages speculation against such policies or insulation against the associated inevitable political crises. Under these conditions, models which seek to determine optimal policies become extremely complex, even given an appropriate definition of the executive function which itself poses severe problems. An approach to the determination of policy objectives through revealed preference estimation of policy weights (e.g. Rausser and Freebairn, 1974) shows that the estimates are subject to the serious problem that weights are not stable over time, and are themselves products of market conditions and policy circumstances.

Usually policy modellers avoid this problem by stopping short of specifying policy optimization models. However, there is a growing literature on the endogenous treatment of government behaviour. A useful, if now somewhat dated review of this work is provided by Rausser *et al.* (1982). Further developments of this line of enquiry result in the economics of public choice, in which policy-makers are seen as responding to pressures for new policies from interest groups and which pictures policy-making as the outcome of a political market place in which the currency is votes, as outlined above. Most agricultural applications to date attempt to explain government intervention in agriculture in general rather than identify specific policy developments (e.g. Olson, 1985; de Gorter and Tsur, 1989).

(ii) *Verification, validation and objectivity:* These issues raise questions of philosophy which are beyond the scope of this paper. The issue is treated in some considerable detail in an agricultural context by Johnson and Rausser (1977). However, some comments are necessary since the issue is one which bedevils policy models in particular, where the objective of such models is to provide information to policy-makers on which to base public decisions. The validity of the advice is therefore of public concern. The essential message is that policy models can be independently verified (i.e. the internal consistency and logic of the model is checked) but are difficult to validate independently from the body of theory and associated evidence which underlies their structure in general as either the only, or even the best explanation of the policy system, where validity refers to the conformity of the model's behaviour with the observed world. Although the performance of the model against previous observations outside the period used in estimating and calibrating the model can provide evidence for or against the explanatory power of the model, the validity of the particular transformation mechanisms employed by the model are the product of the paradigm within which the model was developed. Since it is these mechanisms which are crucial to the controllability of the resulting system in policy terms, this is a serious problem. However, given the costs associated with renouncing predictively useful theories in the absence of adequate replacements, the problem is one which can easily be blown out of proportion. In the final analysis, an existing policy model is useful relative to the alternatives rather than in absolute terms.

Related to the problem of validation is the extent to which a model can capture all the relevant detail in which policy-makers are interested. In a sense, there is a social science equivalent to the Heisenberg Uncertainty Principle: the more accurately one describes the existing system the less certain one can be about how the system will change through space or time. Any model of an open system is a compromise between accurate description and ability to project the consequences of changes in the system. More detailed analysis of the process of change will necessarily mean less detail about the current system, given current methodologies and understanding.

While there is a superficial resemblance between the uncertainty principle and the problems in social science of moving from a comparative static to a dynamic methodology, the underlying logic is quite different. Heisenberg's principle appears to be a universal truth, based on the means by which the position of a particle is determined by shining light at it. This measurement itself disturbs the particle and alters its velocity in an unpredictable fashion, hence distorting measurements of speed. More accurate measurement of position involves reducing the wavelength of light used, which increases the energy of a quantum of light striking the particle and thus increases the disturbance (see, for example, Hawking (1988) for a lay-person's guide to this principle and its consequences). While there are similarities in social science—since measurement, especially by surveys and questionnaires, and reporting of subsequent analysis can alter the knowledge and understanding of those observed and thus perhaps alter their responses—the argument here is that representation of dynamic processes is more difficult than comparative static representations and thus the simplifications have to be more severe and greater limitations have to be imposed on the causal factors which can be considered. This may not be an unalterable truth, since improvements in knowledge (data), understanding, analytical techniques and computational power may one day resolve this uncertainty. Nevertheless, the very act of modelling involves an essential process of simplification, abstraction and generalisation of the actual system under study. A completely accurate and fully comprehensive model of a system would be indistinguishable from the real counterpart and would serve little useful purpose unless as a test-bed against which to examine the consequences of alternative strategies.

However, the latter requires that the model be capable of tracing both cause and effect, i.e. exhibiting controllability and of completing cycles of interactions, in less real time than the real system to be of use. Given the complexity of socio-economic systems, these requirements condemn models to practically eternal simplification with all the dangers that entails.

REFERENCES

Abbott, P. C. (1987). Discussion. Alternative agricultural programs, *American Journal of Agricultural Economics,* **69,** 988–9.

Adelman, I. and Robinson, S. (1978). *Income Distribution Policy in Developing Countries.* Oxford: Oxford University Press.

Agriculture Canada (1978). *Commodity Forecasting Models for Canadian Agriculture, Vols 1 and 2.* Ottawa: Information Services, Agriculture Canada.

Agriculture Canada (1980). *FARM: Food and Agriculture Regional Model.* Ottawa: Information Services, Agriculture Canada.

Allanson, P. F. (1988). *The Manchester Policy Simulation Model of UK Agriculture.* Manchester: Department of Agricultural Economics, University of Manchester, Bulletin no. 217.

Allen, G. R. (1975). Systems analysis in relation to agricultural policy and marketing, pp. 415–30 in: *Study of Agricultural Systems,* ed. G. E. Dalton. London: Applied Science Publishers.

Asimov, I. (1951–68). *The Foundation Saga, Vols 1–6.* London: Grafton.

Bauer, S. (1989*a*). Historical review, experiences and perspectives in sector modelling, pp. 3–22 in: *Agricultural Sector Modelling,* eds S. Bauer and W. Henrichsmeyer. Kiel: Vauk.

Bauer, S. (1989*b*). Some lessons from the Dynamic Analysis and Prognosis System (DAPS), pp. 325–44 in: *Agricultural Sector Modelling,* eds S. Bauer and W. Henrichsmeyer. Kiel: Vauk.

Bauer, S. and Henrichsmeyer, W. (Editors) (1989). *Agricultural Sector Modelling.* Kiel: Vauk.

Bennett, R. J. and Chorley, R. J. (1978). *Environmental Systems: Philosophy, Analysis and Control.* London: Methuen.

Blaug, M. (1980). *The Methodology of Economics.* Cambridge Surveys of Economic Literature. Cambridge University Press.

Buchannan, J. M. and Tollison, R. D. (Editors) (1984). *The Theory of Public Choice—II.* Ann Arbor: University of Michigan Press.

Buckwell, A. E., Harvey, D. R., Thomson, K. T. and Parton, K. (1982). *The Costs of the Common Agricultural Policy.* London: Croom Helm.

CARD (1988). *Annual Report, Centre for Agricultural and Rural Development.* Iowa State University.

CAS (1989). *Countryside Implications of Changes in the Common Agricultural Policy.* Report to the Department of the Environment, Ministry of Agriculture, Fisheries and Food, the Countryside Commission and the Nature Conservancy Council by the Centre for Agricultural Strategy, University of Reading.

Chaveau, J.-M. and Gordon, K. M. (1988). World price uncertainty and agricultural policy formulation under rational expectations, *European Review of Agricultural Economics,* **15,** 437–56.

Chen, D. T. (1977). The Wharton Agricultural Model: structure, specifications and some simulation results. *American Journal of Agricultural Economics,* **59,** 107–16.

Clayton, K. C. (1987). Discussion on alternative agricultural programs, *American Journal of Agricultural Economics,* **69,** 990–1.

Colman, D. R. (1983). A review of the arts in supply response analysis, *Review of Marketing and Agricultural Economics,* **53,** 201–302.

Dosi, C. M. and Stellin, G. (1989). Influencing land use patterns to reduce nitrate pollution from fertilizers and animal manure, pp. 217–30 in:

Economic Aspects of Environmental Regulations in Agriculture, eds A. Dubgaard and A. H. Nielsen. Kiel: Vauk.

Drake, L. (1989). A regionalized non-linear dynamic model of the Swedish agriculture sector with environmental considerations, pp. 247–56 in: *Economic Aspects of Environmental Regulations in Agriculture,* eds A. Dubgaard and A. H. Nielsen. Kiel: Vauk.

Dubgaard, A. and Nielsen, A. H. (Editors) (1989). *Economic Aspects of Environmental Regulations in Agriculture.* Kiel: Vauk.

European Commission (1988). *Disharmonies in EC and US Agricultural Policies.* Luxembourg.

Fox, K. A. (1965). A sub-model of the agricultural sector, pp. 409–61 in: *The Brookings Quarterly Econometric Model of the United States,* eds J. S. Desenberrg *et al.* Chicago: Rand McNally.

Gardner, B. L. (1981). *The Governing of Agriculture.* University of Kansas Press.

de Gorter, H. and Tsur, Y. (1989). *The Political Economy of Price Bias in World Agriculture.* Unpublished mimeograph, Department of Agricultural Economics, Cornell University, Ithaca, NY.

de Gorter, H. and Harvey, D. R. (1990). *Agricultural Policies and the GATT: Reconciling Protection, Support and Distortion.* Unpublished mimeograph, Department of Agricultural Economics and Food Marketing, University of Newcastle upon Tyne, UK.

Grainger, C. W. J. (1969). Investigating causal relationships by econometric models and cross-spectral analysis, *Econometrica,* **37,** 424–38.

de Haen, H. (1979). The use of quantitative sector analysis in agricultural policy: potentials and limitations, pp. 26.1–26.13 in: *Proceedings XVII International Conference of Agricultural Economists,* Banff: International Association of Agricultural Economists.

Hanf, C. (1989). *Agricultural Sector Analysis by Linear Programming Models.* Forum no. 20: Kiel: Vauk.

Harley, F. (1989). Overview of the OECD trade mandate model, pp. 171–4 in: *Agricultural Sector Modelling,* eds S. Bauer and W. Henrichsmeyer. Kiel: Vauk.

Harvey, D. R. (1990). The economics of the farmland market. *Proceedings Agricultural Economics Society Conference on the Farmland Market.* Department of Agricultural Economics and Food Marketing, University of Newcastle upon Tyne.

Harvey, D. R. and Rehman, T. (1989). Environmental change and the countryside: the development and use of a policy model for England and Wales, pp. 231–46 in: *Economic Aspects of Environmental Regulations in Agriculture,* eds A. Dubgaard and A. H. Nielsen. Kiel: Vauk.

Harvey, D. R. and Whitby, M. C. (1988). Issues and policies, pp. 143–77, in: *Land Use and the European Environment,* eds M. C. Whitby and J. Ollerenshaw. London: Belhaven.

Hawking, S. W. (1988). *A Brief History of Time.* London: Bantam Press.

Haxsen, G. (1989). Decreased intensity in crop production and its effects on food supply in emergency situations, pp. 197–206 in: *Economic Aspects of*

Environmental Regulations in Agriculture, eds A. Dubgaard and A. H. Nielsen. Kiel: Vauk.

Hazell, P. B. R. and Norton, R. D. (1986). *Mathematical Programming for Economic Analysis in Agriculture.* New York: Macmillan.

Henrichsmeyer, W. (1989). SPEL-System: concept and overview, pp. 359–64 in: *Agricultural Sector Modelling,* eds S. Bauer and W. Henrichsmeyer. Kiel: Vauk.

Hertel, T. W., Preckel, P. V. and Huang, W. (1989). The CARD linear programming model of US agriculture, *Journal of Agricultural Economics Research,* **40**(2), 20–3.

Johnson, S. R. (1977). Discussion on agriculture sector models and their interface with the general economy. *American Journal of Agricultural Economics,* **59,** 133–6.

Johnson, S. R. and Rausser, G. C. (1977). Systems analysis and simulation: a survey of applications in agricultural and resource economics, pp. 157–304 in: *Quantitative Methods in Agricultural Economics, 1940s to 1970s,* vol. 2, *A Survey of Agricultural Economics Literature,* eds G. J. Judge, R. H. Day, S. R. Johnson, G. C. Rausser and L. R. Martin. Minneapolis: University of Minnesota Press.

Johnson, S. R., Huff, H. B. and Rausser, G. C. (1982). Institutionalizing a large-scale econometric model: the case of Agriculture Canada, pp. 801–30 in: *New Directions in Econometric Modelling and Forecasting in US Agriculture,* ed. G. C. Rausser. New York: North-Holland.

Josling, T. E. (1969). A formal approach to agricultural policy, *Journal of Agricultural Economics,* **20,** 175–95.

Josling, T. E. and Hamway, H. D. (1972). Distribution of costs and benefits of farm policy, pp. 50–85 in: *Burdens and Benefits of Farm Support Policies,* eds T. E. Josling, B. Davey, A. McFarquhar, A. C. Hannak and D. Hammway. London: Trade Policy Research Centre.

Judge, G. J., Day, R. H., Johnson, S. R., Rausser, G. C. and Martin, L. R. (Editors) (1977). *Quantitative Methods in Agricultural Economics, 1940s to 1970s, Vol. 2, A Survey of Agricultural Economics Literature.* Minneapolis: University of Minnesota Press.

Just, R. E. (1990). Modelling the interactive effect of foreign and domestic policies on the agricultural prices of developing countries, pp. 105–29 in *Primary Commodity Prices: Economic Models and Economic Policy,* eds L. A. Winters and D. Sapsford. Cambridge: Cambridge University Press.

Keyzer, M. A. (1986). *An Applied General Equilibrium Model with Price Rigidities.* Staff Working Paper SOW-86-10. Amsterdam: Centre for World Food Studies.

Keyzer, M. A. (1989). Some views on agricultural sector modelling, pp. 23–30 in: *Agricultural Sector Modelling,* eds S. Bauer and W. Henrichsmeyer. Kiel: Vauk.

King, G. A. (1975). Economic models of the agricultural sector. *American Journal of Agricultural Economics,* **57,** 163–71.

Leontief, W. W. (1941). *The Structure of the American Economy 1919–1939.* Oxford: Oxford University Press.

Lipsey, R. G. (1983). *An Introduction to Positive Economics* (6th edition). London: Weidenfeld and Nicolson.
Lowe, P. and Harvey, D. R. (1989). *The Countryside in Question: A Research Strategy*. ESRC Countryside Change Initiative Working Paper No. 1. Countryside Change Unit, Department of Agricultural Economics and Food Marketing, University of Newcastle upon Tyne.
MAFF (annual) *Farm Incomes*. London: HMSO.
Martin, L. and Zwart, A. C. (1975). A spatial and temporal model of the North American pork sector for the evaluation of policy alternatives, *American Journal of Agricultural Economics*, **57**, 55–66.
McCalla, A. F. and Josling, T. E. (1985). *Agricultural Policies and World Markets*. London: Macmillan.
McFarquhar, A. M. M., Mitter, S. and Evans, M. C. (1971). *A Computable Model for Projecting UK Food and Agriculture, Europe's Future Food and Agriculture*. Amsterdam: North-Holland.
Meilke, K. D. and Zwart, A. C. (1979). Effects of alternative specifications on the policy implications of a model: An illustration, *Canadian Journal of Agricultural Economics*, **27**, 15–22.
Meyers, W. H., Womak, A. W., Johnson, S. R., Brandt, J. and Young, R. E. (1987). Impacts of alternative programs indicated by the FAPRI analysis, *American Journal of Agricultural Economics*, **69**, 972–91.
Mueller, D. C. (1979). *Public Choice*. Cambridge Surveys of Economic Literature. Cambridge University Press.
Munk, K. J. (1988). The structure of an agricultural sector model: an alternative approach. Working document in *Disharmonies in EC and US Agricultural Policies*. Luxembourg: European Commission.
Norton, R. D. and Schiefer, G. W. (1980). Agricultural sector programming models: a review, *European Journal of Agricultural Economics*, **7**, 229–64.
Olson, M. (1985). Space, agriculture and organisation, *American Journal of Agricultural Economics*, **67**, 928–37.
Organization for Economic Cooperation and Development (1987). *National Policies and Agricultural Trade*. Paris: OECD.
Parikh, K. S. and Rabar, F. (1981). *Food for All in a Sustainable World: the IIASA Food and Agriculture Program*. Laxenburg: International Institute for Applied Systems Analysis.
Parikh, K. S., Fischer, G., Frohberg, K. and Gulbrandson, O. (1988). *Towards Free Trade in Agriculture*. The Hague: Nijhoff.
Pearce, D., Markandya, A. and Barbier, E. B. (1989). *Blueprint for a Green Economy*. Report for the UK Department of the Environment. London: Earthscan.
Preckel, P. V. and Hertel, T. W. (1988). Approximating linear programmes with differentiable functions: pseudo data with an infinite sample, *American Journal of Agricultural Economics*, **70**, 397–402.
Rausser, G. C. (Editor) (1982). *New Directions in Econometric Modelling and Forecasting in US Agriculture*. New York: North-Holland.
Rausser, G. C. and Freebairn, J. W. (1974). Estimation of policy preference functions: an application to US beef import quotas, *Review of Economics and Statistics*, **56**, 437–49.

Rausser, G. C., Lichtenburg, E. and Lattimore, R. (1982). Developments in theory and empirical applications of endogenous governmental behaviour, pp. 547–614 in: *New Directions in Econometric Modelling and Forecasting in US Agriculture,* ed. G. C. Rausser. New York: North-Holland.

Rawls, J. (1971). *A Theory of Justice.* Cambridge, MA: Harvard University Press.

Rehman, T. and Romero, C. (1984). Goal programming and multiple criteria decision making in farm planning: an expository analysis, *Journal of Agricultural Economics,* **35,** 177–90.

Romero, C. and Rehman, T. (1985). Goal programming and multiple criteria decision making in farm planning: some extensions, *Journal of Agricultural Economics,* **36,** 171–85.

Roningen, V. O. (1986). *A Static World Policy Simulation (SWOPSIM) Modelling Framework.* Staff Report AGES860625. Washington, DC: US Department of Agriculture, Economic Research Service.

Roningen, V. O. and Dixit, P. M. (1989). *Economic Implications of Agricultural Policy Reforms in Industrialised Market Economies.* Staff Report AGES 89-36. Washington, DC: US Department of Agriculture, Economic Research Service.

Roth, A. E. (1988). Laboratory experimentation in economics: a methodological overview, *Economic Journal,* **98,** 974–1031.

Schatzer, R. J. and Heady, E. O. (1982). *National Results for Demand and Supply Equilibrium for Some US Crops in 2000: Theory and Application of Tatonnement Modelling.* CARD Report 106. Ames: Center for Agricultural and Rural Development.

Scherr, B. A. (1978). *The 1978 Version of the DRI Agriculture Model.* Working Paper No. 5. Lexington, Mass.: Data Resources Inc.

Spedding, C. R. W. (1975). Foreword, p. v in: *Study of Agricultural Systems,* ed. G. E. Dalton. London: Applied Science Publishers.

Stoeckel, A. (1985). *Intersectoral Effects of the CAP: Growth, Trade and Unemployment.* BAE Occasional Paper No. 95. Canberra: Bureau of Agricultural Economics, Australian Government Publishing Service.

Tamminga, G. (1989). Integrating agriculture and the environment: a Dutch agricultural model, pp. 257–66 in *Economic Aspects of Environmental Regulations in Agriculture,* eds A. Dubgaard and A. H. Nielson. Kiel: Vauk.

Thompson, R. L. (1981). *A Survey of Recent US Developments in Agricultural Trade Models.* Bibliographies and Literature of Agriculture no. 21. Washington, DC: US Department of Agriculture, Economic Research Service.

Thompson, R. L. (1985). *A Survey of Recent US Developments in International Agricultural Trade Models.* Bibliographies and Literature of Agriculture, Washington, DC: US Department of Agriculture, Economic Research Service.

Thomson, K. J. (1987). A model of the Common Agricultural Policy, *Journal of Agricultural Economics,* **38,** 193–210.

Thomson, K. J. and Buckwell, A. E. (1979). A microeconomic agricultural supply model, *Journal of Agricultural Economics,* **30,** 1–11.

Thomson, K. J. and Harvey, D. R. (1981). The efficiency of the Common Agricultural Policy, *European Review of Agricultural Economics,* **8,** 57–83.
Thomson, K. J. and Rayner, A. J. (1984). Quantitative policy modelling in agricultural economics, *Journal of Agricultural Economics,* **35,** 161–76.
Tweeten, L. (1988). Agriculture in the macro economy, *American Journal of Agricultural Economics,* **70,** 1006–26.
Tyers, R. and Anderson, K. (1988). Liberalizing OECD agricultural policies in the Uruguay round: effects on trade and welfare, *Journal of Agricultural Economics,* **39,** 197–216.
Wallis, K. F. (Editor) (1987). *Models of the UK Economy, a Fourth Review by the ESRC Macroeconomic Modelling Bureau.* Oxford: Oxford University Press.
Walker, N. and Dillon, J. L. (1976). Development of an aggregative programming model of Australian agriculture, *Journal of Agricultural Economics,* **27,** 243–8.
Whitby, M. C. (1990). *Economics and Policies in the Countryside.* ESRC Countryside Change Initiative Working Paper no. 2. Countryside Change Unit, Department of Agricultural Economics and Food Marketing, University of Newcastle upon Tyne.
Wicks, J. A., Mueller, R. A. E. and Crellin, I. R. (1978). *APMAA7—Project Development, Model Structure and Applications of the Aggregative Programming Model of Australian Agriculture.* APMAA Report no 11. Armidale, Australia: Department of Agricultural Economics and Business Management, and Department of Statistics, University of New England.
Winters, L. A. (1987). The economic consequences of agricultural support: a survey, *OECD Economic Studies,* **9,** 7–54.
Zwart, A. C. (1977). *An Empirical Analysis of Alternative Stabilisation Policies for the World Wheat Sector.* Unpublished PhD Thesis, University of Guelph.
Zwart, A. C. and Meilke, K. D. (1976). *Economic Implications of International Wheat Reserves.* Discussion Paper no. 1. School of Agricultural Economics and Extension Education, University of Guelph.
Zwart, A. C. and Meilke, K. D. (1979). The influence of domestic pricing policies and buffer stocks on price stability in the world wheat industry, *American Journal of Agricultural Economics,* **61,** 434–47.

12

Of Agricultural Systems and Systems Agriculture: Systems Methodologies in Agricultural Education

R. J. BAWDEN

Faculty of Agriculture and Rural Development, University of Western Sydney, Australia

INTRODUCTION

In his editorial in the foundation issue of *Agricultural Systems* in 1976, Colin Spedding, with characteristic confident assertion, posited that 'it is now generally recognized that whole agricultural systems deserve to be studied in their own right' (Spedding, 1976*a*). In later statements he warmed to his theme: 'The study of agricultural systems is a proper concern of many different sorts of people' (Spedding, 1976*b*), and that this is not surprising, given that 'agricultural research and development are generally aimed at the improvement of systems' (Spedding, 1978). Given the inherent seduction of this logic and despite significant calls for change framed with similar sentiment (Coombs and Ahmed, 1974; Bawden *et al.*, 1984; Chambers, 1987; Stansbury and Kunkel, 1987), it is disappointing to note how rarely the twin foci of situation improving and systems understanding are central to debates about reform in agricultural education.

Contrary to the claim of Holt and Schoorl (1985), there is little evidence, at least by examination of college and university curricula across the world, that 'systems methodologies' do appear to be well entrenched in agriculture let alone represent the current orthodoxy. It is much easier to support the contention that most intellectual maps of agriculture fail to portray it as 'the basic interface between human societies and their environments' (Dahlberg, 1979)—a complex of human transactions with biotic and abiotic relationships.

It can be argued that it is a singular lack of methodological competence at being able to deal with such complexity that lies at the heart of an increasing disenchantment with the conventional practices of contemporary agricultural professionals and of their institutions. In the USA for instance, Buttel (1983) has drawn attention to a 'Greek chorus of criticism' of the performance of agricultural institutions in that country in their failure to address 'growing externalities of agricultural technology and public policy'. Amongst the issues he cites in this regard are: environmental degradation; concerns for animal welfare; impacts on health and safety of farmers; adverse nutritional effects of production and processing technologies; the extrusion of smaller family farmers from agriculture, the erosion of rural communities and the concentration of agricultural production and economic wealth; and inadequate conservation and commercial exploitation of fragile lands that should not be in cultivation.

Embedded in this problematique, which is but one of many published about similar situations across the globe, is a host of dimensions which enormously complicates the constitution and thus the attainment of improvements. The implication is that both the domain and scope of competencies of professional agriculturalists deserve urgent and fundamental review to embrace what has been referred to in a different context as 'the science and praxis of complexity' (Soedjatmoko, 1985). As used here, praxis is the recursive process by which a practitioner uses theories and practices in ways where each continually improves the other in the overall quest for improvements in the world. It is a dynamic flux between 'reflection and action in the world in order to transform it' (Freire, 1974). For the agriculturalist, this means not only knowing how to actually transform problematic, complex agricultural situations in practice, but also knowing how to transform his or her own practice in such a manner that improvements in the one also informs the other.

Education for praxis is therefore aimed at helping the neophyte agriculturalist to learn ways of learning about agriculture, as well as about agriculture itself, in a dynamic critical manner. To take this further and heed Dahlberg's critique, the agriculturalist must learn how to deal in a dynamic, recursive and critical manner with the complex issues at the 'interface between the natural and social worlds' through the interdependent use of theories about, and practices in both worlds. This complexity lends strong support to the need for new, systemic ways of thinking about agricultural education, and for systemic methods and methodologies for improving it.

In this chapter the focus will be on *learning how to learn how to improve complex situations in agriculture, through systemic praxis.* This is somewhat different from most other systems initiatives in agricultural education which appear to focus on variations on the theme of 'learning about agricultural systems'. In reflecting what Checkland (1984) refers to as 'the shift in systemicity from reality to the process of inquiry into reality', I want to present the case for a change in focus from inquiry into *agricultural systems,* to inquiry through *systems agriculture* (Bawden *et al.*, 1984), from learning *systems* to *learning* systems.

The passion for the case for such a shift is inspired by the comment of Churchman (1971) that 'the most important feature of the systems approach is that it is committed to ascertaining not simply whether the decision makers' choices lead to his desired ends, but whether they lead to ends which are ethically defensible'. It is mainly because of this, and other philosophical matters such as aesthetics, ontology and epistemology, which all lie at the heart of what we value as 'improvements'—what we are trying to improve, and for whom—that systems methodologies deserve any attention at all, either in or for agricultural education.

The section which follows will expand on this assertion and highlight its significance in the development of systemic methods in agricultural education. This will be followed by an exploration of a number of theories of learning which can be considered useful in informing the praxis of the educator. From learning theories, the attention will be turned to learning systems as a key metaphor both for processes and organization in agricultural education. Particular emphasis will be placed on systems agriculturalists, and especially those who are educators, as systemic action researchers. The notion of researching versus researched systems will be briefly elaborated with reference to a number of different systems approaches to agriculture. Finally, the application of systemic praxis in guiding the strategic development of 'whole learning systems' like entire agricultural faculties, colleges and universities will be briefly discussed.

OF METHODS AND METHODOLOGIES

Professions are characterized by the domain of the problems they address and the methods that are used in addressing them. In agriculture this is a complicated matter. In the first place, there are so many different problem domains that the 'profession' is really a whole mosaic of different vocations each with their own constitutive methods. Agriculture educators, concerned with providing strategies which enable students to master methods appropriate to their preferred career options, are thus presented with a spectrum of foci from the operation and management of farms, through the supply and marketing of commodities, through extension and education, to technological developments and scientific research. This situation is further confounded by the increasing recognition of the need for change in the conventional problem domains and in the methods for addressing them within each of these vocations. Therefore, in addition to being competent at using appropriate existing methods, agriculturalists must also develop the capability of designing new methods and this is as true for education as it is for agriculture.

For the systems agriculturalist concerned with dealing with such complexity this suggests at least two levels of involvement in which the use of systems thinking and practices are appropriate to the development of methods. The first of these involves the use of systems methods to improve problematic agricultural situations whilst the second involves the use of systems methods to improve systems methods. This is even more important in the case of the educator of systems agriculturalists—the systems agricultural educator (SAE)! In addition to the complexity of the above situation, they also face a plurality of educational methods which range from the 'transmission' of propositional natural knowledge, the 'demonstration' of practical knowledge and the 'facilitation' of experiential knowledge.

To return to the notion of *praxis,* it follows that, in addressing systems (systemic) methods and methods of methods, the SAE is involved in a dynamic, recursive flux between theories and practices at each of these two levels as well as between them. Following the notion of Banathy (1984), there is a third level of involvement yet in which the educator addresses some of the philosophical and metatheoretical assumptions upon which the other two levels of method use are based.

It is this third, metatheoretical and philosophical level that, Oliga (1988) submits, constitutes the domain of methodology. At this level, the educator specifically addresses ontological assumptions about the 'actual nature of the world' as well as epistemological assumptions about the 'ways by which such knowledge of the nature of the world, is known'. Far from being mere whimsy, these issues are central to any discussion about inquiry methods in agriculture and how one learns to use methodology as the basis for the validation of changes in method and so on. Indeed the very concept of 'wholeness as the essence of systems thinking and practice' is embedded here and it is thus only at this level that discussions about 'the systemicity of the process of inquiry into reality' makes any sense.

MULTIPLE REALITIES

It is often disturbing to agriculturalists born of a tradition in the physical or life sciences that there is no unity about ontological assumptions about the world—that there are those who strongly dispute the positivist assumptions that there is only one reality 'out there' and that it can be known objectively and empirically (Reason and Heron, 1986). For the non-positivist, reality is created through 'the transaction of the observer and the object' (Plas, 1986), echoing the dependency of the observed on the observer as mooted in his principle of uncertainty by the physicist Heisenberg (1959). In stating the non-positivist case for constructivism, Kelly (1955) posits 'that whatever nature may be, or however the quest for truth will turn out in the end, the events we face today are subject to as great a variety of construction as our wits will enable us to contrive'.

Perhaps most shocking of all is the submission that these are not just obscure philosophical issues but that evidence from the natural sciences supports the contention that the way we think can affect what we see; that the physiological aspects of perception cannot be separated from the psychological aspects of interpretation (Maturana and Varela, 1980). Perhaps to add final insult to injury, our view of reality is also highly coloured by our experiences, values, beliefs, intentions, and even emotions (Maturana, 1988) and that we are at once both servants and masters of our own, highly idiosyncratic world view—our *weltanschauung*.

Just as the case for 'multiple realities' or ontological assumptions can be made, so too can the case be made for multiple theories about 'how we know about realities' or epistemological assumptions. In elaborating on the theme of 'the design of inquiring systems', Churchman (1971) illustrates the importance of the lack of comprehensiveness of any one way of knowing and thus the advantages of what Reason and Heron (1986) refer to as 'epistemological heterogeneity'. The central issue here is that epistemological differences can be expected to play a part in any conversation, even between scientists of apparent like minds! As the praxis of the systems agriculturalist is fundamentally a form of social intervention based on dialogue around 'issues to be improved' and on the function of helping others to learn new ways of knowing, the use of theories and philosophies about knowledge and knowing to inform the practice is clearly vital.

This discussion about views of reality and knowledge about reality, is of primary importance when discussing systems methodology in agricultural education for a number of extremely important reasons. Not the least of these is the submission that methodological analysis indeed 'represents a bridging activity that aims at forging a correspondence between paradigmatic ontological assumptions and the particular epistemological positions taken' (Oliga, 1988). In other words, different paradigms of inquiry reflect different assumptions related to ontology and epistemology, as well as to value-based aesthetic and ethical dimensions of human nature. Drawing on the work of Burrell and Morgan (1979) and Habermas (1972), Oliga (1988) elaborates on the thesis that social theories can be classified into three underlying paradigms (functionalist, interpretive and radical/critical) which thus allows the distinction to be made between three methodological approaches (empiricism, hermeneutics and critique or critical heuristics).

Such frameworks for methodological analysis do not appear to have been used to investigate systems approaches to agricultural education or thus to inform the methods that have been adopted. In their absence, and like the situation in the systems movement at large, systems approaches to agriculture have been confounded in no small way through the use of the word '*system*' as both 'an abstract epistemological device and for assumed ontological entities in perceived reality' (Checkland, 1988). In many instances, agricultural researchers, along with many ecologists, have reified the systems they study into ontological entities with but little reference to the methodological context in which their assumptions about reality or about ways of knowing that reality are set. Any curricula that are based on initiatives such as those of cropping systems and farming systems research will therefore be limiting as vehicles for the development of praxis. In the absence of critical debates about the methodological foundations of such methods, and of discussions about the 'correspondence between the ontological assumptions and the particular epistemological positions taken', these educational practices severely restrict the learning potential offered by what Laszlo (1972) has referred to as 'the systems philosophy'.

To illustrate this point, let us view agricultural education from a systemic, epistemological *viewpoint*, examining it as if it were a learning system, in order to gain insights into how it might be improved.

AGRICULTURAL EDUCATION AS A LEARNING SYSTEM

The systems philosophy embraces the essence of wholeness, the concept of emergent properties, the notion of connectivity and of wholes within wholes, and, of particular importance in education, the nature of recursive thinking. As applied to education then, one can imagine learning as a system of learning or inquiring systems with the immanent characteristics of such a philosophy pervasive. The system will be different from the sum of its parts (Bateson, 1972), the community psychology will be a systems psychology (Plas, 1986), and there will be a dynamic born, in the graphic phrase of Bertalanffy (1981), of a 'glorious unity of opposites'.

Key to the development of systemic praxis by the agricultural educator is an exploration of theories of learning, knowledge and knowing. Reason and Heron (1986) provide a useful entry point into this discussion with their submission that it is useful to distinguish between three different kinds of knowledge. Propositional knowledge is knowledge about something, practical knowledge is knowledge of how to do something, and experiential knowledge is knowledge of transformed experience. Restated in educational terms, propositional learning is knowing for knowledge, practical learning is knowing for doing, and experiential learning is knowing for being. As each type of learning is interdependent with the others, they can be seen as together comprising a learning system, with each being a learning system itself at a 'lower level' of resolution. The process of experiential learning can be used to illustrate this contention.

Incorporating key ideas of John Dewey, Jean Piaget, Carl Jung, Paulo Freire and Kurt Lewin, the organizational psychologist David Kolb (1984) has developed a model of the experiential learning process which illustrates how we 'transform our experiences in the world into knowledge about it'. As envisaged, the experiential process involves a dynamic and recursive flux between the concrete and the abstract, and between reflection and action. It thus integrates into a system of learning activities: the feelings that we experience from an event in the world around us, the observations and reflections that we make of these, the abstract patterns that we use to create knowledge through the process of thinking about these reflective observations, and the actions that we take to reflect our new knowledge. No one activity is prime here; each one is vital in informing the others with which it 'interacts' in a recursive, as distinct from a linear, manner.

Indeed it is the very recursiveness of the relationships between the concrete and the abstract, reflection and action, content and process and so on, that enables it to be portrayed as a system—a system of enquiry that has an internal dynamic which is generated through the dialectic that can be assumed to exist between these different states and activities. Drawing on some of the ideas of George Hegel, Kolb (1984) describes these relationships as 'mutually opposed and conflicting processes the results of which cannot be explained by the other, but whose merger through confrontation of the conflict between them results in a higher order process that transcends and encompasses them both'. So reflection and action only differ from each other as different parts of the same whole and it is through their inter-relationship that the whole emerges—thesis and antithesis yielding synthesis! The same can be posited for propositional, practical and experiential learning systems. Instead of the endless and fruitless arguments of dichotomy that persist in the name of curricula reform such as the relative balance between theory and practice in agricultural education, it is much more profitable to systemically explore the nature of their inter-relationship and exploit the 'glorious unity of opposites'—creativity born of the inherent tension of difference.

In addition to this dialectic tension within systems of inquiry, it should also be illuminating to examine a system of inquiring systems at and between different levels of resolution. Already the idea of a systems classification of 'wholes within wholes' has been implied by Banathy's (1984) scheme involving methods and methodologies. Mezirow (1985) has suggested a 'hierarchy' of learning processes which are clearly akin to Banathy's ideas. Thus he discriminates between 'learning *within* meaning schemes', 'learning *new* meaning schemes' and 'learning *through* meaning transformation'. This builds on the concept of 'levels of learning' proposed by Bateson (1972) with its implication that learning how to learn is a high-order level of learning. This assumes particular importance given the submission of Emery (1981), that learning to learn should be the key emphasis of new paradigms for education. As with the notion of a systems dynamic being generated through the recursive flux between opposite states or activities within a system, so too can one assume such a dynamic recursion across a system of systems with each level informing, and thus improving, the others.

Habermas (1972) adds immeasurably to the richness of these notions by exploring the proposition that different systems of inquiry not only use different ways of reasoning about the ways of the world and knowledge about it but also reflect different interests. Revisiting some of the ideas of Arisotle on inquiry, Habermas portrays social theories as reflecting a technical interest for prediction and control (man–nature interaction), a practical interest for understanding (human communicative interaction) or an emancipatory interest for radical change (social relations of power, domination and alienation) (Oliga, 1988). These observations are of profound significance in terms of the purposes of education in agriculture, the access that people have to it, and the nature of that education as it relates to the development of the praxis of the neophyte systems agriculturalist.

When reflecting on the earlier critique of Buttel (1983) of agricultural institutions as translated into a fundamental concern about the praxis of agriculturalists, it becomes obvious that different systems methods of inquiry, conducted for different interests, and reflecting different ontologies and epistemologies, will reveal very different and, indeed, often conflicting notions of improvements. This inherent conflict and contradiction is as applicable to the situation of the institution as an inquiring system as it is to the situations in agriculture that are the subject of its inquiries!

Expansion of the call for greater specificity in the constitution of improvements in agriculture (Spedding, 1978) thus underlines the need for learning systemic ways of managing conflict in complex situations.

LEARNING TO MANAGE CONFLICT

As Kolb (1984) emphasizes, experiential learning is the foundation of our *praxis* and indeed of our very adaptation to the changing events of the world around us. It is a highly personal process and each of us not only develops a unique *weltanschauung,* but an idiosyncratic way of developing it through biases by which we apparently adopt a preference for certain learning activities over others. It is in this way that our own particular way of doing things can be seen to reflect the particular way we go about 'seeing' things. These ways of seeing and doing can be so profoundly different in two individuals that they are a source of conflict so intense that their different viewpoints seem immutable (Cotgrove, 1982).

The challenge of trying to do something about this is clearly central to agricultural education and its very purpose in attempting to improve the quality of learning. Furthermore, the issue of learning is obviously a key focus in the management of conflict. In commenting on the call by Petak (1980) for a synthesis of 'technological' and 'ecological' forms of thinking for improved environmental management, Miller (1983) submits that 'this is easier said than done, for styles of thought are not superficial skills that can be adopted or abandoned at will. . . but complex patterns of behaviour deeply embedded in the individual's personality and held in place by firm conviction'.

In drawing participants into learning systems, these differences are transformed into opposite parts of the same unity where the differences can be used as a foundation for the reconstruction of the issue—a milieu in which mutual negation is resolved through the process of 'bringing forth new worlds together' (Maturana and Varela, 1988).

Ulrich (1988) suggests that the management of conflict represents a high level of systems practice. He distinguishes it, and the characteristics he associates with it, from the lower order practices of the management of complexity and, finally, of scarceness. The criteria of a 'good solution' for each ranges from efficient (for scarceness), effective (for complexity), and ethical (for conflict). Finally, in arguing the case for a synthesis between systems thinking and practical philosophy, he ranks them in terms of 'dimensions of rationalizations', in ascending order as instrumental, strategic and communicative.

We are thus brought recursively back to the centrality of ethics as the critical foundation for all improvements both by and of learning systems. It is unethical for agricultural educators to ignore ethics and in this context it should be revealing to examine some of the initiatives that have been taken in introducing systems thinking and practices into the education of agriculturalists.

SYSTEMS INITIATIVES IN AGRICULTURAL EDUCATION

It would appear from reviews of literature from around the world that the commonest way of introducing students of agriculture to systems concepts is to present agricultural systems for study as modified (and reified!) natural systems. Such agro-ecosystems might be regarded as part of a continuum of systems in nature (Smith and Hill, 1975) or as discrete system types which differ fundamentally from their natural counterparts (Cox and Aitken, 1979). In any event, the concept of the agro-ecosystem has provided a focus for many research studies into the nature of agriculture (Lowrance *et al.*, 1984) as well as a vehicle for educational strategies for learning about it.

However, as Conway (1987) suggests, most of these have tended to concentrate on analyses of flows and cycles of energy and material and have thus had 'little impact on the theory and practice of agricultural development' or the praxis of the developer. More profoundly, the absence of any rigorous methodological contexts seriously weakens their use as efficient, effective or ethical strategies for learning to learn systemically about agriculture.

In Conway's own work, the agro-ecosystem concept is given a different and powerful reorientation through its presentation as a cybernetic system—a system whose behaviour is attributable in part to the way it communicates with its environments. Furthermore, through the participative nature of the inquiring process, the researched agro-ecosystem becomes partially transformed into a purposeful re-searching system with value-laden observers clearly becoming part of the observed and introducing an inevitable 'epistemological hetero-geneity'. It is in this way that the controversy surrounding the ontological assumptions about the 'true nature' of ecosystems (cf. Vayda and McCay, 1975; Engelberg and Boyarsky, 1979) is avoided. Interestingly, Conway (1987) provides a different, if moot perspective on the ontological status of systems by positing that 'there can be little doubt that the transformation of ecosystem to agroecosystem produces well defined systems of cybernetic nature'.

In the language of Checkland (1981), Conway's work presents the study of agricultural systems not as a study of disturbed 'natural systems' but of 'designed physical systems' where the values and goals of those who design and manage them determine in large degree the properties that the systems themselves display. Bawden and Ison (1990) have reviewed the connections that can be drawn between the complexity of value issues and the four properties of productivity, stability, sustainability and equitability which Conway (1983) attributes to agro-ecosystems. They conclude that whilst these are useful guides to improvements in agriculture, each property is replete with ambiguity and contradiction through multiple interpretations as exemplified, for instance, by the observations of Kingma (1985) on productivity and Douglass (1984) on sustainability. Rigorous methodological analysis is vital to allow exploration of the ontological, epistemological, ethical and other 'philosophical heterogeneity' inherent under these circumstances. An example, at least in as much as he refers to rapid rural appraisal, has been provided by Jamieson (1985) with his insights into the importance of methodologies through reference to paradigms, noting that they can differ in at least three distinct ways: in scope, in degrees of specificity and in the degree of conscious awareness with which they are held.

A similar lack of methodological analysis characterizes otherwise prolific writings on the other major systems initiatives in agricultural research which are themselves increasingly used as a focus for both propositional and practical education and training. Where this occurs, curricula, particularly at the postgraduate level, contain courses which are based on cropping systems and farming systems research methods as well as on a host of techniques which are used within these and related methods. As their methodological contexts are rarely explicit and by the nature of the methods of inquiry, it is often more accurate to describe these initiatives as systematic rather than systemic.

Most of the major systems research methods in agriculture present their view of agricultural systems as variations along the theme of 'managed natural systems' which exist in both bio-physical and socio-cultural environments. For all this, the pervasive notion of improvement lies in increasing productivity through the adoption of improved technologies. This is in spite of many accounts of the influence of personal values, beliefs, ethics, and so on, which stress the idiosyncracies of individual farmers, and of the power and ideological influences of the environments in which they must operate. Failure to explore these dimensions really does render such approaches impotent with regard to the development of relevant praxis.

All of these approaches mentioned so far really fit within the rubric of what Checkland (1981) has called the 'hard' tradition in systems approaches. There are underlying positivist and empiricist assumptions about the nature of such systems and how their performance can be optimized through the intervention of 'the expert technologist' or 'management consultant'. These are researched systems with scientists investigating on behalf of, or even on their farmer 'clients'.

An alternative approach sees a major epistemological change from researched systems to researching systems where scientists and farmers and others research 'with' each other in attempting to 'bring forth new worlds together'. The initiatives at Hawkesbury are amongst those that fit within the 'soft' tradition, with its focus on 'human activity systems'. The central strategy is based on the 'soft systems' methods and methodologies first developed at the University of Lancaster and using them to create comprehensive learning systems. Although it has been developed within functionalist and interpretive paradigms, these restrictions are being increasingly recognized and addressed.

The essential educational strategy is based on an experiential action research process which involves students, faculty and 'clients' collaborating together as what Vickers (1984) refers to as appreciative systems. The intentional praxis of the Hawkesbury graduate systems agriculturalist is to be able to bring new methods and methodologies to bear on problematic situations in such a manner that all who are participating in the project learn how to use systemic thinking and practice to improve their situations. Thus the outcomes include improvements in both the situation being examined and in the ways of examining it. Part of their systemic competency is the ability also to use conventional reductionist and 'hard' systemic approaches in such a way that they represent different 'levels' of learning within a 'spiralling hierarchy'—a learning system of recursive learning systems with the tensions of their differences being creatively exploited.

The institution is organized in such a way that it too is envisaged as a learning system, in the sense proposed by Argyris and Schon (1978), as a coherent organization in dynamic co-evolution with its environments. Similarly, each and every action research project also becomes a learning system. Considerable emphasis is placed on the importance of systemic praxis and on the actual use of theories, metatheories and philosophies to inform the practices of inquiry, and *vice versa*. Perhaps somewhat surprisingly, a major source of inspiration for the reconceptualization of the role and praxis of the agriculturalist as a systems agriculturalist has come from work in family therapy (cf. Keeney, 1982; Dell, 1985; Kenny and Gardner, 1988; Anderson *et al.*, 1986).

Details of the evolution of the Hawkesbury approach and of its curricula and supporting theories have been published elsewhere

(Macadam and Bawden, 1985; Bawden and Valentine, 1985; Bawden, 1990). The management of conflict has been an integral part of the whole development and in many senses, the level of conflict continues to increase, particularly with issues concerned with the 'soft systems paradigm'.

A number of writers have been critical of the 'soft systems methodology', particularly as it relates to its inherent functionalism (Jackson, 1982), its idealism (Rosenhead, 1984), and to the lack of inclusion of ethical dimensions (Atkinson, 1989). Each of these critiques have added to the learning process, however, and there have been many responses and further suggestions about improvements (cf. Checkland, 1985; 1986; 1988), with particular reference to an expansion of the 'systems metaphor' (Atkinson and Checkland, 1988) and the integration of ethical dimensions (Atkinson, 1989).

The tension that tends to characterize Hawkesbury, and the source of increasing conflict in field projects, especially those which are being conducted with colleagues in less developed nations, is grounded in the realization that the paradigm must expand to embrace what Ulrich (1988) refers to as 'critical heuristics'—the need to 'free the systems approach from the impossible (and elitist) pretensions of securing a "monological" justification of rational practice'. In the terms of the constitutive theories of Habermas (1972), there is an increasing motivation to refocus the initiatives around issues of emancipation and empowerment. In arguing the case that empowerment is essentially an ecological concept, Rappaport (1987) stresses that it is not only 'an individual psychological construct, (but) it is also organizational, political, sociological, economic, and spiritual'. Without such a critical focus, there are serious doubts that the most significant issues of environmental degradation, rural poverty, restricted land rights and socio-economic inequitability can ever be improved in ways which will be ultimately ethically defensible.

CRITICAL DEVELOPMENT OF AGRICULTURAL UNIVERSITIES AS LEARNING SYSTEMS

In recent years there have been a spate of studies of agricultural universities in many countries across the world (Hansen, 1989). In most instances these have revealed a sense of mismatch between conditions in society and the ways by which the universities are going about their business (Bawden and Busch, 1989). Within the problematique is an increasing sense of alienation between the academy and its rural constituencies, and a degradation of relationships in its political affiliations. At a more fundamental level, there are also many concerns about the relevance and rigour of prevailing paradigms in education, research and extension, and an increasing recognition of the advantages of plurality in types of learning systems (Jackson, 1987; Flood, 1989). The important issue is the capability of converting this awareness into systemic strategies for change based on 'applications of systems dialectics to integrated development' (Delgado, 1988).

This is, perforce, a critical exercise where 'to be critical' does not mean the adoption and defence of a particular position but 'to keep reflecting on the premises of one's position regardless of what that position is' (Ulrich, 1981). Yet there is also an acute awareness of the paradox of critique. The greater the criticism the greater the conflict both within the system itself and in its relationships with key elements in the environment. Thus whilst a society might well applaud the development of the ability to critique by its educated youth, the same society invariably does not tolerate criticism well—and this is particularly true of the cohort within the society that 'governs' it!

From the logic developed above, it is apparent that systemic methodologies can guide the reconstruction of agricultural education institutions as dynamic and critical learning systems which themselves are in co-evolution with the environments which surround them. As the philosophies and metatheories of these methodologies are transformed through practice into praxis, the immanent characteristics of systemicity will begin to pervade the entire community and the things that it does. Patterns of recursiveness will begin to characterize the forms of inquiry, including those used to generate new forms of inquiry! Conventional research, teaching and extension functions will all come under systemic scrutiny as new notions about knowing and knowledge and reality become emergent. New patterns of organization will also take shape both within the institution and, as network linkages, between the institution and its environments. Improvements will be assessed in terms of efficiency and effectiveness set within an all-pervading *weltanschauung* of an ultimately defensible ethic of what is good for the 'whole system'.

And the level of discomfort will inevitably rise; but then, radical changes in the social ecology of universities are, like the paradigmatic shifts they represent, usually only made with 'fierce controversies, international name-calling, and the dissolution of old friendships' (Kuhn, 1970).

ACKNOWLEDGEMENTS

As always it is a great pleasure to acknowledge the inspiration provided by my colleagues in the Hawkesbury learning system. In the particular context of this chapter, I would make special mention of David Russell, Joe Zarb, Graham Bird, and Bob Macadam. I am indebted to Linda Mohammed for all her help in the preparation of this manuscript. And to Colin Spedding, who first encouraged me into this systemic fray three decades ago, I pay my most appreciative respects. May he long enjoy the emergent properties of retiring systems!

REFERENCES

Anderson, H., Goolishian, H., Pulliam, G. and Winderman, L. (1986). The Galveston Family Institute: some personal and historical perspectives, in: *Journeys*: *Expansions of the Strategic and Systemic Therapies,* ed. D. Efron. New York: Brunner/Mazel.

Atkinson, C. J. (1989). Ethics: a lost dimension in soft systems practice, *Journal of Applied Systems Analysis,* **16,** 43–53.

Atkinson, C. J. and Checkland, P. B. (1988). Extending the metaphor, 'system', *Human Relations,* **41,** 709–25.

Argyris, C. and Schon, D. (1978). *Organisational Learning.* Reading, Mass.: Addison-Wesley.

Banathy, B. H. (1984). Systems enquiry in education, *Systems Practice,* **1,** 193–213.

Bateson, G. (1972). *Steps to an Ecology of Mind.* New York: Balentine.

Bawden, R. J. (1990). Towards action researching systems, in: *Proceedings of the First International Action Research Conference,* ed. O. Zuber-Skerritt. Sydney, Australia: John Wiley.

Bawden, R. J. and Busch, L. (1989). Agricultural universities for the 21st century. Paper prepared for an International Workshop on Strategic Development in Agricultural Universities, USAID, Reston, Virginia.

Bawden, R. J. and Ison, R. L. (1990). The purposes of field-crop ecosystems: social and economic aspects, in: *Field Crop Ecosystems of the World,* ed. J. C. Pearson. London: Elsevier.

Bawden, R. J. and Valentine, I. (1985). Learning to be a capable systems agriculturalist, *Program Learning and Educational Technology,* **21,** 273–87.

Bawden, R. J., Macadam, R. D., Packham, R. J. and Valentine, I. (1984). Systems thinking and practices in the education of agriculturalists, *Agricultural Systems,* **13,** 205–25.

Bertalanffy, L. von (1981). In: *A Systems View of Man,* ed. P. A. La Violette. Boulder, CO: Westview Press.

Burrell, G. and Morgan, G. (1979). *Sociological Paradigms and Organizational Analysis*. London: Heinemann.
Buttel, F. H. (1983). The Land Grant System: a sociological perspective on value conflicts and ethical issues, *Agriculture and Human Values,* **2,** 78–95.
Chambers, R. (1987). *Rural Development: Putting the Last First*. New York: Longman Scientific and Technical.
Checkland, P. B. (1981). *Systems Thinking, Systems Practice*. New York: John Wiley & Sons.
Checkland, P. B. (1983). OR and the systems movement: mappings and conflicts, *Journal of the Operational Research Society,* **36,** 821–31.
Checkland, P. B. (1984). Systems thinking in management: The development of soft systems methodology and its implications for social science, pp. 94–104 in: *Self Organisation and Management of Social Systems,* eds H. Ulrich and G. J. B. Probst. Berlin: Springer.
Checkland, P. B. (1985). From optimizing to learning: a development of systems thinking for the 1990s, *Journal of the Operational Research Society,* **36,** 757–67.
Checkland, P. B. (1986). The politics of practice. Paper presented to IIASA International Round Table, November 1986.
Checkland, P. B. (1988). Images of systems and the systems image: presidential address to ISGSR, *Journal of Applied Systems Analysis,* **15,** 37–42.
Churchman, C. W. (1971). *The Way of Inquiring Systems: The Design of Inquiring Systems*. New York: Basic Books.
Conway, G. R. (1983). Applying ecology. Inaugural lecture presented at the Imperial College of Science and Technology Centre for Environmental Technology, London, 7 December 1982.
Conway, G. R. (1985). Agroecosystems analysis, *Agricultural Administration,* **20,** 31–55.
Conway, G. R. (1987). The properties of agroecosystems, *Agricultural Systems,* **24,** 95–118.
Coombs, P. H. and Ahmed, M. (1974). *Attacking Rural Poverty: How Nonformal Education Can Help*. Washington, DC: The Johns Hopkins University Press.
Cotgrove, S. (1982). *Catastrophe or Cornucopia*. New York: Wiley.
Cox, G. W. and Aitken, M. D. (1979). *Agricultural Ecology: An Analysis of World Food Production Systems*. Oxford: W. H. Freeman and Company.
Dahlberg, K. A. (1979). *After the Green Revolution*. New York: Plenum Press.
Delgado, R. R. (1988). Applications of systems dialectics to integrated development, *Systems Practice,* **1,** 259–78.
Dell, P. F. (1985). Understanding Bateson and Maturana, *Journal of Marital and Family Therapy,* **11,** 1–20.
Douglass, G. K. (1984). *Agricultural Sustainability in a Changing World Order*. Boulder, CO: Westview Press.
Emery, F. (1981). Educational paradigms, *Human Futures,* Spring, 2–17.
Engelberg, J. and Boyarsky, L. L. (1979). The noncybernetic nature of ecosystems, *American Journal,* **114,** 317–24.

Flood, R. L. (1989). Six scenarios for the future of systems 'problem solving', *Systems Practice*, **2,** 75–99.
Freire, P. (1974). *Pedagogy of the Oppressed*. New York: Continuum.
Habermas, J. (1972). *Knowledge and Human Interests*. London: Heinemann.
Hansen, G. (1989). *Universities for Development: Lessons for Enhancing the Role of Agricultural Universities in Developing Countries*. Washington, DC: AID. Evaluation Occasional Paper no. 31.
Heisenberg, W. (1959). *Physics and Philosophy*. New York: Harper.
Holt, J. E. and Schoorl, D. (1985). Technological change in agriculture—the systems movement and power, *Agricultural Systems*, **18,** 69–80.
Jackson, M. C. (1982). The nature of 'soft' systems thinking: the work of Churchman, Ackoff and Checkland, *Journal of Applied Systems Analysis*, **9,** 17–28.
Jackson, M. C. (1987). Present positions and future prospects in management science, *Omega*, **15,** 455–66.
Jamieson, N. (1985). *The Paradigmatic Significance of Rapid Rural Appraisal*. Paper presented to International Conference on Rapid Rural Appraisal, Khon Kaen University.
Keeney, B. (1982). What is an epistemology of family therapy? *Family Process*, **21,** 153–68.
Kelly, G. (1955). *The Psychology of Personal Constructs, Vols I and II*. New York: W. W. Norton.
Kenny, V. and Gardner, G. (1988). Constructions of self-organizing systems, *The Irish Journal of Psychology*, **9,** 1–24.
Kingma, O. (1985). p. 67 in: *Agribusiness, Productivity, Growth and Development in Australian Agriculture*, Research Monograph 22. Australia: Transnational Corporations Research Project, University of Sydney.
Kolb, D. (1984). *Experiential Learning: Experience as the Source of Learning and Development*. NJ: Prentice-Hall.
Kuhn, T. S. (1970). *The Structure of Scientific Revolutions*, 2nd edn. Chicago: The University Press.
Laszlo, E. (1972). Basic constructs of systems philosophy, *Systematics*, **10,** 40–54.
Lowrance, R., Stinner, B. R. and House, G. J. (Editors) (1984). *Agricultural Ecosystems: Unifying Concepts*. New York: John Wiley & Sons.
Macadam, R. D. and Bawden, R. J. (1985). Challenge and response: developing a system for educating more effective agriculturalists, *Prometheus*, **3,** 125–38.
Maturana, H. R. (1988). Reality: the search for objectivity or the quest for a compelling argument, *The Irish Journal of Psychology*, **9,** 25–82.
Maturana, H. R. and Varela, F. J. (1980). *Autopoeisis and Cognition—The Realisation of the Living*. Boston, MA: Reidel Publishing.
Maturana, H. R. and Varela, F. J. (1988). *The Tree of Knowledge*. London: Shambhala Press.
Mezirow, J. (1985). A critical theory of self-directed learning. In *Self-Directed Learning: From Theory in Practice*. New Directions for Continuing Education, no. 25. San Francisco: Jossey-Bass.
Miller, A. (1983). The influence of personal biases on environmental problem-solving, *Journal of Environmental Management*, **17,** 133–42.

Odum, H. T. (1971). *Fundamentals of Ecology*. Philadelphia: W. B. Saunders.
Oliga, J. C. (1988). Methodological foundations of systems methodologies, *Systems Practice,* **1,** 88–112.
Petak, W. (1980). Environmental planning and management: the need for an integrative perspective, *Environmental Management,* **4,** 287–95.
Plas, J. (1986). *Systems Psychology in the Schools*. New York: Pergamon Press.
Rappaport, J. (1987). Terms of empowerment/exemplars of prevention: toward a theory for community psychology, *American Journal of Community Psychology,* **15,** 121–45.
Reason, P. and Heron, J. (1986). Research with people: the paradigm of cooperative experiential inquiry, *Person-Centred Review,* **1,** 457–76.
Rosenhead, J. (1984). Debating systems methodology: conflicting ideas about conflict and ideas, *Journal of Applied Systems Analysis,* **11,** 79–84.
Smith, D. F. and Hill, D. M. (1975). Natural and agricultural ecosystems, *Journal of Environmental Quality,* **4,** 143–5.
Soedjatmoko (1985). The science and praxis of complexity. In *Proceedings of United Nations University Symposium at Montpelier*. Multiple Authors. Tokyo, Japan: The United Nations University.
Spedding, C. R. W. (1976*a*). Editorial, *Agricultural Systems,* **1,** 1.
Spedding, C. R. W. (1976*b*). Editorial, *Agricultural Systems,* **1,** 85.
Spedding, C. R. W. (1978). Editorial, *Agricultural Systems,* **3,** 167.
Stansbury, D. L. and Kunkel, H. O. (1987). *Higher Education in Agriculture in Transition: New Delivery Systems and Organizations, Procedures and Changes*. Eighth Working Conference of Representatives of Higher Education in Agriculture, Paris.
Ulrich, W. (1981). Counterpoint to Christenson's critique—a dialogue. C. West Churchman, Thomas A. Cowan and Werner Ulrich, *Journal of Enterprise Management,* **3,** 200–2.
Ulrich, W. (1988). Systems thinking, systems practice, and practical philosophy: a program of research, *Systems Practice,* **1,** 137–63.
Vayda, A. P. and McKay, B. J. (1975). New directions in ecology and ecological anthropology, *Annual Review of Anthropology,* **4,** 293–306.
Vickers, G. (1984). *The Vickers Papers*. London: Harper and Row.

13
Extension Education: Top(s) Down, Bottom(s) Up and Other Things

J. A. GARTNER
Asian Institute of Technology, Bangkok, Thailand

INTRODUCTION

You may wonder at the title of this last chapter. It *is* serious but it jokes with you. It is meant to; a humorist could run amok with it. In keeping with the purpose of this symposium, I want to celebrate, and have recorded, the mercurial sense of humour, sometimes of Rabelaisian proportions, which Colin Spedding brought to the serious business of systems thinking in agriculture. If nothing else in the chapter is discussed, I shall be satisfied if the title leads to a major bout of verbal revelry among old colleagues and friends during coffee time in the Department of Agriculture at the University of Reading.

Sometimes I think of such spontaneous outbursts of humour among these scholars while I live and work with the villagers in northeast Thailand. One can present a serious idea for agricultural development and the villagers will nod wisely and politely during discussion, but, unless the implementation of the idea has an element of fun in it, very little will be done. And yet these people are cash-poor (I spend more on one beer in a bar in Bangkok than they earn from one day's hard work in the fields) and the spectre of hunger, at least in some months of the year, conditions their view of the world. This is a good starting point for any debate on extension education (and for any person being educated institutionally to carry out such work) for it highlights the human, non-science dimension of agriculture. As Spedding (1979) wrote:

> *Agriculture as a subject, therefore, is concerned with an activity of fundamental importance to all communities and consists of a purposeful blend of science and non-science.*

Perhaps professors and peasants have in common this sense of serious fun. Establishing 'something in common' is essential to the process of communication in extension education.

TOP(S) DOWN

Extension education in agriculture before 1950 arose first from a general belief, stimulated by the scientific revolution, that education was a good thing and that it should be accessible to people where they live and work rather than being confined within the boundaries of the institutions responsible for it. Second, it arose in response to social crises due to biological disasters in existing farming systems; and third, it arose because the generation and application of scientific knowledge to agriculture directly, and through derived technology, was seen to be a powerful means of improving production, productivity and profit.

Some History

The historical roots of extension education go back to the Renaissance of Learning when there was a movement to relate education to the needs of human life and to the application of science to practical affairs (FAO, 1984*a*). This period produced the scientific revolution which replaced the Aristotelian view of the world with the Newtonian view which, in spite of some sophisticated modifications, is still recognizably our own (Checkland, 1984). True (1929) detailed the early activities of this movement as it related to agriculture in Europe and North America starting with the influence of Rabelais in the sixteenth century who 'would have pupils study nature as well as books and use their knowledge in their daily occupations'. These activities covered schools in Hungary and Switzerland where agricultural work was an important part of the curriculum, agricultural publications in France and Great Britain, and the establishment of the early agricultural societies in France, Germany, Russia, Scotland and the United States in the eighteenth century.

Extension-type programmes emerged in the United States about the middle of the nineteenth century with the use of 'competent individuals' as itinerant teachers who gave public lectures on the use of scientific knowledge in agriculture (True, 1928). This coincided with what Jones (1982) describes as the first modern agricultural advisory and instructional service which was established in Ireland in 1847 as a response to the social crisis caused by the great potato famine. In this institutionalized service, itinerant instructors worked among small-scale, peasant farmers to reduce their dependence on potato monoculture. They sought to create a system of farming that was less vulnerable to disease epidemics (in this case the potato blight fungus) by stimulating changes in cropping patterns and associated husbandry practices. A similar response to crisis occurred with the outbreak of the cotton boll weevil in the southern states of America at the beginning of the twentieth century. Such responses to the growth of scientific and technical knowledge and to crisis, either biological or economic and thus social, continue to be part of the agricultural scene today.

The actual use of the term 'extension' occurred first in the British university system. It referred to teaching activities away from the main university campus, the objective being to take the educational advantages of the universities to ordinary people where they lived and worked (Farquhar, 1961). 'Extension education', the term coined by Cambridge University in 1873 to describe this educational innovation, was intended to help people outside the universities adjust to the many social and technical changes taking place in nineteenth century Great Britain (Hawkins, 1982). It is interesting to note that the first grants to the university extension movement from public funds came from English county councils for lectures in agricultural science (FAO, 1973).

The idea of extension education soon spread to the United States where it was introduced through city libraries and universities (FAO, 1984*a*). In agriculture, the Land Grant Colleges quickly became involved as did the United States Department of Agriculture (USDA) and the private sector in fostering and funding extension activities, the purpose being to inform farmers about improved methods of farming. All of these efforts culminated in the passage of the Smith–Lever Cooperative Extension Act in 1914 that provided for a combination of federal, state and local funding of agricultural and home economics extension work to be carried out under USDA approval. This led to the formal adoption of the term 'agricultural extension'; the purpose of the activity which it described was to lift farmers, their families and rural communities into the mainstream of American life.

The spread of extension education in agriculture in Europe, Australia, Canada and New Zealand tended to parallel that in the United States but evolved in different ways in different countries

(Hawkins, 1982). For example, the Agricultural Development and Advisory Service of England and Wales is part of the Ministry of Agriculture, Food and Fisheries. In Scotland, on the other hand, the advisory services, established around the beginning of the twentieth century, are part of the agricultural colleges/university system even though they are funded by government. In contrast, extension services for agriculture in Australia were established by the state departments of agriculture which were also responsible for regulation and investigation; the term extension was first used in Queensland in 1889 to describe two trains used as mobile dairy extension units in a programme to foster milk production.

Language and Communication

The countries so far mentioned derive their roots from Western civilization which is characterized by the Judaeo-Christian religious tradition, by forms of art and crafts which are its own and by technologies which have been developed primarily from the knowledge generated by the human activity called science; this activity is unique to Western civilization (Checkland, 1984). A world view, as implied by an international symposium, requires that the approach to the dissemination of knowledge should be studied in other civilizations which have other religious traditions, other forms of art and crafts, and other technologies generated by activities other than science. Regrettably I have not been able to do this, for which I apologize to people who have their origins in these other civilizations. There are complications, however, of which the primary language of communication and the secondary language of the human activity 'farming' are paramount. For example, any Western-style extension activity is brought disconcertingly to a halt when one finds in a language that there are neither words for the elements in the Atomic Table nor words for the 'unseeable' microflora and microfauna in a man-managed fish pond. One must backtrack and begin at the beginning by trying to communicate what goes on in the 'unseen' world, which scientists take for granted, and try to establish the 'images of life' that exist in the brains of the people with whom one is trying to communicate. Even within the Western culture, an American can express apprehension about 'the language problem' when addressing an Australasian Agricultural Extension Conference (Bloome, 1987); and as for an Englishman understanding an Australian, that is another matter to which Colin

Spedding has contributed a great deal of good humour in conjunction with his old friend and sparring partner 'Fearsome Fred' Morley.

Nevertheless, it is possible to communicate without language if one has been an observer of the same phenomenon. I found this in Ethiopia when 'conversing' with a farmer about the effect of improved soil fertility on the botanical composition of his pasture. In his case the vestigial sward of kikuyu grass (*Pennisetum clandestinum* Hochst), was regenerated by the application of cattle dung from the cowyard; in my case the same effect was obtained on degenerate kikuyu pastures in the tropical highlands of Australia by using inorganic nitrogen fertilizer (Gartner, 1969). The natural reserve that exists between strangers was suddenly broken when we found we had this 'scientific' experience in common; further communication was then easy because a bond of mutual interest had been established. Similarly, having 'operational' experiences in common is helpful in communication as I found when talking with a Kenyan farmer about the difficulties of getting his son to do the chaff-cutting before and after school because of other attractions; my father had the same problem!

Civilizations and Colonies

There is another complication. During much of the period covered so far, Western empires held sway over most of the other civilizations in the world and used and abused them according to their purposes, which were concerned primarily with enriching the economy and thus the power of the 'home' country; in some cases the spread of religion added zeal and justification to the primary purpose. Where research and extension services to agriculture were established by colonial governments they were usually associated with commodity improvement programmes for export crops such as groundnuts, oil palm, rubber, sugar and tea (FAO, 1984*a*). These crops were produced by large public 'state' or private 'estate' corporations. Little attention was paid to the improvement of indigenous farming systems and their output of the traditional food crops of the local people; even less attention was paid to indigenous technology. These farming systems remained largely autonomous in their operation and thus not dependent on imported knowledge and technology.

During these times an uneasy relationship existed between the West and the Eastern civilizations which it did not colonize, embodied by China and Japan. To understand the differences between the East and the West one could start by reading Needham (1966) as cited by

Checkland (1984). For the purpose of this paper suffice to say that the traditional approach to nature and to agriculture in Eastern civilizations needs to be studied anew in view of our modern concern with another of the key properties of an agricultural system, sustainability.

This situation continued from 1914 up to about 1950, during which time two World Wars were fought and settled, a political revolution to challenge the West got underway, the Great Depression shattered the world economy and independence movements to challenge colonial power gathered momentum in countries deriving their roots from other civilizations. These world-scale upheavals changed the way agriculture was practised in that national food security became a political issue and protectionism became a government practice. This led to new perceptions of the role of agricultural extension as an instrument of government policy.

Scientist and Farmer

No doubt the exchange of information and technology among farmers continued throughout this time as it had in the centuries before the scientific revolution; in fact many changes in agricultural practice were farmer-led (Bunting, 1986). Furthermore, farmer innovation was particularly important in a country like Australia where a hostile environment, quite different from that faced previously by the European colonists, required an 'innovate or perish' mentality; this was enhanced by a colonial status quite different to that in Africa, Asia and Latin America. However, it seems reasonable to suggest that the prevailing paradigm in agricultural research and extension throughout the world up to the middle of the twentieth century involved what has become known in popular language as the 'top-down' approach, wherein knowledge and techniques are generated in, and transferred from, centres of learning and centres of administration to the centres of production.

BOTTOM(S) UP

Modern times are those that may reside in the professional memory of people attending this special symposium, i.e. from about 1950 onwards.

Western Countries

At the beginning of this period catch-cries like 'farming is a business not a way of life' and 'get big or get out' were heard in Western countries. Agricultural economics became a recognized subject of study (Heady, 1952). In Queensland, Australia, some agricultural extension officers were calling for a 'whole-farm approach' which included not only the management of biology in farming but also the management of money; others were involved in on-farm trials and experiments. In 1962 rural sociology was added to the formal presentation of these ideas and activities in coursework when the University of Queensland commenced the first postgraduate diploma in agricultural extension in Australia.

Agriculture was perceived as a dynamic economic biological complex which, when driven by educational and material inputs based on ever-expanding scientific knowledge, could be turned into a highly productive activity. It would also be possible and profitable for farmers if these inputs were subsidized and if prices were managed to allow a sufficient margin for an income that maintained farmers' livelihoods in line with their urban brothers. This scenario was very attractive to governments concerned with food security after the World Wars, with exports to generate foreign exchange and with bringing rural communities into the mainstream of national life. Furthermore, it was essential for some countries to increase yield per unit area as the area-expansion phase was already over. In small countries like Denmark and the Netherlands an increase in yield per unit area probably began in a modest way at the beginning of the twentieth century (Brown and Finstersbusch, 1972).

Agricultural extension, like agricultural research, was given strong government support in terms of grants (e.g. Commonwealth Extension Services Grant in Australia). In the United States, agricultural extension activities continued to be centred in the universities (Bloome, 1987) while in other countries they were sponsored by government ministries. Salaries were increased and qualifications of extension officers raised to equal those of their brothers in research but professional status was still considered inferior. Output continued to climb, leading to booms and busts in those countries depending on rural exports which in turn depended on international trade policies. All of this led to restructuring in farming communities and an increase in farm size which brought into close focus the social framework within which agriculture is practised. Thus social subsidies were born to assist people to give up farming or to continue farming, particularly in marginal industries and areas, or even at a later stage in the period to reduce production for exports due to imbalances and distortions in world trade. In this regard, 'rural counselling' was added to the responsibilities of some agricultural extension officers in distressed rural communities (Lees, 1987) where the negative aspects of protectionism caused bankruptcies and human hardship.

Public Knowledge

There was another social cost to be borne, however, by the spectacular increases in output per unit area of land which were obtained by the application of scientific knowledge and new technology to agriculture. The first public disquiet came from Rachel Carson's book, *Silent Spring* in 1962. This disquiet in relation to pesticide use, when linked with the older problem of environmental degradation due to expansion of the area used for agriculture, has risen to a crescendo almost 30 years later. The interlocking problems of environmental pollution and environmental degradation due to agriculture are now public knowledge; they are no longer confined within the suite of scientific disciplines that serve agriculture nor are they confined within the subject agriculture. Rather, when merged with the problems of industrial and urban environments they have become part of a larger social and political movement concerned with preservation and conservation of the physical world in which we live and work. And yet we must eat! It is easy to be passionately concerned with these issues when one has money in one's pocket, an abundance of food from which to choose and easily accessible energy with which to cook it, as well as comfortable shelter and suitable clothing. Thus I should like to add, for the purpose of this paper, the phrase 'in a productive state' to the generalized concern of this movement with preservation and conservation of the environment.

There was a social backlash in the community to industrial agriculture. In contrast to the admonishment of 'get big or get out' in the 1950s, in the 1970s 'small is beautiful' became the catch-cry of a cult which drew its inspiration from the book with this title by Schumacher (1973). In the United Kingdom, a Smallfarmers' Association was formed (Centre for Agriculture Strategy, 1981, 1982) which complemented the small-scale structure of farming in continental Europe, the maintenance of which lay at the foundation of the Common Agricultural Policy of the European Economic Community.

In the United States only 3 per cent of Americans remain on farms which are becoming increasingly bimodal in their distribution, with growing numbers of large 'super' farms and small part-time farms. At both ends of the size scale the life style is indistinguishable from that in towns and cities and the educational opportunities and the costs of family living are comparable (Bloome, 1987). It has become popular to declare that extension 'tries to be all things to all people'. Federal and state government officials believe that extension is not sufficiently directed at important federal and state issues. Local groups find extension insufficiently responsive and relevant to local needs. Commercial agriculture believes extension is spreading itself too thinly with urban programmes and programmes for part-time farmers. Urban clientele question the large share of the extension effort devoted to the population on small farms. Bloome sums this up in terms of the modern syndrome of parochialism that says 'extension should be concentrating more resources on issues that are important to me' and cites Breimyer (1987) who points out that 'much of the rebuke, or disagreement, about extension is a contest for preferential treatment'.

Much the same situation has occurred in Australia where a vocal and educated clientele on small farms has grown up demanding new services from an already dwindling set of resources allocated to agricultural extension due to budget cuts; the main source of financial support, the Commonwealth Extension Services Grant, has been terminated. At the University of Queensland, funding for teaching and research in extension has risen and fallen in 30 years (Crouch, 1987). In New Zealand, agricultural research, extension, and regulatory services have been 'privatized' and this idea is being championed for some public functions, including extension education, in other countries.

Thus, within the space of 75 years, the work of traditional agricultural extension in Western countries is finished, if one views it 'through the narrow prism of its original mission' (Bloome, 1987) first enunciated in 1914. 'Extension is dead—long live extension' affirms Crouch (1987) in the title of his paper to the Australasian Agricultural Extension Conference! Or is it? It seems to me, as we approach the end of the twentieth century, that the need for extension education, in which agricultural extension plays a major role, is greater now than it was at the end of the nineteenth century when it evolved formally 'to help people outside the universities adjust to the many social and technical changes taking place in Great Britain' (Hawkins, 1982). All that we need to do is replace Great Britain with 'the world', rewrite the mandate in terms of agriculture and in terms of the environment within which its activities take place, and strengthen the cybernetic loop that encircles a university (the psyche of society) and the 'ordinary' people (the soma of society) within its sphere of influence!

Recent events on the world stage reinforce this view. The refutation of communism, a 100-year-old hypothesis for an improved social, economic and political system, has added a new dimension to the question of how agriculture may be practised in the twenty-first century since it affects all countries in the world. The one human activity system which it failed to manage was agriculture which led to distortions in world trade; recovery in the agricultural output of those countries which tested the hypothesis may cause new distortions in world trade in agricultural products which will create new problems in major exporting countries.

Non-Western Countries

The development of agricultural extension organizations in non-Western countries during modern times depended largely upon whether they had been colonized or not.

Japan

Of the countries not colonized, Japan is unique in that in the last decade of the nineteenth century it had already brought into production most of its cultivable land. According to Brown and Finsterbusch (1972), it was probably the first country in the world to set the conditions for an increase in yield per unit area. The story is familiar in Western countries but with one particular twist that is of great significance to what happened in colonial countries about 50 years later, or rather what did not happen. The policy of generating a yield increase per unit area in rice production was supported by government at national, provincial and local levels. The national government, recognizing that some individual farmers and communities were advanced agriculturally and that others were quite backward, actively sought and supported the spread of the more advanced indigenous farming practices. Japanese officials were also sent abroad to Europe and the United States to search for improved farming techniques that could be applied to Japanese agriculture. The transfer of technology (TOT) was not always smooth because technologies designed for large-scale Western farms had to be adapted to Japan's small-scale, garden-size farms. Japan's greatest gain came from learning about chemical fertilizers and how to apply them in rice culture, but they had to be used in conjunction with other inputs and practices in order to be effective, e.g. a controlled and reliable water supply; varieties that would respond to fertilizer; innovative cultural practices which allowed plants to achieve their photosynthetic potential and which facilitated pest management; and the development of small-scale machinery. Most significantly, however, to make investment in fertilizer and other technologies possible the Japanese devised price supports and protection in whole or in part against imports. Korea successfully followed the Japanese model in the 1970s using loans from the World Bank. In the systems sense 'no man is an island' and so both countries have been under pressure for some years now to open their markets to foreign competition according to GATT (General Agreement on Tariffs and Trade).

Former colonies

The establishment of agricultural extension organizations in colonial countries was, to a very great extent, a post-independence phenomenon, occurring mainly after World War II; this was brought about by international aid, particularly from the United States in Latin America and some countries in South-East Asia (FAO, 1984*a*), and from the British in Africa (Adams, 1982). In nearly every country, agricultural extension was the responsibility of ministries of agriculture because, at independence, few countries had well-established colleges of agriculture or agricultural universities. The general experience of aid agencies with these organizations was not encouraging; the assumption that non-adoption of technology was due to farmer ignorance was not borne out by the extension education prescribed. In this regard their frustration was well expressed by a former Chairman of BIFAD (Board for International Food and Agricultural Development), the entity that provides policy direction for the United States Agency for International Development (USAID):

> *if there is an area where we have been most unsuccessful, it has been the development of cost-effective and program-efficient models for the delivery of new scientific and technical knowledge to the millions upon millions of farm producers of the Third World. We know how to harness the creative and inventive forces of science and technology in the war on hunger, but I submit that we still have not been fully successful in technology diffusion.*
>
> (Wharton, 1983, cited by FAO, 1984*a*)

This statement is most enlightening in that it continues to reflect the top-down TOT approach. Regardless of how successful it had been in Western countries, it certainly did not work when applied, mostly by Western aid agencies, to the previously neglected farming systems and rural communities and cultures of the newly independent countries. With all the goodwill in the world, it doesn't work without the full support of governments, a strong research and extension system operating from within the existing farming systems and other vital conditions and attitudes which have been eloquently described by Bunting (1986). Japan is the classic example of the successful combination and interaction of these factors in a non-Western country. Furthermore, the food aid policies of Western governments, no matter how well meant, exacerbated the problem by destroying local market forces. What politician would waste his energy and popularity in devising and implementing policies to assist indigenous agricultural development if he can get 'free' food from an external source? What farmer will bother to produce a surplus for sale if the reward he gets bears no relation to the risk he takes and the effort he makes? The 1950s and early 1960s were barren years for agricultural development in newly independent countries and for those people involved with agricultural research and extension.

Improved Varieties, Water and Fertilizer

By the mid-1960s, food aid policies were modified, political support for crop production campaigns was mobilized, budget and foreign exchange allocations materialized for the purchase of inputs, particularly fertilizer, and, most importantly, 'country after country adopted guaranteed prices at the farm level for wheat and rice' (Brown and Finsterbusch, 1972). Into this improved environment for agricultural development, by fortuitous timing, came the new high-yielding varieties of wheat and rice. Again by fortuitous circumstance, a reformed version of earlier attempts at top-down extension services, based on systematizing and intensifying old principles, evolved from an irrigated cotton project in Turkey. This was the Training and Visit (T and V) System (Benor and Harrison, 1977) which has been extensively promoted by the World Bank. This system and some similar package extension programmes (Adams, 1982), were well-suited to the recipe-type farming systems required for the introduction of the high-yielding varieties of wheat which did best under high rainfall or irrigated conditions or the high-yielding varieties of rice which required reliable and controlled water supplies. The convergence of these events and their interaction led to what has now gone down in history as the Green Revolution which started dramatically in the late 1960s. After 20 years its benefits have been substantially exploited (Byerlee, 1988), a plateau of improvement has been reached and 'technical assistance can depart once technology has become a habit' (Felix Maramba, 1983, personal communication). Or can it? Byerlee (1988) argues that in many of the post-Green Revolution areas, knowledge and skills of farmers have become critically limiting factors in maintaining increases in productivity and that a new and more complex second generation of inputs and management practices will be needed to maintain the momentum of change generated by the primary revolution; this will require changes in attitudes and in approaches to education, research and extension. This recent paper brings into focus one set of views of agricultural development that has been building for 20 years.

Before continuing with the above story, it is necessary to say that the rural landscapes wherein the Green Revolution took place now resemble in uniformity those of Europe and America and other countries where high input–high output monoculture of cereals and other energy crops is practised. The trees and the hedgerows have gone and so have the diverse habitats of previous agro-ecosystems from which terrestrial and aquatic fauna were harvested by villagers as valuable sources of animal protein in their diet; in this respect fish were often dominant. Furthermore, the indiscriminate use of chemicals to control pest outbreaks, due to the large-scale plantings of uniform varieties (Conway and McCauley, 1983), caused a fear of pollution that is now near the level of public knowledge in Western countries. The rapid change in agricultural practices and crop yields caused changes in the social landscape too; for many people the Green Revolution was a 'poisoned gift' (George, 1977).

The Farming Systems Approach

Like Newton's famous axiom, failure of the traditional top-down TOT approach to research and extension in the 1950s and early 1960s generated an 'opposed and equal reaction' among some Western aid workers which has been aptly described as the bottom-up approach. The farming systems research/extension/development approach (FSR, FSR/E and FSR/D) grew from this reaction based on the argument that constraints at the farm level limited the adoption of technology coming from outside the system. It was enhanced in the 1970s when it became apparent that the Green Revolution was restricted to most favoured areas in terms of water supply and other resources, including personal wealth. Furthermore, the T and V extension system is not suited to more complex, multicultural farming systems operated under ever-changing physical and economic environments. Its strengths and weaknesses, and conditions for its successful application, are discussed by Phongprapai and Setty (1988), and in Adams (1982), Jones (1986) and Howell (1988).

The farming systems approach was expounded by agricultural economists and agronomists working in Latin America, Asia and Africa, some of whom were Hildebrand (1976), Perrin *et al.* (1976), Norman (1978), Harwood (1979), Zandstra *et al.* (1979), Byerlee and Collinson (1980), Gilbert *et al.* (1980), Shaner *et al.* (1982), Zandstra *et al.* (1981), Byerlee *et al.* (1982), and Collinson (1982). The theory and practice have been discussed recently by Norman and Collinson (1985). Studies on farming systems in the 1960s and early 1970s revealed that the previously neglected, small-scale, indigenous farmers were economically rational but risk-averse and sharply constrained by uncertain environments and shortage of cash; they were ready enough, however, to adopt innovations that they themselves perceived to be economically attractive. Accordingly, the doctrine grew that research should be determined by explicitly identified farmer's needs rather than by the preconceptions of researchers. The Consultative Group on International Agricultural Research (CGIAR) therefore introduced economists into all its institutes at an early stage and explicitly espoused FSR ideas with the object of linking research with perceived farmers' needs (Simmonds, 1986). The programmes in FSR at these institutes have been reviewed by Dillon *et al.* (1978) and Simmonds (1985) and discussed further by Anderson and Dillon (1985) in relation to links made with National Agricultural Research Systems (NARS) and the likelihood of their adopting the farming systems approach; in this context Chandrapanya *et al.* (1985) have revealed some of the difficulties in practice.

Activities focused initially on annual crops, particularly the staple food crops, at the Centro Internacional de Mejoramiento de Maiz y Trigo (CIMMYT) and the International Rice Research Institute (IRRI). The fact that the domain of research has shifted in part to the farm means, in theory, that extension workers as well as farmers are brought into the problem identification process and consequent planning and implementation of trials and experiments; once useful technology has been generated and verified, extension workers associated with it, as well as the farmers themselves, are expected to be more effective in its diffusion. Jintrawet *et al.* (1985) have reported a positive result of this process in action during the investigation and transfer of a technique of growing peanuts after rice from a village in one province in Thailand to a village in another province.

This working hypothesis, which links researcher to extension officer to farmer, has been tested in more difficult crop environments by the International Crops Research Institute for the Semi-Arid Tropics (ICRISAT) and the International Centre for Agricultural Research in Dry Areas (ICARDA); and it has been tested in more complex farming systems involving livestock, and perennial trees and shrubs in mixed or integrated farming systems (International Livestock Centre for Africa, ILCA, 1980; International Council for Research in Agro-forestry, ICRAF, 1983). Furthermore, the International Centre for Living Aquatic Resources (ICLARM) has recently appointed a Farming Systems Specialist to test the hypothesis in aquaculture systems; the way was prepared by Edwards *et al.* (1988).

As with any scientific hypothesis, the more severe the test the more likely it will be refuted. In this case, the method evolved for annual crops in resource-rich environments was not directly applicable to semi-arid rainfed farming systems or to livestock production systems in Africa which operate across a whole range of physical environments from arid to humid and highland tropical. New methods appropriate to the needs of the different systems and different environments had to be devised within the context of the same approach. Furthermore, in the development of a new approach or subject, formulating and testing hypotheses about the subject matter, and how to go about investigating it, often get confounded; questions of content and questions of method are inevitably mixed. In this regard, CIMMYT and IRRI had 15 years of prior research experience in their mandate areas and thus considerable subject matter to relay to their collaborators for testing in known environments and systems. The new centres, ICRISAT, ICARDA and ILCA, did not (Anderson and Dillon, 1985); these centres had first to allocate substantial resources to identify, classify and describe their target systems and define their problems before the business of the design and testing of innovations could begin.

Farmer First and Last

From these experiences and others in Africa, Asia and Latin America where agricultural development efforts continued in the 1980s, two new working hypotheses have been proposed to overcome some of the problems inherent in the FSR model and its variants which were reactions to the TOT model. The first, Rapid Rural Appraisal (RRA), concerns method and the second, Farmer First and Last (FFL), concerns content.

First, the socio-economic or baseline survey, which is one of the information-gathering techniques of social scientists and agricultural economists, when used as the starting point for FSR tended to provide a 'Chinese banquet' when all that was required was a 'Continental breakfast'. RRA has emerged because 'decision-makers need information that is relevant, timely, accurate, and usable' and because 'in rural development, a great deal of the information that is generated is, in various combinations, irrelevant, late, wrong and/or unusable; it also costs a lot to obtain, process, analyse, and digest' (Chambers, 1987), if it ever is. A complete rendition of this reaction to the origins of FSR (its methods, tools and techniques, and some case studies) can be found in Khon Kaen University (1987).

The other reaction is in response to the neglect of indigenous farmer technology mentioned above, and the rationale for why farmers do what they do with what they have; inherent in it is a strong sense of social justice and the idea of equitability, the last of the properties of an agricultural system identified by Conway (1987). This brings into focus the third set of farmers in the classification of agriculture proposed by the Brundtland Commission (1987) as cited by Chambers *et al.* (1989). These farmers are engaged in resource-poor agriculture which is identified with difficult areas and characterized by farming systems that are complex, diverse and risk-prone. They are contrasted with farmers engaged in industrial agriculture found mainly in the rich world of the West and in Japan, and with farmers who participated in Green Revolution agriculture. The FFL protagonists argue that the problem is neither the 'ignorant farmer' nor the 'farm constraints', but the technology; and that the faults of the technology can be traced to the priorities and the processes which generate it (Chambers *et al.*, 1989). They feel that a taint of the top-down, TOT paradigm still resides in the FSR model in that the generation and flow of knowledge is still in the researcher-back-to-researcher mode. FFL entails reversals of explanation, learning and location (Chambers and Ghildyal, 1985) resulting in a farmer-back-to-farmer mode (Rhoades and Booth, 1982). In this mode, instead of starting with the knowledge, problems, analysis and priorities of scientists, it starts with the knowledge, problems, analysis and priorities of farmers and farm families. Instead of the research station as the main locus of action, it is now the farm. Instead of the scientist as the central experimenter, it is now the farmer, whether woman or man, and other members of the farm family (Chambers *et al.*, 1989). The danger here is that 'the baby might be thrown out with the bathwater' as in many movements that seek to correct an imbalance in outlook and action. And it appears that the role of research and extension has been reduced to symbolic dashes between the poles! Byerlee (1988) moderates this polarization of views when he says 'the challenge is to combine the knowledge and insights of farmers of their own environment with the information and skills generated by research, extension and formal education that are needed for effective management of a science-based agricultural technology'.

A Man For All Seasons

I want to go further. I hope that this latest and very important hypothesis will be tested to destruction in the Popperian sense and that out of it will come 'a public servant' who is not constrained by instructions from the 'top' nor bent by vested interests at the 'bottom', whose sole mandate is *to do what is necessary* to improve the productivity and profitability of agriculture in his area of influence in order to generate the primary wealth that leads to lasting rural development. The primary quality that this person must possess is an ability to appreciate the totality of a farming system, from within and without, in the light of its potential for change, and then to distinguish between those factors or forces that will yield to change and those factors or forces that will not yield to change within the foreseeable future. Lee (1989) calls this the 'Tao of Development' because of the strong parallel he found between the ancient Chinese philosophy of Taoism and what is called 'balanced development' in modern times. This 'systems' viewpoint matches very well with the 'scientific' viewpoint of the Western philosopher, Karl Popper: an idea which is at one time untestable, and therefore metaphysical, may with changed circumstances become testable and therefore scientific (Magee, 1973). Extending the categories of Mosher (1978), a person with this quality will be more than a facilitator according to the 'Rural Vacuum' theory; and he will be more than just a 'Research Distributor' for he will have the technical competence to conduct his own research and to interpret the results of others, including farmers, for use in specified farming systems. He will continue to be the 'Encouraging Companion' and he will engage in old-fashioned 'Education and Training' according to need. He will be aware of the concerns of rural people and of the need to maintain the physical environment in a productive and unpolluted state. Above all he will be an opportunity maker and taker, and a problem solver. He will certainly not be a 'Promoter' of some particular point of view nor the 'Local Errand Boy' for government. Finally, he will be experienced, an educated citizen in the sense proposed by Spedding (1988) and one of the highest paid public servants in the community he is meant to serve. In this regard how shall his performance be measured in view of our modern obsession with monitoring and evaluation? A not very erudite but immensely important and inexpensive measure of performance is the size of the smile on a farmer's face and the warmth of his welcome to those who advise and work with him.

Is this just a pipe-dream, another flight of fancy by a foolish fellow? Is it just an old-fashioned 'generalist' popping up again amongst the transient vanities of fashion in agricultural development? I think not. But to produce such a person, to create a position for him in 'public service' and to put him in place with a minimum tenure of 5 years might take as long to achieve as it will take for 'a systems approach' to catch up with its elder twin, 'a scientific approach', to the point where they walk together again as equal partners in development. Here I use the word 'development' to imply change to an existing situation but whether it is 'good' or 'bad' or 'in the right direction' depends on one's view of the world.

Colin Spedding foreshadowed the need for this type of person early in the 1970s (Spedding, 1970, 1971*a*, 1973) and thus the need for a new paradigm in institutional education in agriculture. In many ways, in thought and action, he is this type of person.

OTHER THINGS

While working through the literature on agricultural extension and the farming systems approach, which seeks to close the gaps between research, extension and practice, I discovered some curious things. First, there was not much cross-referencing between the two streams of literature although the language used and the causes espoused were not that different. Second, I could count the references to Colin

Spedding's work on less than the fingers of one hand! Third, there was no mention of the literature dealing with the philosophy, theory and practice of the 'systems movement' which, coincident with modern times as defined above, was formally established in 1954 with the formation of the Society for General Systems Research; those concerned were the biologist L. von Bertalanffy, regarded as the founder of the movement, an economist K. E. Boulding, a physiologist R. W. Gerard, and a mathematician A. Rapoport (Checkland, 1984). This situation is symptomatic of the way in which professional agriculture is organized in compartments and highlights the dilemma of language and of communication between people, which is the essence of extension education. Is there a lesson to be learned from this?

This symposium celebrates the career of an extraordinary man. I hope it will also honour him by entertaining the possibility of a new paradigm in the science *and* systems of agriculture which will unite the several streams of thought and action which have evolved more or less in parallel over the last 30 years.

Old Schools and New Schools

Colin Spedding's books (Spedding, 1975, 1979, 1988) evolved quite naturally from his early work on sheep production and grassland ecology (Spedding, 1965, 1971*b*), and collaboration with other systems thinkers in agriculture (Morley and Spedding, 1968; Spedding and Brockington, 1976*a,b*). They provide the 'matter' and conjectures on the 'method' for the new subject, agricultural systems, and place it firmly within the context of the larger systems movement and the philosophy which underpins it, the evolution of which has been clearly expounded by Checkland (1984). The connection between subject matter and subject methodology (methods and recipes, tools and techniques) was made, in a fashion, by other authors from the 'old school', but their interests were mainly descriptive (Duckham and Masefield, 1970; Ruthenberg, 1971; Grigg, 1974) or methodological, i.e. systems simulation (Jones, 1970; Dent and Anderson, 1971; de Wit and Goudriaan, 1974; Brockington, 1979).

Conway (1985) adapted and extended ideas from the old school to develop agro-ecosystems analysis, a technique which brought rigour and focus to the activities of multidisciplinary research and development programmes such as FSR/E/D and IRD (Integrated Rural Development) which arose in the 1970s and 1980s. Within the spectrum of methods that range from Chinese banquets to Continental breakfasts, this technique represents a splendid English Sunday lunch—and its after effects! FAO (1984*b*) reports on an attempt to bring representatives from all the schools together, including institutional education and aquaculture, to establish the common ground of a systems approach to agricultural development.

A Singular Seminal Influence

To the best of my knowledge, Colin Spedding wrote very little about traditional agricultural extension but embedded within the following passage from *The Biology of Agricultural Systems* (pp. 215–6) published in 1975 lies the essence of his singular, seminal influence on how professional agriculture may be practised in the future:

> *The fact is that subjects do not simply have central cores of certainty and peripheral uncertainties. The centre itself is in a state of continuous change, but in any case is not a clearly worked-out, unambiguous, well-organized body of knowledge. The nature of a subject is really much more hazy and the subject content consists of facts and relationships, some of them hypothetical. The result is that research and re-thinking permeate all parts of a subject and the centre is as much the subject of reclarification and restatement as are any of the more peripheral areas.*
>
> *It follows quite naturally from this that there cannot be a sharp separation between the subject as taught and that part of it that is the concern of research workers. Any such separation leads to an overdogmatic view of teaching, the material of which becomes arid and lifeless, and it tends to lead to a restriction of research activity to the less important parts of a subject. If this situation develops, it is bound to be followed at some time by a major upheaval, during which the subject appears to undergo a violent change brought about by a revolution external to it.*
>
> *It is this essential integrity of a subject that is the basis of arguments that teaching, research and* ***advice*** *(or* ***extension****) [my emphasis] should not be separated. Agriculture has been treated in a variety of different ways in different parts of the world and it is inevitable that some people should specialize in one branch or another. The problem is largely an organizational one of maintaining contact between them but the importance of this will only be clear if the essential integrity of a subject is recognized.*

The seeds of the revolution Colin Spedding foresaw, and to which he gave articulate voice, were sown in the 1960s by the *verifiable* success of the Western paradigm for agricultural development within its own environment and its *refutation* when tested in environments which were quite different to those for which it was created, as attested by a multitude of 'voices from the field'. It seems to me, however, that any model for change within the subject of agriculture, how it is taught and how its knowledge may be increased and used in a new paradigm will only be able to progress when it is placed on some philosophical foundation with a body of theory from which to derive new hypotheses for testing. This is the legacy which Colin Spedding leaves us. It is encapsulated in Popper's formula for continuous development as stated by Magee (1973):

$$P1 \rightarrow TS \rightarrow EE \rightarrow P2$$

where P1 is the initial problem, TS the trial solution proposed, EE the process of error elimination applied to the trial solution, and P2 the resulting situation, with new problems. It is essentially a feedback process and not cyclic, for P2 is always different from P1; even complete failure to solve a problem teaches us something new about where its difficulties lie and what the minimum conditions are which any solution for it must meet, and therefore alters the problem situation. Farmers and ordinary people (including you and I when we take off our professional hats) go through this process many times over every day of their lives.

About 400 years ago Cervantes said it another way: 'better the road than the inn'. But sometimes it is wise to rest and reflect awhile at an inn, so why don't we, and leave the serious matters for another time.

ACKNOWLEDGEMENT

The author's thanks are due to Professor Peter Edwards for helpful comments on the first draft of this paper.

REFERENCES

Adams, M. E. (1982). *Agricultural Extension in Developing Countries*. London: Longman.

Anderson, F. M. and Dillon, J. L. (1985). Farming systems at the International Agricultural Research Centres and other international groups, pp. 141–7 in: *Agricultural Systems Research for Developing Countries* (Proceedings of an International Workshop held at Hawkesbury Agricultural College, Richmond, NSW), ed. J. V. Remenyi. Canberra: ACIAR.

Brockington, N. R. (1979). *Computer Modelling in Agriculture.* Oxford: Clarendon Press.

Benor, D. and Harrison, J. Q. (1977). *The Training and Visit System.* Washington, DC: World Bank.

Bloome, P. D. (1987). Review of the USA situation, 1987, in: *Proceedings of Australasian Agricultural Extension Conference—1987,* ed. M. D. Littman. Brisbane, Australia: Queensland Department of Primary Industries.

Brown, L. R. and Finsterbusch, G. W. (1972). *Man and His Environment: Food.* New York: Harper & Row.

Bunting, A. H. (1986). Extension and technical change in agriculture, pp. 37–50 in: *Investing in Rural Extension: Strategies and Goals,* ed. G. E. Jones. London: Elsevier Applied Science Publishers.

Byerlee, D. K. (1988). Agricultural extension and the development of farmer's management skills, pp. 8–27 in: *Training and Visit Extension in Practice,* ed. J. Howell. Agricultural Administration Unit, Occasional Paper 8. London: Overseas Development Institute.

Byerlee, D. K. and Collinson, M. P. (1980). *Planning Technologies Appropriate to Farmers: Concepts and Procedures.* Mexico: CIMMYT.

Byerlee, D. K., Harrington, L. and Winkelman, D. L. (1982). Farming systems research: issues in research strategy and technology design, *American Journal of Agricultural Economics,* **64,** 897–904.

Carson, R. (1962). *Silent Spring.* Boston, MA: Houghton Mifflin.

Centre for Agricultural Strategy (1981). *Smallfarming and the Nation,* ed. R. B. Tranter. CAS Paper no. 9. Reading: CAS.

Centre for Agricultural Strategy (1982). *Smallfarming and the Rural Community,* ed. R. B. Tranter. CAS Paper 11. Reading: CAS.

Chambers, R. (1987). Shortcut methods in social information gathering for rural development projects, pp. 33–46 in: *Proceedings of the 1985 International Conference on Rapid Rural Appraisal.* Khon Kaen University, Khon Kaen, Thailand.

Chambers, R. and Ghildyal, B. P. (1985). Agricultural research for resource-poor farmers: the farmer-first-and-last model, *Agricultural Administration,* **20,** 1–30.

Chambers, R., Pacey, A. and Thrupp, L. A. (Editors) (1989). *Farmer First: Farmer Innovation and Agricultural Research.* London: Intermediate Technology Publications.

Chandrapanya, D., Pantastico, E. B., Suwanjinda, P. and Thaipanich, N. (1985). Farming systems research in Thailand, pp. 130–3 in: *Agricultural Systems Research for Developing Countries* (Proceedings of an International Workshop held at Hawkesbury Agricultural College, Richmond, NSW), ed. J. V. Remenyi. Canberra: ACIAR.

Checkland, P. (1984). *Systems Thinking, Systems Practice.* Chichester, UK: John Wiley & Sons.

Collinson, M. P. (1982). *Farming Systems Research in Eastern Africa: The*

Experience of CIMMYT and Some National Agricultural Research Services, 1976–81. MSU International Development Paper no. 3. Michigan: Department of Agricultural Economics, Michigan State University.

Conway, G. R. (1985). Agroecosystem analysis, *Agricultural Administration,* **20,** 31–5.

Conway, G. R. (1987). The properties of agroecosystems, *Agricultural Systems,* **24,** 95–117.

Conway, G. R. and McCauley, D. S. (1983). Intensifying tropical agriculture: the Indonesian experience, *Nature,* **302,** 288–9.

Crouch, B. R. (1987). Extension is dead—long live extension! Postgraduate extension training and research in extension, in: *Proceedings Australasian Extension Conference—1987,* ed. M. D. Littman. Brisbane: Queensland Department of Primary Industries.

Dent, J. B. and Anderson, J. R. (1971). *Systems Analysis in Agricultural Management.* Sydney: John Wiley & Sons Australasia.

De Wit, C. T. and Goudriaan, J. (1974). *Simulation of Ecological Processes.* Wageningen: PUDOC.

Dillon, J. L., Plucknett, D. L. and Vallaeys, G. J. (1978). *Farming Systems Research at the International Agricultural Research Centres.* Report to Technical Advisory Committee of CGIAR, Rome.

Duckham, A. N. and Masefield, G. B. (1970). *Farming Systems of the World.* London: Chatto & Windus.

Edwards, P., Pullin, R. S. V. and Gartner, J. A. (1988). *Research and Education for the Development of Integrated Crop–Livestock–Fish Farming Systems in the Tropics.* ICLARM Studies and Reviews no. 16. Manila: ICLARM.

FAO (1973). *Agricultural Extension: A Reference Manual,* abridged edition, ed. A. H. Maunder. Rome: Food and Agriculture Organization.

FAO (1984*a*). *Agricultural Extension: A Reference Manual,* 2nd edition, ed. B. E. Swanson. Rome: Food and Agriculture Organization.

FAO (1984*b*). *Proceedings of the Expert Consultation on Improving the Efficiency of Small-scale Livestock Production in Asia; a Systems Approach held 6–10 December, 1983, Bangkok, Thailand,* ed. J. A. Gartner. Rome: Food and Agriculture Organization.

Farquhar, R. N. (1961). Comparative extension education in Anglo-Saxon countries with particular reference to Australia. Mimeo. Melbourne: CSIRO.

Gartner, J. A. (1969). Effect of fertilizer nitrogen on a dense sward of kikuyu, paspalum and carpet grass. 1. Botanical composition, growth and nitrogen uptake, *Queensland Journal of Agricultural and Animal Sciences,* **26,** 21–32.

George, S. (1977). *How the Other Half Dies.* Harmondsworth: Penguin Books.

Gilbert, E. H., Norman, D. W. and Winch, F. E. (1980). *Farming Systems Research: A Critical Evaluation.* MSU Rural Development Paper no. 6. Michigan: Michigan State University.

Grigg, D. B. (1974). *The Agricultural Systems of the World.* Cambridge: Cambridge University Press.

Harwood, R. R. (1979). *Small Farm Development. Understanding and Improving Farming Systems in the Tropics.* Boulder, CO: Westview Press.

Hawkins, H. S. (1982). Agricultural and livestock extension. In *A Course Manual in Agricultural and Livestock Extension,* vol. 2, *The Extension Process,* eds H. S. Hawkins, A. M. Dunn and J. W. Cary: Canberra: Australian Universities' International Development Program.
Heady, E. O. (1952). *Economics of Agricultural Production and Resource Use.* New Jersey: Prentice Hall.
Hildebrand, P. E. (1976). *Generating Technology for Traditional Farmers: A Multidisciplinary Methodology.* Guatemala City: ICTA.
Howell, J. (Editor) (1988). *Training and Visit System in Practice.* Agricultural Administration Unit, Occasional Paper 8. London: Overseas Development Institute.
ICRAF (1983). *Guidelines for Agroforestry Diagnosis and Design.* ICRAF Working Paper no. 6. Nairobi: International Council for Research in Agro-forestry.
ILCA (1980). *ILCA. The First Years.* Addis Ababa: International Livestock Centre for Africa.
Jones, J. G. W. (1970). The use of models in agricultural and biological research, in: *Proceedings of a Symposium—1969, Grassland Research Institute, Hurley, Berkshire,* ed. J. G. W. Jones.
Jones, G. E. (1982). The Clarendon Letter, in: *Progress in Rural Extension and Community Development,* vol. 1, *Extension and Relative Advantage in Rural Development,* eds G. E. Jones and M. J. Rolls. Chichester: John Wiley & Sons.
Jones, G. E. (Editor) (1986). *Investing in Rural Extension: Strategies and Goals.* London: Elsevier Applied Science Publishers.
Jintrawet, A., Smutkupt, S., Wongsamun, C., Katawetin, R., and Kerdsuk, V. (1985). *Extension Activities for Peanuts after Rice in Ban Sum Jan, Northeast Thailand: A Case Study in Farmer-to-Farmer Extension Methodology.* Khon Kaen, Thailand: Khon Kaen University.
Khon Kaen University (1987). *Proceedings of the 1985 International Conference on Rapid Rural Appraisal.* Khon Kaen, Thailand: Khon Kaen University.
Lees, J. W. (1987). Towards a rationale for rural counselling services, pp. 139–49 in: *Proceedings of Australasian Agricultural Extension Conference—1987,* ed. M. D. Littman. Brisbane: Queensland Department of Primary Industries.
Lee, C. (1989). Rice–fish extension in Tung Kula Ronghai. Paper presented at the 2nd Rice–Fish Farming Systems Research and Development Workshop, Munoz, Philippines.
Magee, B. (1973). *Popper.* Glasgow: Fontana/Collins.
Morley, F. H. W. and Spedding, C. R. W. (1968). Agricultural systems and grazing experiments, *Herbage Abstracts,* **38,** 279–87.
Mosher, A. T. (1978). *An Introduction to Agricultural Extension.* New York: Agricultural Development Council.
Norman, D. W. (1978). Farming systems research to improve the livelihood of the small farmer, *American Journal of Agricultural Economics,* **60,** 813–18.
Norman, D. W. and Collinson, M. P. (1985). Farming systems research in theory and practice, pp. 16–30 in: *Agricultural Systems Research for*

Developing Countries (Proceedings of an International Workshop, held 12–15 May, Richmond, NSW), ed. J. V. Remenyi. Canberra: ACIAR.

Perrin, R. K., Winkelman, D. L., Moscardi, E. R. and Anderson, J. R. (1976). *From Agronomic Data to Farmer Recommendations: an Economics Training Manual.* Information Bulletin 27. Mexico: CIMMYT.

Phongprapai, S. and Setty, E. D. (1988). *Agricultural Extension in Thailand: A Study of the Training and Visit System.* HSD Monograph 15. Bangkok: Asian Institute of Technology.

Rhoades, R. E. and Booth, R. H. (1982). Farmer-back-to-farmer: A model for generating acceptable agricultural technology, *Agricultural Administration,* **11,** 127–37.

Ruthenberg, H. (1971). *Farming Systems in the Tropics,* 1st edition. Oxford: Clarendon Press.

Schumacher, E. F. (1973). *Small is Beautiful.* London: Blond & Briggs.

Shaner, W. W., Philipp, P. F. and Schmehl, W. R. (1982). *Farming Systems Research and Development: Guidelines for Developing Countries.* Boulder: Westview Press.

Simmonds, N. W. (1985). *Farming Systems Research, a Review.* Washington, DC: World Bank.

Simmonds, N. W. (1986). A short review of farming systems research in the tropics, *Experimental Agriculture,* **22,** 1–13.

Spedding, C. R. W. (1965). *Sheep Production and Grazing Management,* 1st edition. London: Bailliere, Tindall & Cox.

Spedding, C. R. W. (1970). The relative complexity of grassland systems, pp. A126–A131 in: *Proceedings of the XI International Grassland Congress, Surfers Paradise, Queensland,* ed. M. J. T. Norman. University of Queensland Press.

Spedding, C. R. W. (1971*a*). Agricultural ecosystems, *Outlook on Agriculture,* **6,** 242–7.

Spedding, C. R. W. (1971*b*). *Grassland Ecology.* Oxford: Clarendon Press.

Spedding, C. R. W. (1973). The future of development in agriculture, *Agricultural Progress,* **48.**

Spedding, C. R. W. (1975). *The Biology of Agricultural Systems.* London: Academic Press.

Spedding, C. R. W. (1979). *An Introduction to Agricultural Systems,* 1st edition. London: Applied Science Publishers.

Spedding, C. R. W. (1988). *An Introduction to Agricultural Systems,* 2nd edition. London: Elsevier Applied Science.

Spedding, C. R. W. and Brockington, N. R. (1976*a*). The study of ecosystems. *Agro-Ecosystems,* **2,** 165–72.

Spedding, C. R. W. and Brockington, N. R. (1976*b*). Experimentation in agricultural systems, *Agricultural Systems,* **1,** 47–56.

True, A. C. (1928). *A History of Agricultural Extension Work in the United States 1785–1923,* USDA Miscellaneous publication no. 15. Washington, DC: US Department of Agriculture.

True, A. C. (1929). *A History of Agricultural Education in the United States 1785–1925.* USDA Miscellaneous publication no. 36. Washington, DC: US Department of Agriculture.

Zandstra, H. G., Swanberg, K., Zulberti, C. and Nestel, B. (1979). *Caqueza: Living Rural Development*. Ottawa: IDRC.
Zandstra, H. G., Price, E. C., Litsinger, J. A. and Morris, R. A. (1981). *A Methodology for On-Farm Cropping Systems Research*. Los Baños, Philippines: International Rice Research Institute.

Index